T0213866

Lecture Notes in Computer Science   12071

More information about this series at http://www.springer.com/series/7407

Alexander Raschke · Dominique Méry ·
Frank Houdek (Eds.)

Rigorous State-Based Methods

7th International Conference, ABZ 2020
Ulm, Germany, May 27–29, 2020
Proceedings

 Springer

Editors
Alexander Raschke (iD)
Institute of Software Engineering
and Programming Languages
Ulm University
Ulm, Germany

Dominique Méry (iD)
LORIA, Campus Scientifique
Université de Lorraine
Vandoeuvre-les-Nancy, France

Frank Houdek (iD)
Research and Development
Mercedes-Benz AG
Sindelfingen, Germany

ISSN 0302-9743   ISSN 1611-3349   (electronic)
Lecture Notes in Computer Science
ISBN 978-3-030-48076-9   ISBN 978-3-030-48077-6   (eBook)
https://doi.org/10.1007/978-3-030-48077-6

LNCS Sublibrary: SL1 – Theoretical Computer Science and General Issues

This Springer imprint is published by the registered company Springer Nature Switzerland AG
The registered company address is: Gewerbestrasse 11, 6330 Cham, Switzerland

Preface

The International Conference on Rigorous State-Based Methods (ABZ 2020) is an international forum for the cross-fertilization of related state-based and machine-based formal methods, mainly Abstract State Machines (ASM), Alloy, B, TLA+, VDM, and Z. Rigorous state-based methods share a common conceptual foundation and are widely used in both academia and industry for the design and analysis of hardware and software systems.

The name ABZ was invented at the first conference held in London in 2008, where the ASM, B, and Z conference series merged into a single event. The second ABZ 2010 conference was held in Orford, Canada, where the Alloy community joined the event; ABZ 2012 was held in Pisa, Italy, which saw the inclusion of the VDM community (but not in the title); ABZ 2014 was held in Toulouse, France, which brought the inclusion of the TLA+ community into the ABZ conference series. Lastly, the ABZ 2016 conference was held in Linz, Austria, and ABZ 2018 in Southampton, UK. In 2018 the Steering Committee decided to retain the (well-known) acronym ABZ and add the subtitle "International Conference on Rigorous State-Based Methods" to make more explicit the intention to include all state-based formal methods. Consequently, the title of the proceedings was also modified to "Rigorous State-Based Methods".

Started 2014 in Toulouse, each ABZ asked for the application of formal specifications on industrial case studies. This year, we extend the previous areas (aerospace, medical equipment, rails) with the automotive domain. A specification of an adaptive light and speed control system similar to the available real systems was provided by Frank Houdek, who also answered almost a hundred questions and gave clarifying explanations, for which we would like to thank him. The objective of these case studies is to provide an opportunity to demonstrate the applicability of the ABZ methods to real examples and also to allow for a better comparison of them. These proceedings include the case study as well as several accepted papers outlining solutions to it.

ABZ 2020 received 55 submissions from 21 countries around the world. The selection process was rigorous, where each paper received at least three reviews. The Program Committee (PC), after careful discussions, decided to accept 12 full research papers, 6 case study papers, and 9 short research papers. One extended abstract of one of the keynote speakers and one invited research paper are also included in the proceedings. All accepted papers cover broad research areas on both theoretical systems and practical aspects of state-based methods.

For the first time in the conference's history, ABZ 2020 organized a doctoral symposium and PhD students had to submit a short paper presenting their PhD topics; those six submissions were evaluated by a separate PC including the two chairs of ABZ 2020.

The conference was to be held during May 27–29, 2020, in Ulm, Germany, but due to the historical crisis caused by the corona virus with an unprecedented international lock-down, travel-restrictions, and sadly many deaths around the world, we had to cancel the conference and postpone it to next year (2021). At ABZ 2021, all authors of accepted papers of ABZ 2020 are requested to present their research in addition to the new accepted papers.

We are honored that all three distinguished guests as keynote speakers agreed to give their keynotes next year: Ana Cavalcanti, University of York, UK, will give a talk entitled "RoStar technology—a roboticist's toolbox for combined proof and sound simulation;" Uwe Glässer, Simon Fraser University, Canada, will give a talk entitled "Quantifying Uncertainty in ASM Models with Markov Processes;" and we will hear from Gilles Dowek, INRIA/ENS Paris-Saclay, France.

The EasyChair conference management system was set up for ABZ 2020, supporting submission, review, and volume edition processes. We acknowledge it as an outstanding tool for the academic community.

We would like to thank all the authors who submitted their work to ABZ 2020. We are grateful to the PC members and external reviewers for their high-quality reviews and discussions. Finally, we wish to thank the Organizing Committee members for their continuous support.

We hope the corona crisis will be over within the next weeks or months and that the enormous economic consequences of this crisis will be outweighed by more humanity in the world. We look forward to welcoming many conference attendants in Ulm next year and hope they will enjoy the technical program, informal meetings, and interactions with colleagues from all over the world; and of course, we are confident they will like the city of Ulm, Germany. For readers of these proceedings, we hope these papers are interesting and they inspire ideas for future research.

March 2020   Dominique Méry
                                                          Alexander Raschke
                                                              Frank Houdek

Organization

Program Committee

Yamine Ait Ameur	IRIT/INPT-ENSEEIHT, France
Paolo Arcaini	National Institute of Informatics, Japan
Richard Banach	The University of Manchester, UK
Egon Boerger	University of Pisa, Italy
Eerke Boiten	De Montfort University, UK
Michael Butler	University of Southampton, UK
Andrew Butterfield	Trinity College Dublin, Ireland
David Deharbe	ClearSy System Engineering, France
Juergen Dingel	Queen's University, Canada
Flavio Ferrarotti	Software Competence Centre Hagenberg, Austria
Mamoun Filali-Amine	IRIT, France
Marc Frappier	Université de Sherbrooke, Canada
Leo Freitas	Newcastle University, UK
Angelo Gargantini	University of Bergamo, Italy
Vincenzo Gervasi	University of Pisa, Italy
Uwe Glässer	Simon Fraser University, Canada
Gudmund Grov	Norwegian Defence Research Establishment (FFI), Norway
Stefan Hallerstede	Aarhus University, Denmark
Klaus Havelund	Jet Propulsion Laboratory, USA
Ian J. Hayes	The University of Queensland, Australia
Thai Son Hoang	University of Southampton, UK
Frank Houdek	Daimler AG, Germany
Alexei Iliasov	Newcastle University, UK
Jeremy Jacob	University of York, UK
Felix Kossak	Software Competence Center Hagenberg, Austria
Regine Laleau	Paris-Est Créteil University, France
Thierry Lecomte	ClearSy, France
Michael Leuschel	University of Düsseldorf, Germany
Alexei Lisitsa	The University of Liverpool, UK
Amel Mammar	Télécom SudParis, France
Atif Mashkoor	Johannes Kepler University, Austria
Jackson Mayo	Sandia National Laboratories, USA
Stephan Merz	Inria Nancy, France
Stefan Mitsch	Carnegie Mellon University, USA
Rosemary Monahan	Maynooth University, Ireland
Mohamed Mosbah	LaBRI, University of Bordeaux, France
Dominique Méry	Université de Lorraine, LORIA, France

Shin Nakajima	National Institute of Informatics, Japan
Uwe Nestmann	TU Berlin, Germany
Jose Oliveira	University of Minho, Portugal
Philipp Paulweber	University of Vienna, Austria
Luigia Petre	Åbo Akademi University, Finland
Andreas Prinz	University of Agder, Norway
Shengchao Qin	Teesside University, UK
Philippe Queinnec	IRIT, Université de Toulouse, France
Alexander Raschke	Ulm University, Germany
Elvinia Riccobene	University of Milan, Italy
Victor Rivera	The Australian National University, Australia
Thomas Santen	TU Berlin, Germany
Patrizia Scandurra	University of Bergamo, Italy
Gerhard Schellhorn	Universitaet Augsburg, Germany
Klaus-Dieter Schewe	Zhejiang University, China
Steve Schneider	University of Surrey, UK
Colin Snook	University of Southampton, UK
Michael Stegmaier	Ulm University, Germany
Maurice H. ter Beek	ISTI-CNR, Italy
Laurent Voisin	Systerel, France
Alan Wassyng	McMaster University, Canada
Virginie Wiels	ONERA/DTIM, France
Frank Zeyda	University of York, UK
Wolf Zimmermann	Martin Luther University Halle-Wittenberg, Germany

Additional Reviewers

Bannister, Callum
Bonfanti, Silvia
Charalampous, Tilemachos
Dghaym, Dana
Fantechi, Alessandro
Mazzanti, Franco
Pollitt, Alastair
Salehi Fathabadi, Asieh
Tayebi, Mohammad
Tounsi, Mohamed
Winter, Kirsten

Contents

Short Articles

Articles Contributing to the Case Study

Short Articles of the PhD-Symposium (Work in Progress)

Keynotes and Invited Papers

Modelling and Verification of Robotic Platforms for Simulation Using RoboStar Technology

Ana Cavalcanti[✉]

Department of Computer Science, University of York, York YO105GH, UK
Ana.Cavalcanti@york.ac.uk

The RoboStar framework[1] supports model-based engineering of robotic applications. Modelling is carried out using diagrammatic domain-specific languages: RoboChart [13] and RoboSim [3]. Verification and generation of artefacts is justified by a formal semantics given using a state-rich hybrid version of a process algebra for refinement [7]. It is inspired by CSP [19] and cast in Hoare and He's Unifying Theories of Programming (UTP) [10] formalised in Isabelle [6].

RoboChart is an event-based language for design, while RoboSim is a cycle-based language for simulation. Tool support is provided by RoboTool, which includes facilities for graphical modelling, validation, and automatic generation of CSP (for analysis with the model checker FDR [9]) and PRISM [11] scripts (for verification of probabilistic controllers), and simulations. RoboChart and RoboSim are based on the use of state machines to specify behaviour, akin to notations already in widespread use [2,5,16,20], but RoboChart and RoboSim are enriched with facilities for verification and traceability of artefacts.

Recent work has focussed on enriching RoboSim for physical modelling. Current practice in robotics often uses simulation to understand the behaviour of a robotic controller for a particular robotic platform and environment. A wide variety of simulators for robotics use different tool-dependent or even proprietary programming languages and API [8,12,14,17,18]. Physical modelling of the platforms are encoded by programs in customised notations, generated from graphical tools, or in C++, Java, Python, or C#, for example.

RoboSim, on the other hand, is a tool-independent notation. For physical modelling, we have defined a notation based on SysML block diagrams [15]. Our profile is inspired by XML-based notations used by robotics simulators[2]. It defines a physical model by a diagram that captures the physical components of a platform as links (rigid bodies), joints, sensors, and actuators. Properties of these blocks capture their attributes that are relevant for simulation and for capturing behaviour: movement and use of sensors and actuators.

In contrast with XML-based notations in current use, RoboSim *block diagrams* encourage readability and support modularisation via several mechanisms. Models can be parametrised by constants that represent, for example, key measures of physical bodies. The pose of an element is defined always in reference to

---

[1] www.cs.york.ac.uk/robostar/.

[2] sdformat.org.

© Springer Nature Switzerland AG 2020
A. Raschke et al. (Eds.): ABZ 2020, LNCS 12071, pp. 3–5, 2020.
https://doi.org/10.1007/978-3-030-48077-6_1

the element that contains it. A richer notion of connection captures flexible and fixed compositions. A library fosters reuse by the possibility of defining parts and fragments that can be instantiated or simply included to define a complete model. Finally, well-formedness rules ensure validity of models.

The most distinctive feature of RoboSim block diagrams, however, is the possibility of defining systems of differential algebraic equations that capture behaviour of the platform. For sensors, these equations define how inputs (from the environment) are reflected in sensor outputs for use with the software. For actuators, the equations define how inputs from the software affect the outputs of the actuators, and therefore, affect the platform itself (in the case of motors, for example), or the environment. For joints, the equations define how their movement induces movement on the links connected to them.

A system view is provided by connecting a RoboSim block diagram that specifies a physical model for a robotic platform, to a RoboSim module that specifies a control software. This is achieved by a *platform mapping*, which specifies how software elements that abstract services of the platform are defined. In specifying these services, we can use outputs of sensors and inputs of actuators.

Ongoing work, provides support to translate RoboSim block diagrams to XML for use in simulation (using Coppelia, formerly, v-rep). For mathematical modelling, the UTP semantics constructs a hybrid model, with constructs inspired by those of *Circus* [4], combining Z [1,21] and CSP.

Acknowledgements. The work mentioned is a collaboration with colleagues at the RoboStar group, in particular, Alvaro Miyazawa and Sharar Ahmadi. The author's work is funded by the Royal Academy of Engineering grant CiET1718/45, and UK EPSRC grants EP/M025756/1 and EP/R025479/1. No new primary data was created as part of the study reported here.

References

1. ISO/IEC 13568:2002. Information technology - Z formal specification notation - syntax, type system and semantics. International Standard
2. Brunner, S.G., Steinmetz, F., Belder, R., Domel, A.: Rafcon: a graphical tool for engineering complex, robotic tasks. In: IEEE/RSJ International Conference on Intelligent Robots and Systems, pp. 3283–3290 (2016)
3. Cavalcanti, A.L.C., et al.: Verified simulation for robotics. Sci. Comput. Program. **174**, 1–37 (2019)
4. Cavalcanti, A.L.C., Sampaio, A.C.A., Woodcock, J.C.P.: A refinement strategy for Circus. Formal Aspects Comput. **15**(2–3), 146–181 (2003)
5. Dhouib, S., Kchir, S., Stinckwich, S., Ziadi, T., Ziane, M.: RobotML, a domain-specific language to design, simulate and deploy robotic applications. In: Noda, I., Ando, N., Brugali, D., Kuffner, J.J. (eds.) SIMPAR 2012. LNCS (LNAI), vol. 7628, pp. 149–160. Springer, Heidelberg (2012). https://doi.org/10.1007/978-3-642-34327-8_16
6. Foster, S., Baxter, J., Cavalcanti, A., Miyazawa, A., Woodcock, J.: Automating verification of state machines with reactive designs and Isabelle/UTP. In: Bae, K., Ölveczky, P.C. (eds.) FACS 2018. LNCS, vol. 11222, pp. 137–155. Springer, Cham (2018). https://doi.org/10.1007/978-3-030-02146-7_7

7. Foster, S., Cavalcanti, A.L.C., Canham, S., Woodcock, J.C.P., Zeyda, F.: Unifying theories of reactive design contracts. Theoret. Comput. Sci. **802**, 105–140 (2020)
8. Gerkey, B., Vaughan, R.T., Andrew, H.: The player/stage project: tools for multi-robot and distributed sensor systems. In: 11th International Conference on Advanced Robotics, pp. 317–323 (2003)
9. Gibson-Robinson, T., Armstrong, P., Boulgakov, A., Roscoe, A.W.: FDR3—a modern refinement checker for CSP. In: Ábrahám, E., Havelund, K. (eds.) TACAS 2014. LNCS, vol. 8413, pp. 187–201. Springer, Heidelberg (2014). https://doi.org/10.1007/978-3-642-54862-8_13
10. Hoare, C.A.R., Jifeng, H.: Unifying Theories of Programming. Prentice-Hall, Upper Saddle River (1998)
11. Kwiatkowska, M., Norman, G., Parker, D.: Probabilistic symbolic model checking with PRISM: a hybrid approach. Int. J. Softw. Tools Technol. Transf. **6**(2), 128–142 (2004). https://doi.org/10.1007/s10009-004-0140-2
12. Luke, S., Cioffi-Revilla, C., Panait, L., Sullivan, K., Balan, G.: Mason: a multiagent simulation environment. Simulation **81**(7), 517–527 (2005)
13. Miyazawa, A., Ribeiro, P., Li, W., Cavalcanti, A., Timmis, J., Woodcock, J.: RoboChart: modelling and verification of the functional behaviour of robotic applications. Softw. Syst. Modeling **18**(5), 3097–3149 (2019). https://doi.org/10.1007/s10270-018-00710-z
14. Olivier, M.: Webots^{TM}: professional mobile robot simulation. Int. J. Adv. Robot. Syst. **1**(1), 39–42 (2004)
15. OMG. OMG Systems Modeling Language (OMG SysML), Version 1.3 (2012)
16. Pembeci, I., Nilsson, H., Hager, G.: Functional reactive robotics: an exercise in principled integration of domain-specific languages. In: 4th ACM SIGPLAN International Conference on Principles and Practice of Declarative Programming, pp. 168–179. ACM (2002)
17. Pinciroli, C., et al.: ARGoS: a modular, parallel, multi-engine simulator for multi-robot systems. Swarm Intell. **6**(4), 271–295 (2012)
18. Rohmer, E., Singh, S.P.N., Freese, M.: V-REP: a versatile and scalable robot simulation framework. In: IEEE International Conference on Intelligent Robots and Systems, vol. 1, pp. 1321–1326. IEEE (2013)
19. Roscoe, A.W.: Understanding Concurrent Systems. Texts in Computer Science. Springer, Heidelberg (2011). https://doi.org/10.1007/978-1-84882-258-0
20. Wachter, M., Ottenhaus, S., Krohnert, M., Vahrenkamp, N., Asfour, T.: The ArmarX statechart concept: graphical programing of robot behavior. Front. Robot. AI **3**, 33 (2016)
21. Woodcock, J.C.P., Davies, J.: Using Z - Specification, Refinement, and Proof. Prentice-Hall, Upper Saddle River (1996)

Adding Concurrency to a Sequential Refinement Tower

Gerhard Schellhorn$^{(\boxtimes)}$, Stefan Bodenmüller, Jörg Pfähler, and Wolfgang Reif

Institute for Software and Systems Engineering,
University of Augsburg, Augsburg, Germany
{schellhorn,stefan.bodenmueller,reif}@informatik.uni-augsburg.de,
joerg.pfaehler@gmx.de

Abstract. This paper defines a concept and a verification methodology for adding concurrency to a sequential refinement tower of abstract state machines, that is based on data refinement and a component structure. We have developed such a refinement tower for the Flashix file system earlier, from which we generate executable (C and Scala) Code.

The question we answer in this paper, is how to add concurrency based on locks to such a refinement tower, without breaking the initial modular structure. We achieve this by just enhancing the relevant components, and adding intermediate atomicity refinements that complement the data refinements that are already there. We also give a verification methodology for such atomicity refinements.

1 Introduction

Development of formally proved software systems using incremental refinement has been successfully used in many case studies. Often the system developed is a sequential system, e.g. a compiler. The standard technique used then is data refinement [8,9,14] or closely related definitions [2].

Our group has developed a verified file system for flash memory [12,13,22,26] using a strategy based on data types specified as abstract state machines (ASMs, [4]), data refinement, and subcomponents. The resulting refinement tower is shown in Fig. 1. It starts with an abstract state machine that specifies the POSIX file system operations. This interface is then refined to an implementation VFS (denoted by VFS $\sqsubseteq$ POSIX), which calls operations of a submachine AFS. This machine acts as an abstract interface to the next implementation. This continues until the MTD layer is reached, which is the generic interface for flash hardware used in Linux.

Scala code for simulations as well as C code integrated into the Linux kernel has been generated from the implementations (shown in grey). The file system so far is strictly sequential, i.e., all operations are called in sequential order. Adding

Supported by the Deutsche Forschungsgemeinschaft (DFG), "Verifikation von Flash-Dateisystemen" (grants RE828/13-1 and RE828/13-2).

A. Raschke et al. (Eds.): ABZ 2020, LNCS 12071, pp. 6–23, 2020.
https://doi.org/10.1007/978-3-030-48077-6_2

concurrency is however relevant for practical usability and efficiency on at least three levels: top-level operations, garbage collection and wear leveling.

Since existing refinement strategies are typically designed to start with an atomic specification that is refined to a concurrent system, this raises the question how to add concurrency a posteriori to intermediate levels of such a refinement tower without losing modularity and without having to start verification from scratch. This paper gives a positive answer to the question, by "shifting" parts of the refinement towers, i.e., by modifying individual specifications and implementations, to make them concurrent.

We will use erase block management (the EBM interface) and the concurrent implementation of wear leveling (WL) based on the interface Blocks as an example to demonstrate how concurrency is added. A specification of the sequential specifications and refinements involved has already been published in [23].

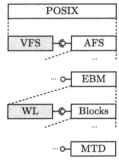

Fig. 1. Flashix refinement tower

The next section will give a simplified version of the relevant sequential specifications and implementation, to demonstrate in Sect. 3 how concurrency using locks is added and how restrictions are encoded as *ownership* constraints. Section 4 informally introduces the well-known concept of *linearizability* as the relevant concept to verify correctness of concurrent implementations, and shows how the proof of linearizability can be split into one of data refinement (that reuses the original proof) and one of *atomicity refinement*. Section 5 will give a proof strategy based on *rely-guarantee* proofs and *reduction*. Both have been implemented in our KIV [11] theorem prover. The specifications and proofs for the case study are available online [18]. Section 6 gives related work, and Sect. 7 concludes.

2 The Refinement for Wear Leveling

Flash hardware is partitioned into erase *blocks*. Blocks can be written sequentially, and erased as a whole. Erasing wears out the block until it becomes unusable. Therefore, for efficient usage of a flash device, blocks must be worn out evenly. In particular if a device is filled to a large part with static data, the blocks with these data must sometimes be swapped with other (currently empty) blocks, that have often been modified and erased. This is called *wear leveling*. Wear leveling is hidden from the more abstract levels of the file system by the erase block manager (EBM) interface. The interface offers access to *logical blocks*. The task of the implementation (WL) is to map them to the *physical blocks* offered by the hardware, and to change the mapping when this is advisable, using an internal operation for wear leveling that has no effect (implements skip) for the interface EBM.

An abstract specification of the erase block manager is given with the ASM EBM. The state consists of a function that maps logical block numbers to actual content and a set of currently used ("mapped") block numbers.

$$\textbf{state} \qquad Contents : nat \rightarrow content \qquad Mapped : set\langle nat \rangle$$
$$\textbf{initial state} \qquad Contents = \lambda\, n.\ \texttt{empty} \wedge Mapped = \emptyset$$

For simplicity, we do not specify *content*, except for a default value `empty`. The interface of EBM shown in Fig. 2 allows to read and to write the content of logical blocks. The operations use a semicolon to separate input and output parameters.

ebm_write(*lnum*, *c*)
  Contents(*lnum*) := *c*
  Mapped := *Mapped* ∪ {*lnum*}

ebm_read(*lnum*; *c*)
  if ¬ *lnum* ∈ *Mapped* **then**
    c := `empty`
  else
    c := *Contents*(*pnum*)

Fig. 2. Sequential specification of the erase block manager (EBM)

The implementation of EBM is given by the ASM WL together with a specification Blocks as a submachine. This refinement introduces the distinction between logical and physical blocks. Blocks allows reading and writing of physical blocks while WL is responsible for the mapping of logical to physical blocks. Furthermore, the wear leveling algorithm is implemented in WL.

To enable wear leveling each physical block in Blocks contains a header. This header stores which logical block is mapped to the physical block or if the block is currently unmapped (⊥).

$$\textbf{data } header = \texttt{mapped}(\texttt{blockno} : nat) \mid \perp$$
$$\textbf{data } block = \texttt{mkb}(\texttt{header} : header, \texttt{content} : content)$$

The state of Blocks is a function that maps physical block numbers to blocks. Initially all blocks are unmapped and empty.

$$\textbf{state } Blocks : nat \rightarrow block \qquad \textbf{initial state } Blocks = \lambda\, m.\ \texttt{mkb}(\perp, \texttt{empty})$$

The interface of Blocks as shown in Fig. 3 provides additional functionality to write and read the header of a physical block. Accessing the content of a block requires it to be mapped, i.e., the header of the block must not be ⊥. For wear leveling the interface also offers an interface operation **blocks_get_wl** that returns two physical blocks *from* and *to*, that are suitable for wear leveling. The actual decision is based on erase counts (also stored in block headers), but we leave the concrete implementation open here. To signal that wear leveling is currently unnecessary, the operation returns a block *from* with an unmapped header.

The operations of WL are depicted in Fig. 4. To avoid scanning the headers of all blocks, the state of WL maintains an in-memory mapping from logical block numbers to headers, which contain the corresponding physical block numbers if the logical block is mapped.

$$\textbf{state } LMap : nat \rightarrow header \qquad \textbf{initial state } LMap = \lambda\, n.\ \perp$$

blocks_write($pnum, c$)
pre $Blocks(pnum)$.header $\neq \bot$
　$Blocks(pnum)$.content $:= c$

blocks_read($pnum; c$)
pre $Blocks(pnum)$.header $\neq \bot$
　$c := Blocks(pnum)$.content

blocks_write_h($pnum, h$)
　$h := Blocks(pnum)$.header

blocks_read_h($pnum; h$)
　$Blocks(pnum)$.header $:= h$

blocks_map($; pnum$)
　choose m **with**
　　$Blocks(m)$.header $= \bot$
　in
　　$pnum := m$

blocks_get_wl($; from, to$)
　choose m_1, m_2 **with**
　　$Blocks(m_2)$.header $= \bot$
　　/* $\wedge\ m_1, m_2$ *are suitable*
　　　　for wear leveling */
　in
　　$from := m_1, to := m_2$

Fig. 3. Sequential specification of the physical block layer (`Blocks`)

Reading and writing of content delegates to the corresponding operations of `Blocks` by following *LMap*. If a logical block is unmapped, the write operation first maps this block to an unused physical block by writing a header and updating *LMap*. Therefore `Blocks` provides an operation **blocks_map** that returns a fresh block that can be mapped.

The wear leveling operation **wl_wear_leveling**, that is not visible to the clients, first requests a pair of blocks to be wear leveled by calling **blocks_get_wl**. If the *from* Block is mapped, its header and content are copied to the *to* Block and *LMap* is updated. We leave away many details here, that ensure, that crashing in the middle of wear leveling will result in a consistent state, see [23].

To prove the refinement WL $\sqsubseteq$ EBM three invariants are established in WL.

$injective(lmap) \leftrightarrow$
　$\forall\ n_1, n_2.\ lmap(n_1) \neq \bot \wedge lmap(n_2) \neq \bot \rightarrow lmap(n_1) \neq lmap(n_2)$
$lmapblocks(lmap, blocks) \leftrightarrow$
　$\forall\ n.\ lmap(n) \neq \bot \rightarrow blocks(lmap(n).\text{blockno}).\text{header} = \text{mapped}(n)$
$blockslmap(blocks, lmap) \leftrightarrow$
　$\forall\ m.\ blocks(m).\text{header} \neq \bot \rightarrow lmap(blocks(m).\text{header.blockno}) = \text{mapped}(m)$

The three predicates guarantee a valid mapping between logical and physical blocks. *injective* prohibits that two logical blocks are mapped to the same physical block, *lmapblocks* ensures that each mapped physical block in *lmap* points to the correct logical block, and *blockslmap* ensures that each mapped physical block also has a matching entry in *lmap*.

The abstraction relation between states of the specification and states of the implementation ensures that mapped blocks in *Mapped* conform with mapped logical blocks in *LMap* and that contents of *Contents* conform to the contents of the mapped physical blocks in *Blocks*.

$(\forall\ n.\ n \in Mapped \leftrightarrow LMap(n) \neq \bot)$
$\wedge\ (\forall\ n.\ n \in Mapped \rightarrow Contents(n) = Blocks(LMap(n).\text{blockno}).\text{content})$

```
wl_write(lnum, c)                              wl_wear_leveling()
  let pnum = 0 in                              internal
    if LMap(lnum) = ⊥ then                       let h = ⊥, c = empty,
      blocks_map(; pnum);                            from = 0, to = 0
      blocks_write_h(pnum, mapped(lnum));        in
      blocks_write(pnum, empty);                   blocks_get_wl(; from, to);
      LMap(lnum) := mapped(pnum);                  blocks_read_h(from; h);
    else                                         if h ≠ ⊥ then
      pnum := LMap(lnum).blockno;                  let lnum = h .blockno in
      blocks_write(pnum, c);                         blocks_read(from; c);
                                                     blocks_write_h(to, h);
  wl_read(lnum; c)                                   blocks_write(to, c);
    if LMap(lnum) = ⊥ then                          LMap(lnum) := mapped(to);
      c := empty;                                    blocks_write_h(from; ⊥) ;
    else
      let pnum = LMap(lnum).blockno in
        blocks_read(pnum; c);
```

Fig. 4. Sequential implementation of the wear leveling layer (WL)

Together with the invariants this is sufficient to prove a data refinement using forward simulation.

3   Adding Concurrency and Ownership

The sequential code calls the wear leveling operation at the end of every other operation. This causes small pauses in between operations. A better solution is to call wear leveling in a separate thread concurrently. This exploits that even the MTD hardware interface is capable of reading and writing different blocks concurrently. This is not possible for individual blocks, since these do not provide random access, but can be written sequentially only.

Adding concurrency implies that interface operations are now called concurrently by several threads, and it is natural to assume that they now have an *atomic* semantics (which is the natural semantics of ASMs, but was not required in a sequential context). We emphasize this, by writing EBM_{At} and $Blocks_{At}$ for EBM and Blocks with atomic semantics, although the machines are the same. Assuming an atomic semantics for the implementation is however unrealistic.

A simple solution that enforces an atomic semantics for an implementation is to use a single global mutex, that is set before each operation and released afterwards. Doing so for the operations of WL would however prevent wear leveling from running concurrent.

An implementation of Blocks that uses such a simple locking strategy would be correct to enforce atomicity, but too restrictive as it would prevent concurrent access to different blocks. It would also not be sufficient for the correctness of WL. To understand this, consider the implementation of **wl_write** in Fig. 4 and a potential interleaving of two concurrent executions of this operation as depicted in Fig. 5. Here two threads tid_1 and tid_2 write two contents to different logical blocks $lnum_1$ resp. $lnum_2$. Both logical blocks are unmapped so by calling **blocks_map** unmapped physical blocks are chosen to be mapped. Although the operation is atomic it is possible that for tid_2 the same physical block $pnum$ is returned as for tid_1 since tid_1 has not written the new header yet. Both threads would then write to the same physical block, first different headers that point to $lnum_1$ resp. $lnum_2$, then different contents c_2 resp. c_1. After both writes finish an inconsistent state is reached to the effect that the written data of tid_2 is lost and the injectivity of the block mapping is violated.

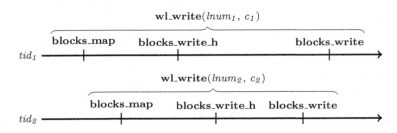

Fig. 5. Critical interleaving of two **wl_write** executions

A concept is needed that enforces on the level of *Blocks* that its implementation can assume that only one thread is writing each block at one time, and that headers are written by a single thread only.

The concept we use is that of threads *owning* data structures.

> **data** $owner$ = **readers**($tids : set\langle threadid\rangle$) | **writer**($tid : threadid$)
>
> **ghoststate** $OBlocks : nat \rightarrow owner$ $OHeaders : owner$

An owner can either own a data structure non-exclusively (typically for reading) or exclusively for writing. That a thread owns all headers or some block for reading or writing is specified as two ghost variables *OHeaders* and *OBlocks*. To ensure, that clients of the extended interface Blocks$_{\mathrm{Owns}}$ shown in Fig. 6 respect the ownership, we add preconditions to the operations, that request read-ownership for reading and write-ownership for writing blocks and headers. A thread that wants to call an operation of Blocks$_{\mathrm{Owns}}$ must now *acquire* ownership before it and can *release* ownership afterwards. For this purpose the interface is extended with two auxiliary acquire and release operations. These acquire and release full ownership, which is sufficient for the concurrent implementation of wear leveling given below. It is possible to add operations that acquire and release read-ownership too. Acquiring full ownership has the precondition that there is no current owner. If two threads now try to write the same block, one of them will violate the precondition of acquire (if it tries to acquire) or it will violate the precondition of writing (if it does not). But this is impossible, since submachine calls in implementations are checked to satisfy their preconditions.

data *mutex* = **free**   **locked**(tid : *threadid*)

blocks_acquire(*pnum*)
pre *OBlocks*(*pnum*) = readers(∅)
atomic ghost
  OBlocks(*pnum*) := writer(*tid*)

blocks_acquire_h()
pre *OHeaders* = readers(∅)
atomic ghost
  OHeaders := writer(*tid*)

blocks_write(*pnum*, c)
pre   *Blocks*(*pnum*).header ≠ ⊥
   ∧ *tid* ∈ *OBlocks*(*pnum*).writers
atomic
  Blocks(*pnum*).content := c

blocks_write_h(*pnum*, h)
pre   *tid* ∈ *OHeaders*.writers
   ∧ *tid* ∈ *OBlocks*(*pnum*).writers
atomic
  h := *Blocks*(*pnum*).header

blocks_get_wl(; *from*, *to*)
pre *tid* ∈ *OHeaders*.readers
atomic
  choose m_1, m_2 **with**
   Blocks(m_2).header = ⊥
   /* ∧ m_1 is good for WL */
  in
   from := m_1, *to* := m_2

blocks_release(*pnum*)
pre *tid* ∈ *OBlocks*(*pnum*).readers
atomic ghost
  OBlocks(*pnum*) := release(*tid*, *OBlocks*(*pnum*))

blocks_release_h()
pre *tid* ∈ *OHeaders*.readers
atomic ghost
  OHeaders := release(*tid*, *OHeaders*)

blocks_read(*pnum*; c)
pre   *Blocks*(*pnum*).header ≠ ⊥
   ∧ *tid* ∈ *OBlocks*(*pnum*).readers
atomic
  c := *Blocks*(*pnum*).content

blocks_read_h(*pnum*; h)
pre *tid* ∈ *OHeaders*.readers
atomic
  Blocks(*pnum*).header := h

blocks_map(; *pnum*)
pre *tid* ∈ *OHeaders*.readers
atomic
  choose m **with**
   Blocks(m).header = ⊥
  in
   pnum := m

Fig. 6. Atomic specification of the physical block layer with ownership (Blocks$_{\text{Owns}}$)

Calls to acquire and release in the augmented code of wear leveling will now ensure, that ownership is properly acquired. They are used for verification, but are "ghost code" that is eliminated when generating executable code.

To make sure, that calls to acquire never violate their precondition, we have to use locks in the extended implementation of WL given in Fig. 8. The simple implementation we give here just uses mutexes.

data *mutex* = **free** | **locked**(tid : *threadid*)

The locking and unlocking operations **mutex_lock** and **mutex_unlock** are specified as the atomic program statements given in Fig. 7. The definition of **mutex_lock** uses the program construct **atomic** φ { α }. The **atomic** construct blocks the current thread until its guard φ is satisfied. Immediately afterwards, the program α is executed in a single, indivisible step.

Figure 8 shows the result of applying sufficient locking and ownership acquisition to WL.
Additionally, each atomic step gets an individual label (W1–W18, R1–R8, and WL1–WL21) to give
assertions for this program point when reasoning about atomicity (see Sect. 5). We refer to
this concurrent implementation as WL$_{\text{Conc}}$. The
state of WL$_{\text{Conc}}$ is enhanced by a lock that protects the headers of all blocks, and locks for each
logical block that protects its contents.

```
mutex_lock(mutex)
   atomic (mutex = free) {
      mutex := locked(tid)
   }

mutex_unlock(mutex)
   pre mutex = locked(tid)
   mutex := free
```

Fig. 7. Mutex locking operations

$$\textbf{state} \quad \ldots \quad Lock : mutex \qquad Locks : nat \rightarrow mutex$$

We use mutexes for all locks, since they match our simplification of acquiring
write-ownership only. The actual Erase-Block-Manager in Flashix employs reader-
writer locks whenever parallel reading is unproblematic. The general locking concept of WL$_{\text{Conc}}$ is to acquire *Lock* only if the mapping from logical to physical blocks
needs to be updated. This is the case when writing to an unmapped block or when
wear leveling is active. Otherwise, locking only one individual *Locks(lnum)* of a
specific logical block *lnum* is sufficient. This lock protects the corresponding entry
LMap(lnum) of the block mapping as well as the content of the physical block
LMap(lnum).blockno. With this strategy multiple reads and writes to different,
mapped logical blocks are possible, even in parallel to wear leveling.

```
       wl_write(lnum, c)
W1       let pnum = 0 in
W2         mutex_lock(Lock);
W3         mutex_lock(Locks(lnum));
W4         if LMap(lnum) = ⊥ then
W5           blocks_acquire_h();
W6           blocks_map(; pnum);
W7           blocks_acquire(pnum);
W8           blocks_write_h(pnum, mapped(lnum));
W9           blocks_write(pnum, empty);
W10          blocks_release(pnum);
W11          LMap(lnum) := mapped(pnum);
W12          blocks_release_h();
         else
W13          pnum := LMap(lnum).blockno;
W14          mutex_unlock(Lock);
W15          blocks_acquire(pnum);
W16          blocks_write(pnum, c);
W17          blocks_release(pnum);
W18          mutex_unlock(Locks(lnum));


       wl_read(lnum; c)
R1       mutex_lock(Locks(lnum));
R2       if LMap(lnum) = ⊥ then
R3         c := empty
         else
R4         let pnum = LMap(lnum).blockno in
R5           blocks_acquire(pnum);
R6           blocks_read(pnum; c);
R7           blocks_release(pnum);
R8       mutex_unlock(Locks(lnum));
```

```
       wl_wear_leveling()
       internal
WL1      let h = ⊥, c = empty,
             from = 0, to = 0
       in
WL2        mutex_lock(Lock);
WL3        blocks_acquire_h();
WL4        blocks_get_wl(; from, to);
WL5        blocks_read_h(from; h);
WL6        if h ≠ ⊥ then
WL7          let lnum = h .blockno in
WL8            mutex_lock(Locks(lnum));
WL9            blocks_acquire(from);
WL10           blocks_read(from; c);
WL11           blocks_acquire(to);
WL12           blocks_write_h(to, h);
WL13           blocks_write(to, c);
WL14           LMap(lnum) := mapped(to);
WL15           blocks_write_h(from; ⊥);
WL16           blocks_release(to);
WL17           blocks_release(from);
WL18           mutex_unlock(Locks(lnum));
WL19       blocks_release_h();
WL20       mutex_unlock(Lock);
```

Fig. 8. Concurrent implementation of the wear leveling layer (WL$_{\text{Conc}}$)

One exception is that the *Lock* has
to be acquired in every **wl_write** execution (W2–W14 in Fig. 8), at least for a
short amount of time. This is due to
the locking hierarchy that is employed to
avoid deadlocks. When running in parallel, it is possible that a **wl_write** and
wl_wear_leveling may both need to
acquire *Lock* and the same *Locks(lnum)*,
so it must be ensured that those opera-

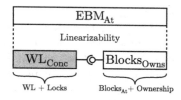

Fig. 9. Concurrency refinement of the
erase-block-manager

tions request the locks in the same order. Because **wl_wear_leveling** needs
to be owner of *OHeaders* to get suitable physical blocks at WL4 before a logical block can be locked, **wl_write** must request *Lock* (W2) ahead of requesting
Locks(lnum) (W3).

Figure 9 shows the resulting refinement of EBM_{At}. Proving $WL_{Conc} \sqsubseteq EBM_{At}$
using linearizability is discussed in detail in the next sections. It remains to
integrate the new "shifted" refinement into the refinement tower. The layers
above EBM_{At} can remain untouched since EBM_{At} is identical to EBM, and sequential
use of EBM_{At} is not problematic. Below $Blocks_{Owns}$ an adjustment is necessary:
a simple one is to use a global lock around the operations of its implementation.
Since the level is already close to the MTD hardware interface, the real solution
propagates ownership down to ownerships at the hardware level (where blocks
store a sequence of bytes instead of a header and content).

4   Linearizabilty and Atomicity Refinement

The standard correctness criterion we use to prove correctness of the refinement
of EBM_{At} to WL_{Conc} from Fig. 9 is *linearizability*. A formal definition can be found
in [15], we only give an informal description here.

A concurrent implementation CASM with nonatomic programs COP_i is linearizable to an atomic specification AASM with atomic operations AOP_i, if the
input/output behaviors of each concurrent run can be explained by mapping
them to the sequential input/output behavior of some sequential run of AASM.

The mapping between a concurrent and a
sequential run is as follows: for each concurrent call of an operation COP_i that is started
at time t_i and returns at time t_i' find some
point in time l_i with $t_i \le l_i \le t_i'$, such that
all l_i are different. The point is called the *linearization point* of the operation call. Then
construct some sequential run of AASM that
executes each corresponding abstract operation AOP_i atomically at time l_i. Note that
even for fixed linearization points this may
give several sequential runs if the abstract
operations are nondeterministic.

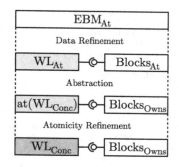

Fig. 10. Splitting the refinement

A refinement from AASM to CASM then is *linearizable*, if for every concurrent run linearization points and an abstract sequential run can be found, such that all operation calls *have the same inputs and outputs*.

The clients of the interface then cannot distinguish the concurrent run from one, where each operation call is delayed until time l_i, executes AOP_i atomically and then is delayed again until time t'_i.

Our proof technique will use an intermediate machine $at(WL_{Conc})$ that is the same as WL_{Conc}, but executes the code of each operation as one atomic step. This splits the refinement problem into three parts as shown in Fig. 10. The data refinement $WL_{At} \sqsubseteq EBM_{At}$, that we have already proved (since the ASMs are the same as WL and EBM). Second, a trivial refinement $at(WL_{Conc}) \sqsubseteq WL_{At}$ that *abstracts* from the locking/unlocking (and acquire/release) instructions in $at(WL_{Conc})$, since the overall effect of locking/unlocking in one atomic step is empty. Finally, the *atomicity refinement* $WL_{Conc} \sqsubseteq at(WL_{Conc})$, where both machines have the same data and operations, but different atomicity. Splitting the refinement from an atomic AASM to a concurrent CASM by using an intermediate $at(CASM)$, which executes the operations of CASM atomically, has the advantage that data refinement is completely decoupled from atomicity refinement.

The next section will describe a proof strategy for proving the atomicity refinement between $at(WL_{Conc})$ and WL_{Conc}, which is the new problem we get from adding concurrency to the refinement tower.

5 Proof Strategy for Atomicity Refinement

The proof strategy we use to prove atomicity refinement consists of two steps. First we prove that the concurrent runs of WL_{Conc} satisfy some assertions at all program points. These proofs use thread-local reasoning with the rely-guarantee calculus. They additionally ensure termination and deadlock-freedom, which are not implied by linearizability alone. Second we prove that based on the assertions, atomic program steps can be *reduced* to larger and larger atomic steps, until we arrive at $at(WL_{Conc})$. We sketch the basic strategy in the first subsection, and give results for the case study in Sect. 5.2.

5.1 Rely-Guarantee Proofs and Reduction

The variant of the rely-guarantee calculus used here is similar to the one given in [30], Section 5. The basic correctness statement[1] is of the form

$$pre \wedge I \rightarrow \langle R, G, I, run, \alpha \rangle \, post$$

---

[1] The notation in [30] is: α *sat* $(pre, R \wedge (I \rightarrow I(\underline{x}')), run, G \wedge (I \rightarrow I(\underline{x}')), post)$.

where program α is assumed to be the sequential program of some thread, that executes atomic steps. These alternate with environment steps, where one environment step is an arbitrary sequence of steps of other threads.

The program is assumed to use the state variables $\underline{x}$. Precondition pre, postcondition $post$, predicate run, and global invariant I are predicates over this state. The rely R and the guarantee G restrict environment and program steps. They are predicates over $\underline{x}$ and $\underline{x}'$ We write arguments in predicates if they differ from the standard ones only.

The formula asserts, that program α, when started in a state that satisfies precondition pre and global invariant I, will execute steps that satisfy G and preserve the invariant I, as long as all previous environment steps satisfy R and preserve I too. No program step will block, when at that time run holds. In addition, when all environment steps satisfy R and preserve I, then the program will either terminate and the final state will satisfy $post$, or it will stop in a blocked state where run is false.

The calculus to prove such formulas in KIV is based on symbolic execution. The basic rule to execute one atomic step at label L, that is annotated with an assertion φ_L is

$$
\frac{
\begin{array}{l}
pre \wedge I \rightarrow \varphi_L \wedge (run \rightarrow \varphi) \\
pre \wedge I \wedge \langle\alpha\rangle\ \underline{x} = \underline{x}' \rightarrow G(\underline{x}, \underline{x}') \wedge I(\underline{x}') \\
pre(\underline{x}_0) \wedge \langle\alpha(\underline{x}_0)\rangle\ \underline{x}_0 = \underline{x}_1 \wedge R(\underline{x}_1, \underline{x}) \wedge I(\underline{x}) \rightarrow \langle R, G, I, run, \beta\rangle\ post
\end{array}
}{
pre \wedge I \rightarrow \langle R, G, I, run, L : / * \varphi_L * / \ \mathbf{atomic}\ \varphi\ \{\alpha\}; \beta\rangle\ post
}
$$

The rule reduces the conclusion at the bottom to premises. The first premise states that before executing α the assertion at the initial label holds, and that the first step does not block (φ holds) whenever the run predicate is true.

The second premise uses the Dynamic Logic formula $\langle\alpha\rangle\ \underline{x} = \underline{x}'$ which asserts that the sequential program α has a terminating run that yields a state $\underline{x}'$. The premise ensures that the first atomic step of the program, which executes α is a step that satisfies G and preserves the invariant I.

The third premise continues symbolic execution with the rest of the program. Its precondition uses two sets $\underline{x}_0$ and $\underline{x}_1$ of fresh variables, to represent the two old states before and after the first atomic program step. The subsequent environment step from $\underline{x}_1$ to the current state $\underline{x}$ is assumed to satisfy R. Since rely steps preserve the invariant, it can be assumed for the current state again.

One common instance of the rule is a parallel assignment $\underline{y} := \underline{t}$, which can be viewed as an abbreviation for $\mathbf{atomic}\ true\ \{\underline{y} := \underline{t}\}$. In this case the formula $\langle\alpha\rangle\ \underline{x} = \underline{x}'$ reduces to $\underline{y}' = \underline{t} \wedge \underline{z}' = \underline{z}$, where $\underline{z}$ are the remaining variables from $\underline{x}$ that are not assigned.

The rules for other constructs like conditionals resemble the usual rules for symbolic execution of programs, except that similar to the rule above they have rely steps in between program steps and side conditions for assertions and guarantee. For loops, a loop invariant (that holds at the start of each iteration) and a variant, that decreases with a wellfounded order are needed. Proofs for recursive routines need wellfounded induction.

Individual rely-guarantee proofs for single threads can be combined to a rely-guarantee property of a concurrent system. The crucial property that needs to hold for this to work, is that the relies and guarantees must be *compatible*: the guarantee of each thread G_{tid} must imply the relies $R_{tid'}$ of other threads $tid' \neq tid$. For our state machines where all threads are known to execute the same operations, the guarantee can be chosen to be $G_{tid} := \bigwedge_{tid' \neq tid} R_{tid'}$, the weakest guarantee possible that is trivially compatible. The system is deadlock-free, if the disjunction of all $\bigvee_{tid} run_{tid}$ holds. When a mutex is used, run_{tid} is chosen to be $lock = locked(tid) \lor lock = Free$ which implies this condition. This easily generalizes to the hierarchy of locks used in the case study.

In summary, to verify assertions for a specification of a concurrent state machine with operations OP_i, the user has to provide an invariant I, a rely R_{tid} and a predicate $idle_{tid}$. The latter describes states, where a thread is not currently executing an operation. From these predicate logic proof obligations (e.g. the R must be reflexive, initial states satisfy the invariant etc.) are generated, together with the following rely guarantee proof obligation for each operation.

$$tid \neq tid', I, idle_{tid}, pre_{tid} \vdash \langle R_{tid}, R_{tid'}, I, run_{tid}, OP_i \rangle \, idle_{tid}$$

Successful verification guarantees that each of the assertions φ_L holds every time a thread reaches label L, that the operations terminate and that the implementation is deadlock-free.

The verified assertions are then used to combine atomic statements to larger ones following Lipton's [19] strategy of reduction. The idea is that a thread executing two atomic steps At_{L1} and At_{L2} (at labels L1 and L2) with an environment step in between is often equivalent to first executing the environment step, then At_{L1} and At_{L2} with no intermediate environment step. In this case the two steps can be merged together to form one atomic step.

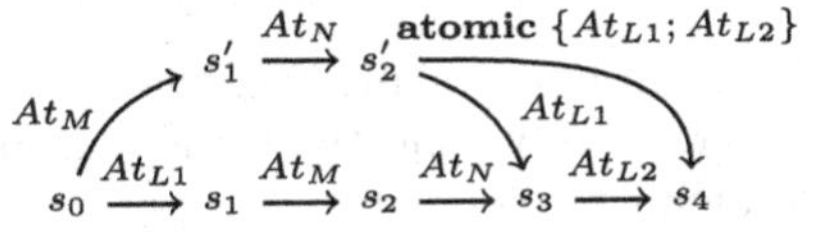

Reverting the order of first executing At_{L1} and then an environment step is possible, if all steps of other threads, that could be a part of the environment step, commute to the right with At_{L1}, in the sense that executing them in both orders gives the same final state. In this case

Fig. 11. At_{L1} commutes to the right of environment step At_M; At_N

At_{L1} is called a *right mover*. Analogous to this, a step that commutes to left with all steps is called a *left mover*. Figure 11 shows an example, where the environment step consists of two steps At_M and At_N of other threads. The original run is shown at the bottom, the alternative run which allows executing At_{L1} and At_{L2} as one atomic step at the top. The intermediate states of the runs are different, but they reach the same final state.

The atomic steps of the programs can all be written in the form

$$At_L \equiv L : / * \varphi_L * / \textbf{ atomic } \varepsilon_L \{\alpha_L\}$$

where L is the label, and φ_L the assertion established. The guard ε_L is **true** for all statements, except locking instructions, cf. Figure 7. Program α_L is either an assignment, or the call of a submachine operation. For a conditional or a while loop with test δ, α_L is defined to be $b := \delta$ using a fresh variable b, while binding a local variable **let** $y = t$ **in** . . . gives $\alpha_L \equiv \{y := t\}$. The formal condition for At_{L1} to commute to the right with At_{L2} executed by another thread is

$$\varphi_L \wedge \varphi'_M \wedge \varepsilon_L \wedge \varepsilon'_M \wedge tid \neq tid' \wedge \langle \alpha_L ; \alpha'_M \rangle \, \underline{x} = \underline{x}_0 \rightarrow \langle \alpha'_M ; \alpha_L \rangle \, \underline{x} = \underline{x}_0 \qquad (1)$$

In the formula, $\varphi'_M, \varepsilon'_M, \alpha'_M$ are variants, that rename thread local variables used in At_M to new, primed variables disjoint from the shared state and the local variables of At_L. The criterion critically uses the assertions at both labels, since they often show that the preconditions of the implication contradict each other, trivializing the proof. If, for example the two steps are both in a region where a common lock is needed, they commute trivially: φ_L implies $lock = locked(tid)$, while φ'_M implies $lock = locked(tid')$, so the proof obligation trivially holds. A general result is that locking is always a right mover, while unlocking is always a left mover.

Combining steps to larger steps can be translated into rules for making statements like sequential composition, conditionals and loops atomic, when their parts are atomic already. We use rules similar to the reduction rules given in [10]. Iterated application gives larger and larger atomic blocks. Ideally, the final result is that the whole concurrent program of one operation has been combined into a single atomic step. If this is possible, then a linearizability proof becomes trivial, as the linearizability point then simply *is* the single atomic step.

5.2   Proving the Case Study

The main task for proving the atomicity refinement of the case study is to find assertions, rely conditions and a global invariant that are strong enough to allow atomicity refinement.

The rely conditions are derived from the crucial ideas what data structures are protected from being changed, when thread tid has a certain lock or ownership. This results in the following clauses.

$$tid \in OHeaders.\textbf{readers} \rightarrow \forall\, m.\ Blocks(m).\textbf{header} = Blocks'(m).\textbf{header}$$
$$tid \in OBlocks(m).\textbf{readers} \rightarrow Blocks(m) = Blocks'(m)$$
$$Lock = \texttt{locked}(tid) \rightarrow LMap' = LMap$$
$$Locks(n) = \texttt{locked}(tid) \rightarrow LMap'(n) = LMap(n)$$
$$Locks(n) = \texttt{locked}(tid)$$
$$\rightarrow \forall\, m.\ Blocks(m).\textbf{header} = \texttt{mapped}(n) \leftrightarrow Blocks'(m).\textbf{header} = \texttt{mapped}(n)$$

The only rely that is somewhat difficult to find is the last one: if a thread locks logical block n, then other threads are not allowed to change the block header to point to or to point away from n.

The global invariant and the assertions are derived from several sources. First, ownership as used in the interface $\text{Blocks}_{\text{Owns}}$ has to be compatible with the use of locks.

$$OHeaders \subseteq Lock.\texttt{owner} \tag{2}$$

$$\forall\ m.\quad Blocks(m).\texttt{header} \neq \bot$$
$$\rightarrow OBlocks(m) \subseteq Locks(Blocks(m).\texttt{header.blockno}).\texttt{owner} \tag{3}$$

$$\forall\ m.\ Blocks(m).\texttt{header} = \bot \rightarrow OBlocks(m) \subseteq Lock.\texttt{owner} \tag{4}$$

The invariant (2) states that headers are owned only if the lock has been taken. Invariant (3) states that a mapped physical block m can be owned (and therefore changed) only if the corresponding logical block that is stored in its header is locked. For unmapped blocks property (4) states that they can be owned only if WL_{Conc} has taken the header lock.

Second, the three global invariants of the sequential code are relevant. Dropping them completely would result in illegal states where e.g. the block mapping is no longer injective. However, the invariants of the sequential verification are only guaranteed to hold in idle states, where no thread is running. So it is necessary to give weaker assertions for intermediate states, that are still sufficient to avoid illegal ones.

For the given case study, it turns out that *lmapblocks* and *injective* are preserved by all steps, but that *blockslmap* does not hold while the headers are locked. As a result the global invariant can include *blockslmap(Blocks, LMap)* only when the headers are currently not owned ($Oheaders = readers(\emptyset)$). To establish this assertion, after a step that releases $OHeaders$, assertions have to be given for all labels, where $OHeaders$ is taken. For writing the predicate is violated between line W9 after the header of block *pnum* has been set to *lnum* and line W11, where $LMap(lnum)$ is set to *pnum*. For all lines in this range *blockslmap(Blocks, LMap(lnum; pnum))* holds: if $LMap$ were already updated, then *blockslmap* would hold. The wear leveling algorithm gives similar assertions for the range WL13–WL15.

Finally, assertions are sometimes necessary for the code after a test or after assignments to a variable. In a purely sequential setting, the test for $LMap(lnum) \neq \bot$ at R2 ensures that this formula holds, until the subsequent **let** binding $pnum = LMap(n).\texttt{blockno}$ at line R4, which will ensure $pnum = LMap(lnum).\texttt{blockno}$ when the variable *pnum* is used later on. However, in the concurrent setting $LMap$ may be assigned by other threads, destroying each of these properties. In the given case, the rely conditions are strong enough to propagate the formulas, so we assert that at line R4 the first formula holds, while for lines R5–R7 the second holds. A number of similar assertions are needed for other local variables.

Proving the rely-guarantee proof obligations for the individual programs requires the main effort in proving the concurrent setting correct. This is in line with case studies we have done for lock-free algorithms [25, 27–29], where proving rely-guarantee assertions caused the main effort too.

After establishing assertions for all program points, the program can then be reduced, combining atomic steps to larger ones. This requires to find out, which steps are left or right movers (or both). The current strategy implemented in KIV does simple syntactic checks to check whether the resulting commutativity requirement (1) is trivial: either the accessed variables are disjoint, or the preconditions of the proof obligation trivially reduce to false. Otherwise it is possible to generate proof obligations, by manually asserting that certain steps (identified by their label) are left or right movers (or both).

For the case study, manual specifications of mover types are currently necessary for the atomic calls **blocks_acquire** (right mover) and **blocks_release** (left mover) of $Blocks_{At}$. The reader may check, that this trivially implies that the other operations of $Blocks_{At}$ are left and right movers. After the mover types have been determined, the reduction rules are then applied automatically, to form maximally large atomic blocks.

This immediately results in a single atomic block for **wl_write** and **wl_read**. Reducing **wl_wear_leveling** creates three atomic blocks. The first ends at the conditional at line WL6 and is a right mover. The second is for the **let**-block WL7–WL19. The third is for the last two lines WL20–WL21, and is a left mover. The conditional cannot be reduced, since its then-branch requires the lock for block *lnum* to be free, while the empty else-branch does not have this guard. With the atomic blocks now being much larger than before, it becomes possible to prove much stronger invariants that just hold in between blocks, but did not hold for the original programs. In particular, since all locking and unlocking of blocks is now within atomic regions, the simple invariant that all *Locks*(*lnum*) are always free can be established using another simple rely-guarantee proof. With the new invariant established, another reduction step finds, that the conditional at line WL6 can now be reduced to an atomic block. Together with the initial and the final block being right resp. left movers already, the wear leveling code is combined by another reduction step into a single step. This implies that the concurrent implementation of wear leveling is indeed linearizable and a correct refinement.

6   Related Work

Related work on wear leveling and the flash file system we have developed has already been given in [23], where the full version of the sequential wear leveling algorithm has been specified.

This paper is based on the PhD of Jörg Pfähler [21], where concurrency was added to the full wear leveling algorithm. The full version needs to add ownership annotations and locks to several refinements. This version is now used in our actual flash file system implementation. The PhD also contains extensions that allow verifying crash-safety, which we could not address in this paper.

The flash file system by Damchoom et al. [7] has concurrent wear leveling. The synchronization between threads is implicitly performed by the semantics of Event-B models, i.e., an event in an Event-B model is always executed atomically, and not explicitly via locks or other synchronization primitives. This makes the step to actual running code more difficult and less straightforward. The full erase block management used in our flash file system is also more general, because it does not use additional bits of out-of-band data of an erase block.

Verification of concurrent, lock-based systems is of course a very broad topic with lots of important contributions, and the proof techniques we use are from this field. We are not aware of other formal methods that specifically address the question of this paper: how to add concurrency a posteriori to an existing modular, sequential system, without having to prove the system from scratch. Adding concurrency to components of an existing software system to increase efficiency is however a recurring software engineering task that should be supported by formal methods.

Refinement and abstraction of atomicity is quite common for concurrent systems, and many refinement definitions for concurrent systems like [1] or [20] address refinements of atomicity. The refinement calculus of Back [3] uses the opposite direction. It starts out with an atomic program and splits it into smaller actions in refinement steps.

The calculus of atomic actions due to Elmas et al. [10] is an extension of Lipton's [19] original approach for highly concurrent, linearizable programs. It provides a more incremental verification methodology than the calculus given here for highly concurrent systems and its implementation is better automated. The assertions and invariants are incrementally validated in [10], whereas here a rely/guarantee proof is used to validate them before applying any reductions. The rules of the calculus in [10] address partial correctness, so termination would have to be proven differently. Nevertheless, many of the reduction rules given in this paper are directly used in our approach too.

Ownership annotations are used in the C verifier VCC [6] and Spec# [16] in order to ensure data-race freedom of the code. They are typically coupled to objects of the programming language, while we decouple the use of ownership from objects. Fractional permissions [5] in concurrent versions of separation logics [24] serve a similar purpose as ownership. These are for example supported by the C code verifier VeriFast [17].

7   Conclusion

We have presented an approach for adding concurrency to an existing refinement tower. The given approach allows to add concurrency by enhancing some of the components of the refinement tower. Abstract interfaces are extended with acquire and release operations, that specify allowed concurrency. In our case study concurrent writes on different blocks are possible, while concurrent writes on the same block are disallowed. Concurrent code using these interfaces is then possible, that enhances the existing sequential code with suitable locking strategies. We have evaluated this strategy of "shifting parts of the refinement" tower by making wear-leveling concurrent in the Flashix file system. Specifications using the same concept have been defined for concurrent garbage collection, with executable code already running. Verification is work in progress. We also work on a allowing concurrent calls for POSIX file system operations.

References

1. Abadi, M., Lamport, L.: The existence of refinement mappings. Theoret. Comput. Sci. **2**, 253–284 (1991). Also appeared as SRC Research Report 29
2. Abrial, J.-R.: Modeling in Event-B - System and Software Engineering. Cambridge University Press, Cambridge (2010)
3. Back, R.J.R.: A method for refining atomicity in parallel algorithms. In: Odijk, E., Rem, M., Syre, J.-C. (eds.) PARLE 1989. LNCS, vol. 366, pp. 199–216. Springer, Heidelberg (1989). https://doi.org/10.1007/3-540-51285-3_42
4. Börger, E., Stärk, R.F.: Abstract State Machines—A Method for High-Level System Design and Analysis. Springer, Heidelberg (2003). https://doi.org/10.1007/978-3-642-18216-7
5. Boyland, J.: Checking interference with fractional permissions. In: Cousot, R. (ed.) SAS 2003. LNCS, vol. 2694, pp. 55–72. Springer, Heidelberg (2003). https://doi.org/10.1007/3-540-44898-5_4
6. Cohen, E., et al.: VCC: a practical system for verifying concurrent C. In: Berghofer, S., Nipkow, T., Urban, C., Wenzel, M. (eds.) TPHOLs 2009. LNCS, vol. 5674, pp. 23–42. Springer, Heidelberg (2009). https://doi.org/10.1007/978-3-642-03359-9_2
7. Damchoom, K., Butler, M.: Applying event and machine decomposition to a flash-based filestore in Event-B. In: Oliveira, M.V.M., Woodcock, J. (eds.) SBMF 2009. LNCS, vol. 5902, pp. 134–152. Springer, Heidelberg (2009). https://doi.org/10.1007/978-3-642-10452-7_10
8. de Roever, W., Engelhardt, K.: Data Refinement: Model-Oriented Proof Methods and their Comparison. Cambridge Tracts in Theoretical Computer Science, vol. 47. Cambridge University Press, Cambridge (1998)
9. Derrick, J., Boiten, E.: Refinement in Z and in Object-Z: Foundations and Advanced Applications. FACIT. Springer, Heidelberg (2001). https://doi.org/10.1007/978-1-4471-5355-9. Second, revised edition 2014
10. Elmas, T., Qadeer, S., Tasiran, S.: A calculus of atomic actions. In: Proceeding POPL 2009, pp. 2–15. ACM (2009)
11. Ernst, G., Pfähler, J., Schellhorn, G., Haneberg, D., Reif, W.: KIV - overview and verifythis competition. Softw. Tools Techn. Transf. **17**(6), 677–694 (2015)
12. Ernst, G., Pfähler, J., Schellhorn, G., Reif, W.: Inside a verified flash file system: transactions & garbage collection. In: Gurfinkel, A., Seshia, S.A. (eds.) VSTTE 2015. LNCS, vol. 9593, pp. 73–93. Springer, Heidelberg (2015). https://doi.org/10.1007/978-3-319-29613-5_5
13. Ernst, G., Pfähler, J., Schellhorn, G., Reif, W.: Modular. Crash-Safe Refinement for ASMs with Submachines. Science of Computer Programming (SCP) (2016)
14. He, J., Hoare, C.A.R., Sanders, J.W.: Data refinement refined resume. In: Robinet, B., Wilhelm, R. (eds.) ESOP 1986. LNCS, vol. 213, pp. 187–196. Springer, Heidelberg (1986). https://doi.org/10.1007/3-540-16442-1_14
15. Herlihy, M.P., Wing, J.M.: Linearizability: a correctness condition for concurrent objects. ACM Trans. Program. Lang. Syst. (TOPLAS) **12**(3), 463–492 (1990)

16. Jacobs, B., Leino, K.R.M., Piessens, F., Schulte, W.: Safe concurrency for aggregate objects with invariants. In: Software Engineering and Formal Methods (SEFM) 2005, pp. 137–146. IEEE (2005)
17. Jacobs, B., Smans, J., Philippaerts, P., Vogels, F., Penninckx, W., Piessens, F.: VeriFast: a powerful, sound, predictable, fast verifier for C and Java. NASA Formal Methods **6617**, 41–55 (2011)
18. KIV proofs for wear leveling (2020). https://kiv.isse.de/projects/WearLeveling. html
19. Lipton, R.J.: Reduction: a method of proving properties of parallel programs. Commun. ACM **18**(12), 717–721 (1975)
20. Lynch, N., Vaandrager, F.: Forward and backward simulations - part i: untimed systems. Inf. Comput. **121**(2), 214–233 (1995). Also: Technical Memo MIT/LCS/TM-486.b, Laboratory for Computer Science, MIT
21. Pfähler, J.: A modular verification methodology for caching and lock-based concurrency in file systems. Ph.D. thesis, Universität Augsburg, Fakultät für Informatik (2018). https://opus.bibliothek.uni-augsburg.de/opus4/frontdoor/index/in dex/docId/41890
22. Pfähler, J., Ernst, G., Bodenmüller, S., Schellhorn, G., Reif, W.: Modular verification of order-preserving write-back caches. In: Polikarpova, N., Schneider, S. (eds.) IFM 2017. LNCS, vol. 10510, pp. 375–390. Springer, Cham (2017). https://doi. org/10.1007/978-3-319-66845-1_25
23. Pfähler, J., Ernst, G., Schellhorn, G., Haneberg, D., Reif, W.: Formal specification of an erase block management layer for flash memory. In: Bertacco, V., Legay, A. (eds.) HVC 2013. LNCS, vol. 8244, pp. 214–229. Springer, Cham (2013). https:// doi.org/10.1007/978-3-319-03077-7_15
24. Reynolds, J.C.: Separation logic: a logic for shared mutable data structures. In: Proceedings of 17th Annual IEEE Symposium on Logic in Computer Science, pp. 55–74. IEEE (2002)
25. Schellhorn, G., Derrick, J., Wehrheim, H.: A sound and complete proof technique for linearizability of concurrent data structures. ACM Trans. Comput. Logic **15**(4), 31:1–31:37 (2014)
26. Schellhorn, G., Ernst, G., Pfähler, J., Haneberg, D., Reif, W.: Development of a verified flash file system. In: Ait Ameur, Y., Schewe, K.D. (eds.) ABZ 2014, vol. 8477. LNCS, pp. 9–24. Springer, Heidelberg (2014). https://doi.org/10.1007/978-3-662-43652-3_2
27. Schellhorn, G., Travkin, O., Wehrheim, H.: Towards a thread-local proof technique for starvation freedom. In: Ábrahám, E., Huisman, M. (eds.) IFM 2016. LNCS, vol. 9681, pp. 193–209. Springer, Cham (2016). https://doi.org/10.1007/978-3-319-33693-0_13
28. Tofan, B., Schellhorn, G., Reif, W.: Formal verification of a lock-free stack with hazard pointers. In: Cerone, A., Pihlajasaari, P. (eds.) ICTAC 2011. LNCS, vol. 6916, pp. 239–255. Springer, Heidelberg (2011). https://doi.org/10.1007/978-3-642-23283-1_16
29. Tofan, B., Travkin, O., Schellhorn, G., Wehrheim, H.: Two approaches for proving linearizability of multiset. Sci. Comput. Program. **96**(P3), 297–314 (2014)
30. Xu, Q., de Roever, W.-P., He, J.: The rely-guarantee method for verifying shared variable concurrent programs. Formal Aspects Comput. **9**(2), 149–174 (1997)

Regular Research Articles

Diverse Scenario Exploration in Model Finders Using Graph Kernels and Clustering

Robert Clarisó[1]([✉])[iD] and Jordi Cabot[2][iD]

[1] Universitat Oberta de Catalunya (UOC), Barcelona, Spain
rclariso@uoc.edu
[2] ICREA, Barcelona, Spain
jordi.cabot@icrea.cat

Abstract. Complex software systems can be described using modeling notations such as UML/OCL or Alloy. Then, some correctness properties of these systems can be checked using model finders, which compute sample scenarios either fulfilling the desired properties or illustrating potential faults. Such scenarios allow designers to validate, verify and test the system under development.

Nevertheless, when asked to produce several scenarios, model finders tend to produce similar solutions. This lack of diversity impairs their effectiveness as testing or validation assets. To solve this problem, we propose the use of *graph kernels*, a family of methods for computing the (dis)similarity among pairs of graphs. With this metric, it is possible to *cluster* scenarios effectively, improving the usability of model finders and making testing and validation more efficient.

Keywords: Model-driven engineering · Verification and validation · Testing · Graph kernels · Clustering · Diversity

1 Introduction

The structure and behavior of a software system can be described by means of *software models*, using notations such as Alloy [10], graph-based formalisms [20] or UML/OCL [17]. These notations describe software systems at a high level of abstraction, hiding implementation details while preserving its salient features. Analysing these models can reveal complex faults in the underlying systems.

In this analysis, the key assets for checking the correctness of software models are *model finders* [8], tools capable of computing *instances* of a model that satisfy a set of constraints and properties of interest. Each model finder targets

This work is partially funded by the H2020 ECSEL Joint Undertaking Project "MegaM@Rt2: MegaModelling at Runtime" (737494) and the Spanish Ministry of Economy and Competitivity through the project "Open Data for All: an API-based infrastructure for exploiting online data sources" (TIN2016-75944-R).

© Springer Nature Switzerland AG 2020
A. Raschke et al. (Eds.): ABZ 2020, LNCS 12071, pp. 27–43, 2020.
https://doi.org/10.1007/978-3-030-48077-6_3

a particular modeling notation and uses a different reasoning engine, like search-based methods [1,24], SAT [10], SMT [24,28] or constraint programming [3].

For verification purposes, it is usually enough to search for one instance, which either proves or disproves the property of interest. However, for testing and validation purposes several instances are usually required to increase our confidence in the correctness of the model. It is highly desirable that those instances exhibit *diversity*, *i.e.*, distinct configurations of the system and interesting corner cases [11]. Lack of diversity may make validation and testing more time consuming, as the analysis includes almost-duplicate instances that do not provide added value; and less effective, as the sample of instances may fail to include relevant scenarios.

Nevertheless, most model finders focus on efficiency and expressiveness of the input modeling notation, so few of them ensure diversity of the generated instances [6,11,20,23,26]. In these few, diversity assurance is integrated into the solver: it guides the search process to look for diverse instances. However, this integration makes it harder to transfer the proposed methods to other solvers and notations. Thus, designers are limited in terms of expressiveness (*e.g.*, no support for integer or string attributes [11,20,26] or dynamic properties [6,11,23,24]) and cannot benefit from additional features provided by others model finders (*e.g.*, computation of minimal instances [16] or support for max-satisfiability [28]).

This paper proposes a method for distilling diverse instances in the model finder output based on the use of *clustering*. Instances are classified into categories according to their *similarity*, which is calculated using information about their *structure* (the existing objects and the links between them), *typing* (the specific type of each object) and *attribute values*. This calculation is based on the use of *graph kernels*, a family of methods for computing distances among graphs. Selecting a representative instance from each category ensures diversity while reducing testing and validation time, as redundant instances can be safely discarded. As a drawback, this method does not *force* the model finder to look for diverse instances, it only distills the most diverse ones.

Compared with related works, our approach offers the following advantages:

- It is independent of the solver used by the model finder (SAT, SMT, ...) and the modeling notation being analyzed (Alloy, UML/OCL, ...).
- It does not require manual intervention from the designer to define what kind of instances are "relevant" or when two instances are "similar".
- The similarity computation can be customized, *e.g.*, by selecting a trade-off between precision and accuracy.

Paper Organization. The remainder of the paper is structured as follows. Section 2 presents an overview of the method illustrated with a simple example. Then, we describe the three steps of our method: the *abstraction* process for transforming instances into graphs (Sect. 3); *graph kernels* (Sect. 4), the framework for computing similarities among graphs; and *clustering* algorithms that can use this similarity to build groups of related instances (Sect. 5). Section 6 presents some experimental results of the application of this method. After that,

Sect. 7 describes previous work on diversity and model finding. Finally, Sect. 8 outlines the conclusions and lines for future work.

2 Method Overview

The overview of our approach for identifying diverse instances in model finder output is depicted in Fig. 1. Our **input** is a set of instances computed by a model finder, and our **output** is a set of clusters grouping those instances according to their similarity. From this output, it is possible to select a representative instance for each cluster, *e.g.*, choosing the smallest instance.

The method can be divided into three steps:

1. **Graph abstraction**: First, each instance is abstracted as a labeled graph, where labels store type and attribute value information and the underlying graph captures the objects and the links among them.
2. **Graph kernel:** Then, the pairwise similarity among the n graphs is computed using a state-of-the-art labeled graph comparison technique. The result of this computation is a $n \times n$ matrix S where each cell S_{ij} provides information about the similarity between graphs i and j.
3. **Clustering:** Finally, the similarity data is used by a clustering procedure to classify instances into groups of similar instances. The most suitable number of groups is determined by using *clustering validity indices*, which measure whether elements in the cluster are similar to each other and different from elements in other clusters.

To illustrate how the method works and the type of results it can achieve, we will use the UML class diagram in Fig. 2(a). This model describes the relationships between employees who work in or lead a department. There are two constraints regarding the salary, defined as OCL invariants: all salaries must be below a salary threshold and also below the salary of the department's director.

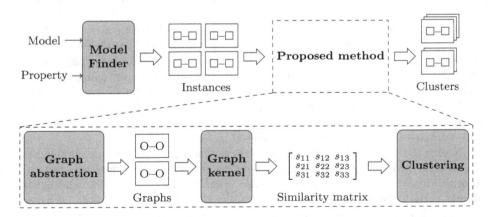

Fig. 1. Overview of the method presented in this paper.

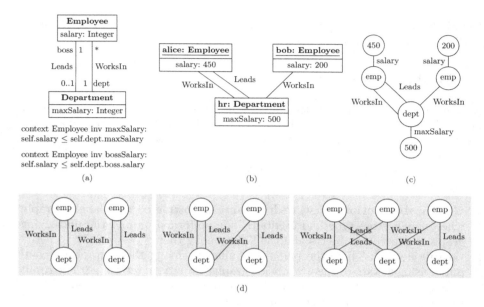

Fig. 2. Motivating example: (a) UML/OCL class diagram; (b) Sample instance; (c) Encoding of the instance as a labeled graph; (d) Graph shapes of the three clusters.

To be usable in practice, this model should be *strongly satisfiable* [3]: it should have some instance where all integrity constraints are satisfied with each class having a non-empty population. In our example, the class diagram is satisfiable and a potential solution is the instance shown in Fig. 2(b). Instances like this can then be used for validating and testing the UML/OCL model.

We have used the USE Model Validator [12] to generate 25 valid instances for this model. By manually inspecting these instances, we can easily realize that most of them are very similar. A designer would be interested in a smaller and more diverse set of instances that gives the same or even more information as the 25 original ones. We explain next how this can be achieved with our method.

Applying our method, each object diagram is abstracted as a labeled graph. As an example, Fig. 2(c) shows the abstraction for the object diagram in Fig. 2(b). We then apply hierarchical clustering to our 25 graphs using the similarity information provided by a graph kernel algorithm. From the results, validity indices recommend choosing 3 clusters. Thus, we have discovered that out of the 25 instances, there are only 3 types of solutions worth considering. The common pattern in each cluster is depicted in Fig. 2(d).

Notice that one cluster identified by our method (the middle one) highlights a potential problem in the model: a department where the director works in another department. This is a corner case worth studying, to decide whether it should actually be allowed or it is a mistake in the model that needs to be fixed.

The following sections describe the different phases of our approach in detail.

3   Graph Abstraction

Depending on the model finder, instances have a different structure, *e.g.*, an object diagram, an enriched graph or a set of tuples. In order to take advantage of off-the-shelf graph comparison algorithms, we translate these instances into labeled undirected graphs. To this end, we define the vertices, edges and labels in the graph in terms of the original instance.

Intuitively, the vertices of the graph will describe the object elements in the instance, while the edges will describe the relationships among them. Labels are integer values assigned to vertices. Labels will be used to describe information such as the type of each element or the values of attributes that can, later on, help to establish whether a pair of vertices from two different graphs can be considered "equivalent".

The complexity of this step depends on the kind of output provided by the model finder. Our approach provides a specific solution for each type of output. As shown in Fig. 2, the abstraction of object diagrams is straightforward according to this pattern: objects and attributes becomes vertices, links become edges, and types and attribute values become labels. Similarly, the mapping from instances in graph-based modeling notations is also trivial: the vertices and edges of the original graph are preserved while the type of each element is used as a label for the corresponding vertex. Nevertheless, the transformation from the relational notation used by Alloy is more involved. Thus, we devote the remainder of this Section to formalize the abstraction of Alloy instances.

Alloy Models. An Alloy specification is defined as a collection of *signatures* and *constraints*, followed by a *command*.

Signatures (**sig**) describe the data in the model. Each signature has a unique name and represents a set of atoms, the base individuals in Alloy's logic. Signatures can have *fields* which take values for each atom of the signature. These values can be basic data types like integers, other signatures or complex values like functions or sets. Internally, these values are managed as *relations*, collections of tuples with the same *arity* (number of elements).

It is possible to define a hierarchy among signatures (**extends**). Moreover, fields and signatures may have *multiplicity constraints* limiting their population, *e.g.*, **one** or **lone** (zero or one). In addition to user-defined signatures, Alloy provides some *built-in signatures* to describe common data types such as booleans, integers, strings or sequences.

Regarding constraints, there are different types of constraint: *facts* (**fact**) describe invariants that should always hold; *assertions* (**assert**) state desired properties that should be checked; and *predicates* (**pred**) are reusable constraints where some elements are passed as parameters. Each constraint can be defined using a mixture of logical operators (*e.g.*, **and**, **not** or **implies**), relational operators (*e.g.*, dot join or transpose) and quantifiers (*e.g.*, **all** or **some**).

Finally, commands instruct the solver which constraint should be analyzed and the *scope* (number of atoms) that should considered for each signature.

Command `check` searches for a counterexample of an assertion, while command `run` searches for an example of a predicate.

Alloy Snapshots. Executing a command with the Alloy Analyzer may yield two outcomes: either no instance within the scope satisfies the constraints or an instance has been found. Instances are called *snapshots* in the Alloy terminology.

An Alloy snapshot is defined by the following elements:

- A list of signatures, including both built-in and user-defined signatures.
- A list of relations, each one with a fixed arity n.
- A list of *free variables* in the model, *e.g.*, parameters of predicates and existentially quantified variables.
- For each signature, a set of atoms.
- For each relation with arity n, a set of tuples of n atoms.
- For each free variable with arity n, a *witness*, *i.e.*, a set of tuples of n atoms.

That is, when checking for a property with existential quantifiers, Alloy not only answers whether it is satisfied or not: if it holds, it also computes for which specific value of the quantified variable (the witness) the property holds.

From Snapshots to Graphs. We need to define how to translate: (1) built-in signatures, (2) user-defined signatures and (3) relations. As witnesses are a special type of relation, we do not need to treat them separately.

Regarding built-in signatures, we need to make sure that each value will be given the same label in different snapshots: an integer like 7 and a string like "John" should be considered equal among different snapshots. Thus, the first step is traversing the set of snapshots being abstracted to construct a *vocabulary of values*. In this way, we compute a *unique label* for each value of a basic type.

1. **Built-in signatures:** We create a vertex for each atom in these signatures, plus a vertex for each built-in value (string, integer or sequence) used in the model. We label each vertex with the unique label for that built-in value.
2. **User-defined signatures:** We create a vertex for each atom. It is labeled with its signature, *i.e.*, the innermost signature in the signature hierarchy where it belongs.
3. **Relations:** We create a vertex v for each tuple, labeled with the name of the relation. Then, for each i-th element in the tuple, we create a vertex[1] labeled with i connected to both v and the vertex of the corresponding value.

Figure 3 shows an example of this abstraction process. The Alloy model in Fig. 3(a) describes a DNS server lookup process. We want to validate the potential scenarios in this process, for instance, whether two names may resolve to the same IP address. To do that, Alloy finds example instances, highlighting the offending names (n1 and n2) and DNS (d). Figure 3(b) and (c) show one sample Alloy instance in textual and graphical format. The corresponding graph

---

[1] The intermediate vertex is omitted when the position i can be inferred: no other position in the relation has a compatible signature, *i.e.*, with a common supertype.

abstraction is depicted in Fig. 3(d). For clarity, vertices are depicted in a different shape according to their origin: circles for atoms; rectangles for relations (white) and positions within relations (grayed); and hexagons for witnesses.

Abstraction and Diversity. Some approaches aimed at achieving diversity use *uniform sampling* [5,14,15,18] as their goal: achieving a uniform distribution among solutions. Nevertheless, the desired notion of diversity may be more complex (a target probability distribution, a partition into meaningful classes), and specific to a domain or even a particular problem [6,24]. In the following, we discuss how this information about the desired type of diversity can be integrated in the graph abstraction process with very few changes.

For example, let us consider the specification of a banking system. From our domain knowledge, it seems reasonable to think that the name of the owner an account is not very relevant: if there are 10 clients in our system, the fact

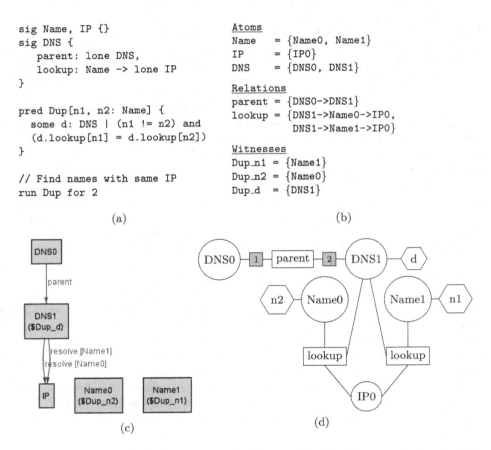

```
sig Name, IP {}
sig DNS {
    parent: lone DNS,
    lookup: Name -> lone IP
}

pred Dup[n1, n2: Name] {
    some d: DNS | (n1 != n2) and
    (d.lookup[n1] = d.lookup[n2])
}

// Find names with same IP
run Dup for 2
```

(a)

```
Atoms
Name  = {Name0, Name1}
IP    = {IP0}
DNS   = {DNS0, DNS1}

Relations
parent = {DNS0->DNS1}
lookup = {DNS1->Name0->IP0,
          DNS1->Name1->IP0}

Witnesses
Dup_n1 = {Name1}
Dup_n2 = {Name0}
Dup_d  = {DNS1}
```

(b)

(c)

(d)

Fig. 3. Example of graph abstraction: (a) Alloy model; (b) Alloy snapshot in textual format; (c) Alloy snapshot depicted graphically; (d) Abstracted graph.

that all of them are called "John Smith" might not be problematic. Thus, the name of the owner could be abstracted away in our graph representation, *i.e.* remove from the graph the vertices related to this particular attribute. On the other hand, focusing on the balance of an account, we might be interested in considering accounts with a positive, negative and zero balance. In this case, we are not interested in specific values for the balance, only if they fit in these three categories. In our graph abstraction, this situation can be modeled by using these categories (instead of the integer value) as the label for the vertex.

4 Graph Kernels

There are different ways to compare a pair of graphs and establish the degree of similarity between them. For instance, the *edit distance* measures the number of atomic changes required to transform one graph into the other. An alternative is checking for *isomorphism*[2] between the whole graphs or their subgraphs. However, these approaches have a high computational complexity and may be unsuitable for comparing large graphs or sizable collections of graphs.

An alternative approach is taken by graph kernels [7,27], a family of methods for measuring the (dis)similarity among pairs of graphs. Rather than computing an exact measure for similarity, kernels aim to provide an efficient approximation that can be computed efficiently but still captures relevant topological information about the graphs. A typical approach is counting the number of matching substructures within the graphs, like paths, subtrees or subgraphs. In this work, we have used the Weisfeiler-Lehman kernel [22], as it has been shown to provide good precision with an efficient computation in a variety of domains [13,22].

Algorithm 1 describes the Weisfeiler-Lehman (WL) kernel. The procedure computes the distance between a pair of graphs G_1 and G_2 by counting the number of common subtrees up to height h. To avoid enumerating subtrees explicitly, a characteristic label is computed for each subtree. This label is constructed iteratively: each iteration i computes the label for the tree of height i rooted in each node v ($\texttt{label}(i, v)$). Iteration 0 (line 11) uses the original labels in the graph. Then, each iteration i (lines 14–21) assigns a label to each vertex v by combining the labels of v and its adjacent vertices in iteration $i - 1$. Finally, the distance between the pair of graphs is computed by counting the original labels (line 12) and the labels for subtrees up to height h (line 22) and comparing their frequencies (lines 4–6). The complexity of this procedure is $O(hm)$, with m being the number of edges in the graphs [22]. The parameter h allows us to control the trade-off between performance and precision.

Notice that thanks to how our graph abstraction process is defined (types and attribute values as labels), the similarity value computed by the kernel is implicitly taking advantage of topological, type and attribute value information from the instance.

---

[2] Graphs $G_1 = (V_1, E_1)$ and $G_2 = (V_2, E_2)$ are called isomorphic if there is a mapping $f : V_1 \rightarrow V_2$ such that $\forall x, y \in V_1 : (x, y) \in E_1$ iff $(f(x), f(y)) \in E_2$.

```
1  Function WLKernel(G₁, G₂, h)                    // Weisfeiler-Lehman graph kernel
      input  : G₁, G₂: a pair of labeled graphs; h : an integer (the tree height)
      output: A distance measure between G₁ and G₂
2  |   freq1 ← WLTest(G₁, h);                 // frequency of each label in G₁
3  |   freq2 ← WLTest(G₂, h);                 // frequency of each label in G₂
4  |   distance ← 0;                      // distance = difference among frequencies
5  |   foreach label lab do
6  |   |   distance ← distance + |freq1[lab] − freq2[lab]|;
7  |   return distance;

8  Function WLTest(G, h)                       // Weisfeiler-Lehman isomorphism test
      input  : G: a labeled graph G = (V, E); h : an integer (the tree height)
      output: A map counting the frequency of labels in G
9  |   // Initially all labels x have frequency[x] = 0
10 |   foreach vertex v ∈ V(G) do
11 |   |   label(0,v) ← label of v in G;
12 |   |   frequency[label(0,v)] ← frequency[label(0,v)] + 1
13 |   for i ← 1 to h do
14 |   |   foreach vertex v ∈ V(G) do
15 |   |   |   adjacentLabels ← labels(i-1, neighbours(v,G));
16 |   |   |   // signature = my label + sorted labels of adjacent vertices
17 |   |   |   signature ← append(label(i-1,v), sort(adjacentLabels));
18 |   |   |   // Assign an integer label that summarizes signature
19 |   |   |   // Two equal signature should always receive the same label
20 |   |   |   // Compressed labels not reused in the next iterations
21 |   |   |   label(i, v) ← compressLabels(signature) ;
22 |   |   |   frequency[label(i,v)] ← frequency[label(i,v)] + 1
23 |   return frequency;
```

Algorithm 1: Pseudocode for the Weisfeiler-Lehman graph kernel [22].

5 Clustering

Clustering is one of the fundamental tasks in the field of Machine Learning (ML). Intuitively, it consists in the analysis of a collection of elements to identify groups of similar individuals, for a given definition of "similarity".

Algorithm Selection. Several algorithms have been proposed for this task [29]. There is no single "best" clustering algorithm: the most suitable one depends on the collection being analyzed. This is because the strategies for finding clusters can be very different. For example, *means* and *medoids* are different definitions of the "center" of a cluster, and algorithms like *K-means* and *K-medoids* aim to find the best location for those centers. On the other hand, methods like *hierarchical clustering* initially consider each element as a cluster and then iteratively merge the two nearest clusters.

In order to select which clustering algorithm should be used, the required input information should be considered:

- **Feature versus Kernel methods**: Some algorithms like K-means require each element to be described by a vector of features (relevant characteristics) of a fixed length. Meanwhile, other algorithms like hierarchical clustering only require a distance (or similarity) measure among pairs of elements.
- **Target number of clusters**: Algorithms like K-means or K-medoids require knowing the target number of clusters a priori. Conversely, algorithms like hierarchical clustering do not require this information beforehand.

In our context, the elements we are trying to cluster are labeled graphs abstracting the outputs of a model finder. The number of target clusters is unknown a priori and, as discussed in the previous section, we will be using a similarity metric. Given this setting, we have chosen hierarchical clustering.

Choice of Number of Clusters. Hierarchical clustering computes a hierarchical structure called *dendogram*, a tree that describes the order in which clusters should be merged according to their similarity. A clustering is obtained when we decide where (in which level of the tree) the merging should stop. In order to decide that, we can use *cluster validity indices*, metrics that measure the quality of a clustering. In a good clustering, elements within a cluster should be very similar and very dissimilar to elements in other clusters. The metric is evaluated in each level of the tree and the clustering providing the optimal value is selected.

In this work, we have used the *silhouette coefficient* [19], a classical metric that measures the average distance to elements in the same cluster compared to the minimum of the average distances to elements in other clusters. It provides a value in the $[-1, 1]$ range (higher is better), where values below 0.5 signal a bad fit in the clustering. As mentioned previously, the clustering achieving the highest average silhouette width is selected as our output.

6   Experimental Results

In order to assess the computational effort of the proposed method and the usefulness of its output, we have performed several experiments. These experiments aim to answer the following research questions:

RQ1. How does the execution time of the method compare to model finding?
RQ2. Do the resulting clusters provide a concise yet diverse summary of the model finder output?

Experiment Design. We have analyzed a collection of Alloy models provided in the Alloy GitHub model repository[3]. Among them, we have chosen examples dealing with the generation of examples or counterexamples, rather than proving their absence. These type of models could be used for validation and testing, and thus they are the target of the proposed method. For these models, we have used the Alloy Analyzer to generate up to 100 instances (less if there are not enough valid instances available). Table 1 provides information about the size and complexity of these models: the number of signatures (**Sig**), fields (**Fields**), facts (**Fact**) and predicates (**Pred**) in each Alloy model.

Implementation. We have implemented our method as two separate components. First, we have developed a Java program that calls the latest version of the Alloy API (5.0.0) to compute a collection of instances and generate their graph abstraction. The output of this tool is stored as a set of files in GML format. Then, a R script reads the GML files, computes the graph kernel and

---

[3] https://github.com/AlloyTools/models.

Table 1. Summary of the models analyzed with the Alloy Analyzer.

Model	Domain	Sig	Field	Fact	Pred
chord-bug-model	Chord distributed hastable lookup protocol	4	8	3	15
file-system	Generic file system	7	4	0	3
firewire	Leader election in the Firewire protocol	15	16	2	15
flip-flop	Flip-flop state machine	6	8	1	2
genealogy	Genealogical relationships	5	2	4	1
grandpa	"I am my own grandfather" puzzle	3	3	3	2
philosophers	Dining philosophers problem	3	5	1	2
railway	Train safety in a railway system	4	5	3	6
reset-flip-flop	Evolution of a flip-flop	7	8	1	2

performs the clustering. This script takes advantage of existing libraries for representing graphs (the `igraph` package[4]), similarity analysis among graphs (the `graphkernels` package[5]) and clustering (the `cluster` package[6]).

The experiments have been performed on a quad-core Intel i5-760 2.8 GHz with 4 GB of RAM. On the software side, we have used Java 9.0.4 64 bits and R 3.50 64 bits. With respect to the settings, Alloy has used MiniSat as the SAT solver back-end with the highest amount of symmetry breaking (symmetry=20). Regarding the graph kernel, the Weisfeiler-Lehman graph kernel has been used with the default number of iterations ($h = 5$).

Execution times have been measured in each step of the computation: the Alloy analysis, the graph abstraction phase and the kernel and clustering phases.

Table 2. Experimental results.

Model	Scope	Inst	Model finding	Execution time				Output	
				Abst	Kern	Clust	Total	# Cl	Sil
chord-bug-model	2	52	498 ms	169 ms	90 ms	30 ms	289 ms	5	0.31
file-system	5	100	825 ms	165 ms	180 ms	30 ms	375 ms	3	0.99
firewire	2–7	100	1474 ms	209 ms	180 ms	40 ms	429 ms	3	0.76
flip-flop	10	100	652 ms	203 ms	180 ms	50 ms	433 ms	2	0.04
genealogy	6	100	830 ms	129 ms	140 ms	50 ms	319 ms	33	0.45
grandpa	4	48	554 ms	88 ms	70 ms	40 ms	198 ms	2	0.96
life	3–6	100	1681 ms	283 ms	180 ms	40 ms	503 ms	14	0.30
philosophers	4	100	1539 ms	157 ms	160 ms	40 ms	357 ms	2	0.30
railway	1–4	100	735 ms	179 ms	170 ms	30 ms	379 ms	50	0.46
reset-flip-flop	10	100	672 ms	250 ms	160 ms	40 ms	450 ms	14	0.48

---

[4] https://igraph.org.
[5] https://cran.r-project.org/package=graphkernels.
[6] https://cran.r-project.org/package=cluster.

Results and Discussion. Table 2 shows, for each experiment, the scope used in the analysis (**Scope**) and the number of computed instances (**Inst**). Notice that for two models there were less than 100 satisfying instances. Then, we describe the time (in milliseconds) required by Alloy to compute the instances (**Model finding**), compared to the time taken by the different steps of our method: graph abstraction (**Abst**), graph kernel (**Kern**) and clustering (**Clust**). The total time for the three steps is reported as well. Finally, we list the optimal number of clusters (**# Cl**) identified by our method and the silhouette coefficient (**Sil**). As mentioned in Sect. 5, the silhouette is a value in the $[-1,+1]$ range that estimates the quality of the clustering (higher is better).

Considering these results, regarding RQ1 (efficiency) the execution time of the method is always below 0.5 seconds and less than the time required by Alloy to compute the instances. This was somewhat expected, as the computational effort of our approach depends on the number of instances and their size, but it is unaffected by the hardness of finding instances, the decisive factor in Alloy's execution time. Therefore, we can conclude that using our approach does not incur in a significant overhead with respect to using the model finder.

With respect to the scalability of our approach, let us consider the computational complexity of our method. We consider two parameters in this analysis: n, the number of instances that will be computed by the model finder; and m, the size (number of atoms, tuples in the relation and witnesses) of an instance. Graph abstraction performs a traversal of the instance, requiring $O(m)$ time. The graph kernel takes $O(m)$ time for each comparison and performs $O(n^2)$ comparisons, so in total it requires $O(m \cdot n^2)$. Finally, clustering requires $O(n^3)$ time, so the overall complexity is $O(m \cdot n^2 + n^3)$. In terms of space complexity, we require $O(m \cdot n)$ to store the n graphs, $O(n^2)$ to store the similarity matrix and perform clustering, that is, $O(m \cdot n + n^2)$ in total.

Regarding RQ2 (quality of the output) we can see that the proposed number of clusters varies significantly from one model to another, and so does the silhouette coefficient:

- Models with a high silhouette (*e.g.*, file-system and grandpa) exhibit some sort of symmetry that is not being detected by the Alloy Analyzer. For instance, in file-system there is a symmetry between directory names, so in practice, it is as if Alloy was only returning the same 3 effective instances all the time. Models like this one are the scenarios where our approach is most effective.
- Models with a low number of clusters and a low silhouette (*e.g.*, flip-flop) highlight scenarios where all instances are very similar. For instance, in flip-flop the instance models 10 steps of a trace in the evolution of a flip-flop. All these traces are very similar, so no salient features can be used to classify them. Diversity can only be slightly improved for these scenarios.
- Models with a high number of clusters (*e.g.*, genealogy or railway) describe scenarios where the instances produced by the solver are already very dissimilar among them. In this case, the output of the solver was already diverse before applying our method.

– The rest of models, with an average silhouette between (0.4–0.7) illustrate a middle ground: some instances share similarities but the boundaries between each group may overlap or be hard to establish. Choosing a representative from each cluster ensures diversity, but there is the risk (higher for lower silhouette values) of discarding relevant instances. To reduce this risk, it would be possible to select a higher number of representatives per cluster.

To sum up, our method can reduce the number of instances to consider while preserving diversity. Furthermore, this method provides an estimate of the quality of its result that helps designers deciding when and how to employ it.

7   Related Work

Several works have considered how to improve the diversity in the output of model finders, *e.g.*, [6,9,11,20,23,26]. We will classify them according to two criteria: (i) how diversity is specified by the designer and (ii) how it is achieved.

We exclude from this discussion all methods designed for general-purpose solvers [5,15,25], as they have not been used within model finders and they consider diversity at a lower level of abstraction (*e.g.*, assignments to a boolean formula) where some model-level similarities may be lost (*e.g.*, isomorphic instances with different bit-vector representations are still equivalent). For instance, a related software engineering problem that relies on low-level constraint solvers is finding valid configurations in a software product line. In this context, it has been shown [18] that SAT solvers designed for *uniform sampling* (*i.e.*, computing satisfying assignments that are distributed as close as possible to a uniform distribution) do not achieve a uniform distribution in the set of computed configurations.

Definition of Diversity. The designer has different ways to specify the desired notion of diversity. Some methods [6,23] need to be given a *probability distribution* that the output instances should follow. Otherwise, the designer can partition the universe of instances by defining predicates called *classifying terms* [9]. For instance, for an attribute the designer may only be interested in its sign (positive, negative or zero), defining 3 partitions. Diversity is then achieved by finding instances that cover each partition.

Meanwhile, other methods such as [11,26] or the one proposed in this paper do not require any input from the designer: diversity is defined implicitly by ensuring non-equivalence or enforcing some distance metric between the output instances. Nevertheless, in our case, the designer has some degree of control over the desired type of diversity by adapting the graph abstraction process, as explained in Sect. 3.

Implementation of Diversity. Most methods operate inside the model finder, reducing the number of instances being computed in different ways.

Some techniques aim to automatically *detect equivalent solutions* during the analysis in order to avoid exploring them. In the context of boolean satisfiability

(SAT), SAT-Modulo Theories (SMT) and Constraint Programming (CP) this notion is called *symmetry breaking* [3,10] and it is achieved by including additional constraints a priori. These constraints can also be added dynamically each time a new instance is found [9,23], to forbid exploring equivalent instances in the future. Another way to avoid exploring equivalent instances is requiring the solution to be *minimal* [2,4,16].

In search-based methods like genetic algorithms [2] or simulated annealing [4], similarity among solutions can be detected through a *distance measure*: neighbors that are too close to previously explored solutions can be ignored. Similarly, in graph solvers *graph shape analysis* [20,21] can detect equivalent or similar graphs. Nevertheless, this approach does not support features like attributes, relations or witnesses like the approach presented in this paper.

Moreover, model finders can introduce *randomness* [6], such as random selection of the next value to be explored or random restarts that can help explore different areas of the search space. Another take on randomness, *randomized partitioning* [11], shares the goal of classifying terms (partitioning the solution space) but generates the partitions by randomly splitting the domains of model elements. While this approach may be successful in problems with simple and local constraints, it is ineffective when dealing with complex constraints.

Finally, the COMODI tool [6] provides several techniques for clustering the object diagrams produced by a UML/OCL model finder. First, it defines a feature vector encoding for object diagrams that captures, for each object, information about attribute values and adjacent objects. And second, it defines a centrality metric (similar to the `pagerank` algorithm of search engines) that measures the importance of each object within the object diagram. Compared to our method, this approach is specific for object diagrams: it cannot deal with features from other modeling notations, such as Alloy's relations or witnesses. Furthermore, the proposed similarity metrics do not consider information about types, structure and attribute values simultaneously: the centrality metric omits attribute values entirely; and the feature vector approach does not consider topological information about the structure of the object diagram.

8   Conclusions

We have presented a method for addressing the lack of diversity among the instances computed by a model finder. Our approach uses clustering to group instances according to their similarity, using information both about topology, types and attribute. The method is solver- and notation-agnostic: it can be applied to model finders using different types of solvers (*e.g.*, SAT, SMT or CP) and even targeting different modeling notations (*e.g.*, UML/OCL or Alloy).

This approach is capable of computing meaningful clusters and has an execution time that is negligible with respect to that of the model finder itself. Still, as our diversity computation is an *a posteriori* procedure, it is intended for validation and testing scenarios where model finders are able to find instance solutions with relative ease. In this sense, our approach does not increase the diversity of the model finder output. However, it maximizes diversity by selecting, on behalf of the user, the widest possible variation among the output set.

As future work, we plan to define custom kernels for comparing instances that take into account specific characteristics of the input model. For instance, the invariants and multiplicities in the model can be used to identify which model elements are more constrained: this is where diversity is most relevant, rather than elements where we are free to choose almost any value. Also, we plan to look into combining graph kernels with topological and label features [13] that can improve the quality of the similarity analysis. Finally, we will consider strategies for tailoring the graph abstraction to particular problems and domains.

References

1. Ali, S., Zohaib Iqbal, M., Arcuri, A., Briand, L.C.: Generating test data from OCL constraints with search techniques. IEEE Trans. Softw. Eng. **39**(10), 1376–1402 (2013). https://doi.org/10.1109/TSE.2013.17
2. Batot, E., Sahraoui, H.: A generic framework for model-set selection for the unification of testing and learning MDE tasks. In: ACM/IEEE International Conference on Model Driven Engineering Languages and Systems (MODELS 2016), pp. 374–384. ACM Press, New York (2016). https://doi.org/10.1145/2976767.2976785
3. Cabot, J., Clarisó, R., Riera, D.: On the verification of UML/OCL class diagrams using constraint programming. J. Syst. Softw. **93**, 1–23 (2014). https://doi.org/10.1016/j.jss.2014.03.023
4. Cadavid, J.J., Baudry, B., Sahraoui, H.: Searching the boundaries of a modeling space to test metamodels. In: IEEE International Conference on Software Testing, Verification and Validation (ICST 2012), pp. 131–140. IEEE (2012). https://doi.org/10.1109/ICST.2012.93
5. Dutra, R., Laeufer, K., Bachrach, J., Sen, K.: Efficient sampling of SAT solutions for testing. In: International Conference on Software Engineering (ICSE 2018), pp. 549–559. ACM (2018). https://doi.org/10.1145/3180155.3180248
6. Ferdjoukh, A., Galinier, F., Bourreau, E., Chateau, A., Nebut, C.: Measurement and generation of diversity and meaningfulness in model driven engineering. Int. J. Adv. Softw. **11**(1/2), 131–146 (2018). https://hal-lirmm.ccsd.cnrs.fr/lirmm-02067506
7. Ghosh, S., Das, N., Gonçalves, T., Quaresma, P., Kundu, M.: The journey of graph kernels through two decades. Comput. Sci. Rev. **27**, 88–111 (2018). https://doi.org/10.1016/J.COSREV.2017.11.002
8. González, C.A., Cabot, J.: Formal verification of static software models in MDE: a systematic review. Inf. Softw. Technol. **56**(8), 821–838 (2014). https://doi.org/10.1016/j.infsof.2014.03.003
9. Hilken, F., Gogolla, M., Burgueño, L., Vallecillo, A.: Testing models and model transformations using classifying terms. Softw. Syst. Modeling **17**(3), 885–912 (2016). https://doi.org/10.1007/s10270-016-0568-3
10. Jackson, D.: Software Abstractions: Logic, Language and Analysis. MIT Press, Cambridge (2006). https://mitpress.mit.edu/books/software-abstractions
11. Jackson, E.K., Simko, G., Sztipanovits, J.: Diversely enumerating system-level architectures. In: International Conference on Embedded Software (EMSOFT 2013), pp. 1–10. IEEE, September 2013. https://doi.org/10.1109/EMSOFT.2013.6658589

12. Kuhlmann, M., Hamann, L., Gogolla, M.: Extensive validation of OCL models by integrating SAT solving into USE. In: Bishop, J., Vallecillo, A. (eds.) TOOLS 2011. LNCS, vol. 6705, pp. 290–306. Springer, Heidelberg (2011). https://doi.org/10.1007/978-3-642-21952-8_21
13. Li, G., Semerci, M., Yener, B., Zaki, M.J.: Effective graph classification based on topological and label attributes. Stat. Anal. Data Mining 5(4), 265–283 (2012). https://doi.org/10.1002/sam.11153
14. Mougenot, A., Darrasse, A., Blanc, X., Soria, M.: Uniform random generation of huge metamodel instances. In: Paige, R.F., Hartman, A., Rensink, A. (eds.) ECMDA-FA 2009. LNCS, vol. 5562, pp. 130–145. Springer, Heidelberg (2009). https://doi.org/10.1007/978-3-642-02674-4_10
15. Nadel, A.: Generating diverse solutions in SAT. In: Sakallah, K.A., Simon, L. (eds.) SAT 2011. LNCS, vol. 6695, pp. 287–301. Springer, Heidelberg (2011). https://doi.org/10.1007/978-3-642-21581-0_23
16. Nelson, T., Saghafi, S., Dougherty, D.J., Fisler, K., Krishnamurthi, S.: Aluminum: principled scenario exploration through minimality. In: International Conference on Software Engineering (ICSE 2013), pp. 232–241. IEEE, May 2013. https://doi.org/10.1109/ICSE.2013.6606569
17. Petre, M.: UML in practice. In: International Conference on Software Engineering (ICSE 2013), pp. 722–731. IEEE Press (2013). https://doi.org/10.1109/ICSE.2013.6606618
18. Plazar, Q., Acher, M., Perrouin, G., Devroey, X., Cordy, M.: Uniform sampling of SAT solutions for configurable systems: are we there yet? In: IEEE Conference on Software Testing, Validation and Verification (ICST 2019), pp. 240–251. IEEE (2019). https://doi.org/10.1109/ICST.2019.00032
19. Rousseeuw, P.J.: Silhouettes: a graphical aid to the interpretation and validation of cluster analysis. J. Comput. Appl. Math. 20(1), 53–65 (1987). https://doi.org/10.1016/0377-0427(87)90125-7
20. Semeráth, O., Nagy, A.S., Varró, D.: A graph solver for the automated generation of consistent domain-specific models. In: International Conference on Software Engineering (ICSE 2018), pp. 969–980. ACM Press (2018). https://doi.org/10.1145/3180155.3180186
21. Semeráth, O., Varró, D.: Iterative generation of diverse models for testing specifications of DSL tools. In: Russo, A., Schürr, A. (eds.) FASE 2018. LNCS, vol. 10802, pp. 227–245. Springer, Cham (2018). https://doi.org/10.1007/978-3-319-89363-1_13
22. Shervashidze, N., Schweitzer, P., van Leeuwen, E.J., Mehlhorn, K., Borgwardt, K.M.: Weisfeiler-Lehman graph kernels. J. Mach. Learn. Res. 12, 2539–2561 (2001). https://dl.acm.org/citation.cfm?id=2078187
23. Soltana, G., Sabetzadeh, M., Briand, L.C.: Synthetic data generation for statistical testing. In: IEEE/ACM International Conference on Automated Software Engineering (ASE 2017), pp. 872–882. IEEE (2017). https://doi.org/10.1109/ASE.2017.8115698
24. Soltana, G., Sabetzadeh, M., Briand, L.C.: Practical model-driven data generation for system testing. ACM Transactions on Software Engineering and Methodology (2020, to appear). http://arxiv.org/abs/1902.00397
25. Vadlamudi, S.G., Kambhampati, S.: A combinatorial search perspective on diverse solution generation. In: AAAI Conference on Artificial Intelligence, pp. 776–783. AAAI Press (2016). https://dl.acm.org/citation.cfm?id=3015927

26. Varró, D., Semeráth, O., Szárnyas, G., Horváth, Á.: Towards the automated generation of consistent, diverse, scalable and realistic graph models. In: Heckel, R., Taentzer, G. (eds.) Graph Transformation, Specifications, and Nets. LNCS, vol. 10800, pp. 285–312. Springer, Cham (2018). https://doi.org/10.1007/978-3-319-75396-6_16
27. Vishwanathan, S., Schraudolph, N.N., Kondor, R., Borgwardt, K.M.: Graph kernels. J. Mach. Learn. Res. **11**(Apr), 1201–1242 (2010). http://www.jmlr.org/papers/v11/vishwanathan10a.html
28. Wu, H.: MaxUSE: a tool for finding achievable constraints and conflicts for inconsistent UML class diagrams. In: Polikarpova, N., Schneider, S. (eds.) IFM 2017. LNCS, vol. 10510, pp. 348–356. Springer, Cham (2017). https://doi.org/10.1007/978-3-319-66845-1_23
29. Xu, R., Wunsch, D.: Survey of clustering algorithms. IEEE Trans. Neural Netw. **16**(3), 645–678 (2005). https://doi.org/10.1109/TNN.2005.845141

Formal Verification of Interoperability Between Future Network Architectures Using Alloy

Mohammad Jahanian[1]([✉]), Jiachen Chen[2], and K. K. Ramakrishnan[1]

[1] University of California, Riverside, CA, USA
mjaha001@ucr.edu, kk@cs.ucr.edu
[2] WINLAB, Rutgers University, North Brunswick, NJ, USA
jiachen@winlab.rutgers.edu

Abstract. The Internet is composed of many interconnected, interoperating networks. With the recent advances in Future Internet design, multiple new network architectures, especially Information-Centric Networks (ICN) have emerged. Given the ubiquity of networks based on the Internet Protocol (IP), it is likely that we will have a number of different interconnecting network domains with different architectures, including ICNs. Their interoperability is important, but at the same time difficult to prove. A formal tool can be helpful for such analysis. ICNs have a number of unique characteristics, warranting formal analysis, establishing properties that go beyond, and are different from, what have been used in the state-of-the-art because ICN operates at the level of content names rather than node addresses. We need to focus on node-to-content reachability, rather than node-to-node reachability. In this paper, we present a formal approach to model and analyze information-centric interoperability (ICI). We use Alloy Analyzer's model finding approach to verify properties expressed as invariants for information-centric services (both pull and push-based models) including content reachability and returnability. We extend our use of Alloy to model counting, to quantitatively analyze failure and mobility properties. We present a formally-verified ICI framework that allows for seamless interoperation among a multitude of network architectures. We also report on the impact of domain types, routing policies, and binding techniques on the probability of content reachability and returnability, under failures and mobility.

1 Introduction

Today's computer networks, the Internet being a dominant example, are heavily used to fulfill users' *information-centric* needs: users primarily seek information over the network without necessarily wanting to focus on its location or the underlying mechanisms used to retrieve it [9]. However, the current way of using "location-based" access in IP networks results in a less convenient and less efficient means for information retrieval and dissemination. *Information-Centric Networks* (ICNs) address this content-oriented networking paradigm by separating content identity from its location [9]. ICN enables access to content based on

A. Raschke et al. (Eds.): ABZ 2020, LNCS 12071, pp. 44–60, 2020.
https://doi.org/10.1007/978-3-030-48077-6_4

its name, from wherever it resides, supporting mobility as well as accessing the named content from the best, any, or all source(s). It also allows for network-wide caching to reduce access latency. There are a variety of ICN architectures which have been proposed in the past decade. Two of the most notable ones, which we primarily focus on in this paper, are Named Data Networks (NDN) [20], and MobilityFirst [16], which have been considered for *Future Internet* designs [3].

Currently, there are two main factors that make the discussion of network interoperability important: 1) Today, IP is ubiquitous and used on a majority of network devices, despite the legacy of end-point address-oriented communication, especially considering new services and demands on today's networks [15]. 2) Research on designing new network architectures radically different from IP, is ongoing, and in many cases has already led to implemented systems; our focus in this paper is on an important class of such architectures, namely ICN. It is anticipated that we may have a number of interconnected networks (domains) using different architectures [15]. To go beyond the *interconnection* (*i.e.*, physical connections between different domains) towards *interoperation* between them (*i.e.*, being able to use a service, or content, provided by one domain in another domain), we need network interoperability. In the past decade, several designs have been proposed for interoperation between an ICN architecture (either NDN or MF) with IP [3]. However, such designs and their requirements were presented informally, describing the primitives and operations. It has been observed that network interoperability is complex [19]; thus, a formal structure for analysis of *information-centric interoperability* (ICI) can be very helpful, as it can provide proofs or expose errors early on, before the universal deployment of ICI frameworks for Future Internet.

Formal methods have been extensively used for designing and analyzing computer networks and protocols (surveyed in [14]). As for interoperability, work in [19] proposed a formal model to analyze interoperation of legacy networks. However, it only deals with *host-centric interoperability* (HCI), and only uses classic *model finding* [17] reasoning techniques. We extend that to support ICI as well as modeling failure and mobility with *model counting* [7] techniques. Network verification tools have also been proposed to analyze network data and control planes. Recently, work in [10] proposed a tool to verify ICN data planes, analyzing properties such as reachability. However, it only deals with a single domain, while our goal here is to cover multiple domains with different architectures coexisting with each other. Also, the symbolic execution nature of works such as [10] is computationally too expensive when expanded across multiple domains, each having its own data plane.

We present an Alloy [8]-based formalization of ICI, to analyze interoperability correctness. We cover both pull-based (request/response) and push-based (publish/subscribe) [6] content retrieval services, and their most essential properties such as content reachability and returnability. To analyze content-oriented services, we distinguish between static and dynamic content, justifying their differences, and specifying no-conflict properties, especially for dynamic content retrieval. For verification of these properties, we use Alloy Analyzer's built-in SAT solver-based model finding engine [2]. We also consider failure and mobility; to analyze them, mere model finding is not sufficient, as failure and mobility, when severe, can cause any network protocol to become "incorrect" (and raise counterexamples). Thus, for such analysis, we resort to model counting (to count and compare the number of satisfying instances and counterexamples) to assess "how well" a particular domain or architecture is doing under failure and mobility.

The major contributions of this paper are: 1) a model finding method to analyze basic properties (mainly reachability and returnability) of information-centric interoperability (ICI); 2) a formally-verified ICI framework; and 3) a model counting method to analyze gateway failure and mobility.

2 Background and Related Work

2.1 Information-Centric Networking (ICN) and Interoperability

ICN enables access to content independent of its location, focusing on the fact that what matters to users is *what* the content is rather than *where* that content is located [9]. An ICN network layer recognizes and makes its forwarding decisions based on content *names* (or IDs) instead of addresses (unlike host-centric networks, as in today's IP networks), achieving efficiency and scalability.

Among many different ICN architectures proposed recently, we focus on the two most popular ones, namely Named Data Networks (NDN) [20] and MobilityFirst (MF) [16]. Both allow users to retrieve content using content names, through pull-based request/response or push-based publish/subscribe methods [6]. In-network content caching in routers is an important feature of ICN, allowing for requests to be satisfied from an intermediate cache on the path to the server/repository [9]. An in-network *namespace* is generally a graphical structure that captures the content names and their relationships in an ICN's content space [12]. Despite both being ICNs, NDN and MF have important differences [16,20]: NDN uses human-readable hierarchically-structured names, with Longest Prefix Matching-based forwarding. NDN content requests (called *Interests*) leave "breadcrumb" state in the routers on their path, which the associated response (called *Data* packets) then follow back, via *Reverse Path Forwarding (RPF)*. MF, on the other hand, uses flat IDs (called *GUIDs*) to identify content. Response packets contain the consumer's ID and do not need to follow the same path as the request. Also, MF inherently supports mobility by *late binding*, which re-directs in-flight packets towards a mobile content repository. Early binding assigns names to locations strictly at the original client, while late binding allows such assignment to be updated on its way in the network [16].

There have been several proposals for interoperability frameworks for ICNs (surveyed in [3]). These frameworks typically consist of *interoperation gateways* between *domains* of different network architectures, performing *translations* between them. All of these proposals allow interoperation of just two domains, IP and one ICN (either NDN or MF), and often require addition of new protocols or modification of existing ones. We generalize these solutions in our model to an interoperability framework of multiple (≥2) domain types (we allow IP, NDN and MF to coexist simultaneously), and do not change any domain-specific protocols.

2.2   Alloy

Alloy is a declarative language based on relations and first order logic [8]. Alloy models a system, M, through the declaration of *signatures* (objects and their relations) and *facts* (constraints and axioms). A *predicate* is defined as a logical formula. An *Assertion* is a logical formula (which can be a combination of predicates) that are required to be always true (*i.e.*, as *invariants*) in the system. Alloy Analyzer [2] allows the automatic analysis of models and their properties through utilizing off-the-shelf SAT solvers. The tool translates Alloy descriptions into *Conjunctive Normal Form (CNF)* expressions. It uses an enumeration of instances, also called *model finding*, within a *bound (scope)*, to prove whether or not a predicate P *ever* holds (by SAT-solving $M \wedge P$), or an assertion A *always* holds as an invariant (by SAT-solving $M \wedge \neg A$, to look for *counterexamples*).

Alloy has been used in modeling and analysis of many systems, including network protocols and architectures [8]. In the particular case of network interoperability, Zave [19] used Alloy to formally analyze host-centric interoperability for legacy networks, with domains of the Public Switched Telephone Network (PSTN), BoxOS and the Session Initiation Protocol (SIP). We extend the approach to model and analyze interoperability of information-centric services and architectures, since we are dealing with radically different network designs (name-based networking *vs.* address-based [9]) and required properties (node-to-content reachability *vs.* node-to-node reachability [10]). Additionally, we extend the classic Alloy-based model finding approach, such as in [19], to a model counting one, to quantitatively analyze the impacts of failure and mobility. An important feature of Alloy is its strength in efficiently handling graph structures and properties [18], a feature that we benefit from, in two ways: 1) the composite network topology, and 2) a graph-based information namespace. Further, Alloy helps provide proofs for properties with a reasonably large scope [18].

3   Modeling Information-Centric Interoperability

We now describe the basics of our formal model[1]. First and foremost, let us define information-centric interoperability (ICI):

Definition 1. *A sequence of interconnected domains in a network are information-centrically interoperable if and only if any client in any of the domains can access information-centric services provided in any other domain.*

Throughout this paper, we use the term "network" to mean "a composition of multiple network domains", each domain being a different type of standalone architecture (*e.g.*, IP, NDN, or MF). An interoperability framework (such as [3]) is a set of protocols and architectural components that allow interconnected networks of different types to interoperate. Information-centric services are broadly sub-categorized as: 1) requesting for and retrieving content (pull-based), and 2)

---

[1] Full source files are available in [1].

Fig. 1. Information-centric interoperability (ICI): request for content

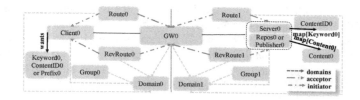

Fig. 2. Example (partial) instance for ICI Alloy model (objects and relations)

subscribing to and receiving content (push-based). Both of these may be based on namespaces defined by content producers. An example 3-domain ICI scenario is depicted in Fig. 1. As shown, ICI accesses content by name, rather than an address. Also, requests can be satisfied at any cache node, not just the original server. As for formal analysis, in ICI, the main property we care about is *node-to-content reachability* [10], while in traditional host-centric interoperability (HCI) analysis [19], the focus is on *node-to-node reachability*.

We model our networked environment using Alloy's relational and logical atoms. We have *Domains* (as abstract signatures), each of which can be an IP, NDN, or MF type (extended signatures) (Listing 3.1). A *Node* is at least in one *Domain* and has at least one *NodeID*. A *Node* can be either a *Client*, *Repos* (repository/server), or *GW* (gateway). A gateway is associated with exactly two *Domains* (constrained using facts), that it is stitching together (Listing 3.2.)

<table>
<tr><td>

Listing 3.1. Domains

```
abstract sig Domain{}
sig IPdomain extends Domain{}
sig NDNdomain extends Domain{}
sig MFdomain extends Domain{}
```
</td><td>

Listing 3.2. Nodes

```
abstract sig Node{domains: some Domain, id:
    some NodeID}
sig Client extends Node{...}{...}
sig Repos extends Node{...}{...}
sig GW extends Node{...}{#domains=2 && ...}
```
</td></tr>
</table>

Our declarations specify a network *meta-model* [8], which maps to a number of instances (models) each being a network configuration (*i.e.*, with their own topology, content, namespace, *etc.*). An example 2-domain instance is depicted in Fig. 2, as a high-level schematic, showing objects and their inter-relations. The *Client* here wishes to retrieve some *Content* using its *ContentID* or a (set of) *Keyword*(s). Objects of type *Route* and *RevRoute* (reverse route) couple the notion of "a series of links" and "packets carried over them", the packet carrying content request and response, respectively. A *Route* has attributes such as *initiator*, *acceptor*, and a request for *ContentID*. We also extend signatures to add more fine-grained, domain-specific characteristics. One of *Route*'s extended object types, namely *IPRoute*, inherits its attributes and constraints, and also has additional attributes such as *srcIPaddress* and *destIPaddress*, and constraints saying that source and destination IP addresses must correctly correspond to initiator and acceptor nodes. Gateways perform translation for forwarding requests (over a composition of *Routes*), and retain state information which they use to forward the content back to the client (over composition of *RevRoutes*). We also add a number of additional facts, such as uniqueness of node ID, absence of self-looping routes, and the existence of one-to-one mapping between NDN's forward and reverse routes (to reflect NDN's RPF policy [20]).

We define a global-state relation C that captures routes to/from gateways. To model connectivity, we use the transitive closure of the route-connections relation C where $(r1, r2) \in C$ if and only if there exists a gateway between two domains that connects routes $r1$ and $r2$. E.g., if we have $C = \{(r1, r2), (r2, r3)\}$, then its transitive closure $C^+ = \{(r1, r2), (r2, r3), (r1, r3)\}$ will represent existing paths of any length (*i.e.*, number of routes). We define object type *Connections* (as a singleton) to capture these connections (*i.e.*, relation C); it has attributes being relations themselves, primarily *connected* and *revconnected*, to capture connection relations of *Routes* and *RevRoutes* respectively. Relation *revconnected* has an additional constraint, which says that for two reverse routes $rr1$ and $rr2$ connected at gateway gw, corresponding state information (associated with the *ContentID* or other multiplexing/demultiplexing values in $rr1$ and $rr2$) must be stored on gw, so that the content can be carried over this cascade of reverse routes towards the consumer (Listing 3.3). Additionally, we define a fact (*path_exists*, Listing 3.4) that ensures any two nodes are connected (through one or multiple *Routes* or *RevRoutes*), to reduce our instance space to only the ones with strongly connected topology.

Listing 3.3. Connections: capture the connectivity of routes and groups

```
one sig Connections{connected: Route->Route, revconnected: RevRoute->RevRoute,
    chain: Group->Group, revchain: Group->Group}
fact connectivity{ -- conditions for two (reverse) routes being ''connected''
    all r1,r2:sRoute, c:Connections |
        (r1->r2) in c.connected <=>
            r1.acceptor = r2.initiator && r1.contentID = r2.contentID &&
                r1.reposdomain = r2.reposdomain -- requests for same content
    -- similar condition for RevRoute paths (with extra criteria: gateway state
        information should match for the two connecting reverse routes) ...
}
```

Listing 3.4. Constraints to ensure that a path exists between any two nodes

```
fact path_exists{
    all co:Connections, disj n1,n2:Node, cid:ContentID, rd: Domain |
        (some r:Repos | rd in r.domains =>
            (some r1,r2:Route | (r1->r2) in ^(co.connected) && r1.initiator = n1
                && r2.acceptor = n2 && r1.contentID = cid && r2.contentID = cid
                && r1.reposdomain = rd && r2.reposdomain = rd))
    -- similar condition for RevRoute paths ...
}
```

While *Routes* represent unicast exchange paths, we define *Groups* to denote multicast groups (one-to-many communication), enabling push-based notification models. Following the principles of ICN, each group is associated with a content name *Prefix* [6] and can be used for publish/subscribe exchanges regarding that prefix. Each group belongs to one domain. To model a connection of groups across multiple domains, we add relation attributes *chain* and *revchain* to *Connections* (Listing 3.3), to capture connectivity of groups (as a chain) for subscription and publication respectively. To ensure strong connectivity, we add a fact that says any two groups serving the same prefix are chained (Listing 3.5).

Listing 3.5. Constraints to ensure that a chain of connectivity exists between groups

```
fact GroupRules{
    all disj g1,g2:Group, co:Connections | -- group chain conditions
        (g1->g2) in co.chain <=>
            (g1.prefix = g2.prefix &&
            (some gw:GW| g1.domain in gw.domains && g2.domain in gw.domains))
    all disj d1,d2:Domain, co:Connections, p:Prefix | -- chains for each prefix
        some disj g1,g2:Group |
            g1.domain = d1 && g2.domain = d2 && (g1->g2) in ^(co.chain) &&
            g1.prefix = p && g2.prefix = p
    -- similiar conditions for revchain ...
}
```

Content naming is integral in ICI. We define names, *i.e.*, *ContentID* objects for each *Content*. Based on domain type, *ContentID* can be either *URL* (in IP), *NDNName* (in NDN) or *ContentGUID* (in MF) (Listing 3.6). Each *ContentID* is a leaf node under a *Prefix* in the prefix tree (*PTree*). An example prefix tree is shown in Fig. 3, which represents the network's content namespace. *PTree* may contain a number of fragmented sub-trees (*i.e.*, as a forest), each sub-tree representing the namespace of a different (set of) content provider(s) in different domains. To represent the structure of hierarchical prefixes, we use binary relations to model the immediate parent-child relationship between prefixes in *PTree*. In Fig. 3, the relation $P = \{(P1, P2), (P1, P3), (P2, P4), (P2, P5)\}$ represents such relationships, and is captured in the prefix-to-prefix relation *map* in *PTree* (Listing 3.6). We also use its transitive closure to model the ancestor-descendant relationships. We add additional facts to ensure basic constraints on the tree, such as the non-existence of loops.

Listing 3.6. Content IDs and Prefix Tree

```
abstract sig ContentID{prefix: Prefix}
sig URL extends ContentID{} -- if in IP
sig NDNName extends ContentID{} -- if in NDN
sig ContentGUID extends ContentID{} --if in MF
sig Prefix{parent: lone Prefix, domains: some
    Domain} -- each Prefix has exactly one
    parent and is at least in one domain
one sig PTree {map: Prefix set -> set Prefix}
```

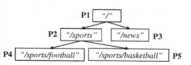

Fig. 3. Prefix tree example

4 Satisfying Information-Centric Service Properties

There are a number of important properties that are required from the framework, to ensure interoperability as defined in Definition 1. We consider properties of two classes of information-centric services here: pull-based (for unicast request/response), and push-based (for multicast publish/subscribe) content retrieval. We further divide the pull-based services into two categories: static content retrieval (SCR) and dynamic content retrieval (DCR). This distinction is important as the nature, protocol for retrieval, and thus formal properties of the two are different: static content is one that does not change in a long time (*e.g.*, a movie) and can be retrieved from its original producer as well as a cache, while dynamic content is created once on demand (*e.g.*, result of a Google search), and must be retrieved from its original server (not from a cache). Additionally, we assume content requests are assumed to be genuine and correct, *i.e.*, false and bogus content requests are not our focus here.

We study essential invariant properties, guaranteed to hold at all times. These properties are primarily associated with content-oriented reachability and returnability. We formally specify these properties, using Alloy predicates and assertions. For verification, Alloy's built-in model finding engine is used to find satisfying instances and counterexamples. Any counterexample found indicates interoperability violations: *e.g.*, a client cannot generate a request native to its domain, or the gateway does not know what to do with a returned response.

4.1   Pull-Based Retrieval: Request/Response

Static Content Retrieval. In the static content retrieval (SCR) service, the request packets carry content IDs which the client requests, and the response packets produced by repositories (can be content producers or router caches) carry the data associated with that content ID. We describe two of SCR's essential content-oriented properties using Alloy (Listings 4.1 and 4.2).

Property 1.1. SCR Reachability: For every client that wants to retrieve content associated with a content ID and has a direct route to a gateway, there is a repository with content having that ID reachable from that gateway.

Property 1.2. SCR Returnability: For every client that reaches a repository with a request, there is a path back to the client for the response with the content.

Listing 4.1. SCR reachability property

```
pred reach[c:Client, cid:ContentID, re:Repos, gw:GW]{ -- reachability predicate
    all co: Connections | cid in c.want => -- if requested
        (some r:Route, con:Content | r.initiator = c && r.acceptor = gw &&
        r.contentID = cid && (cid->con) in re.map =>
            some r1,r2:Route | (r1->r2) in ^(co.connected) && r1.initiator = gw &&
            r2.acceptor = re && r1.contentID = cid && r2.contentID = cid &&
            r1.reposdomain in re.domains && r2.reposdomain in re.domains)}
assert reach{ -- reachability assertion
    all c:Client, cid:ContentID| some re:Repos, gw:GW | reach[c,cid,re,gw]}
```

Listing 4.2. SCR returnability property

```
pred return[c:Client, cid:ContentID, re:Repos, gw:GW]{ -- returnability
    predicate
    all co:Connections | some gw1:GW | reach[c,cid,re,gw1] => -- if reachable
        (some r,r1,r2:RevRoute | (r1->r2) in ^(co.revconnected) &&
        r1.initiator = re && r2.acceptor = gw &&
        r1.content = re.map[cid] && r2.content = re.map[cid] &&
        r.initiator = gw && r.acceptor = c && r.content = re.map[cid])}
assert return{ -- returnability assertion
    all c:Client, cid:ContentID, re:Repos | some gw:GW | return[c,cid,re,gw]}
```

Dynamic Content Retrieval. In DCR, every request has to be mapped to a unique response, as opposed to SCR. To facilitate this, having a *demux* value (for multiplexing/demultiplexing) is essential for DCR, to provide the correct mapping of responses to requests; since every generated response is specific to not just the request's name, but also its input parameters. To access dynamic content from a server, a client generates a query for which the gateway keeps state as <nodeID, demux> of the requesting side and <demux> for the serving side. Reachability and returnability are still important in DCR (Properties 2.1–2.2). However, if the same SCR protocol is used for DCR, there can be *conflicts* between multiple requests, *e.g.*, a cached content may get sent back to multiple distinct clients. Therefore, we define no-conflict properties for DCR (Property 2.3).

Property 2.1–2.2. DCR Reachability and Returnability: These two properties are similar to those of SCR; with the difference being additional constraints regarding elements of DCR requests, *i.e.*, including generation and verification of the correct *demux* values at gateways (*i.e.*, in addition to contentID, *etc.*).

Property 2.3. No-conflict between distinct requests/clients: For every client that searches for two distinct content items (*no-conflict-A*, Listing 4.3), or a dynamic content requested by two different clients (*no-conflict-B*, Listing 4.4), two distinct, appropriately associated responses, should be received back. In *no-conflict-A*, the focus is on the distinction between two *return*-ed contents, associated with two distinct requests made by a given *Client* for distinct *Keywords* $k1$ and $k2$. On the other hand, *no-conflict-B* focuses on the distinction between two *return*-ed contents, associated with requests for a particular *Keyword* initiated by two distinct *Clients* $c1$ and $c2$.

This property shows the importance of having two separate *demux* values in packets, namely both the request ID (required for Property 2.3.a) and client ID (required for Property 2.3.b), to make each dynamic request globally unique, for correct multiplexing/demultiplexing. If we remove either of those two elements, this property will be violated and counterexamples will arise; *i.e.*, the gateway would not know how to demultiplex incoming response data to serve the correct, corresponding requesting client.

Listing 4.3. DCR - No conflict between 2 distinct requests from the same client

```
assert no-conflict-A{ -- Property 2.3.a
    all c:Client, disj k1,k2:Keyword | some s1,s2:Server, gw1,gw2:GW |
      return[c,k1,s1,gw1] && return[c,k2,s2,gw2] => some n1,n2: NodeID,
      d1,d2,d3,d4:Demux |
        (n1->d1->d2) in gw1.state && (n2->d3->d4) in gw2.state &&
        n1 in c.id && d1 in c.demux && d2 in gw1.demux &&
        n2 in c.id && d3 in c.demux && d4 in gw2.demux &&
        !(n1 = n2 && d1 = d3 && d2 = d4) && (some disj r1,r2:RevRoute |
            r1.initiator = gw1 && r1.acceptor = c && r1.contentID = s1.map[k1]
            && r1.demux = d1 && r2.initiator = gw2 && r2.acceptor = c
            && r2.contentID = s2.map[k2] && r2.demux = d3)}
```

Listing 4.4. DCR - No conflict between 2 identical requests from two distinct clients

```
assert no-conflict-B{ -- Property 2.3.b
    all c1,c2:Client, k:Keyword | some s1,s2:Server, gw1,gw2:GW |
      return[c1,k,s1,gw1] && return[c2,k,s2,gw2] => some n1,n2: NodeID,
      d1,d2,d3,d4:Demux |
         (n1->d1->d2) in gw1.state && (n2->d3->d4) in gw2.state &&
         n1 in c1.id && d1 in c1.demux && d2 in gw1.demux &&
         n2 in c2.id && d3 in c2.demux && d4 in gw2.demux &&
         !(n1 = n2 && d1 = d3 && d2 = d4) && (some disj r1,r2:RevRoute |
           r1.initiator = gw1 && r1.acceptor = c1 && r1.contentID = s1.map[k]
           && r1.demux = d1 && r2.initiator = gw2 && r2.acceptor = c2
           && r2.contentID = s2.map[k] && r2.demux = d3)}
```

4.2   Push-Based Retrieval: Publish/Subscribe

In pub/sub, we have domain-specific multicast groups that are associated with prefixes [6]. We want a client to be able to subscribe to and receive all relevant publications in accordance with the prefix tree of the namespace over "chain" of groups across domains. Groups $G1$ and $G2$ form a chain if and only if the publisher of $G1$ can be a subscriber of $G2$, and is then able to relay data received from $G2$ to his subscribers in $G1$.

Property 3.1. Ability to subscribe to any prefix. For every client that wants to retrieve future publications under/associated with an existing prefix and has a direct route to a gateway, if there is some publisher that will publish content under that prefix, then that publisher is accessible through a chain of groups.

Property 3.2. Ability to receive any content published directly associated with the subscribed prefix. For every client who is subscribed to a prefix and can reach the associated publisher, there is a path back to the client to carry any content with a content ID belonging to that prefix. For example, a subscriber of $P2$ in Fig. 3 should receive publications pertaining to $P2$ across domains.

Property 3.3. Ability to receive all content published that is associated with prefixes under the subscribed prefix. This property says that for every client that has subscribed to a prefix and has reached the associated publisher, there is a path back to the client to carry any content with content ID either directly belonging to that prefix or under it in the hierarchy on the prefix tree. For example, a subscriber of $P2$ in Fig. 3 should receive publications pertaining to $P2$ *and also* $P4$ across domains. The assertion *rcvall* in Listing 4.5 depends on how relationships among groups and also between content IDs and prefixes are represented by *Connections* and *PTree*. For a domain with a namespace that does not capture relationships between prefixes, *i.e.*, does not map a prefix to a set of multiple relevant prefixes according to a graph, then *rcvall* would be equivalent to receiving a single content element (Property 3.2). Properties 3.1–3 collectively model and verify properties of a service offering hierarchical pub/sub.

Listing 4.5. Pub/Sub - receiving all relevant publications

```
assert rcvall{ -- all relevant publications in accordance with the prefix tree
    all pub:Publisher, con:Content, cid:ContentID |
        all co:Connections, pt:PTree | (cid->con) in pub.map =>
          ((some c:Client, p:Prefix | (p in c.want || (all p1:Prefix |
          (p1->p) in ^(pt.map) && p1 in c.want)) && cid.prefix = p =>
            (some r1,r2:Route | r1.initiator = pub && r2.acceptor = c &&
            (r1>r2) in ^(co.connected) && some g1,g2:Group |
               g1.domain = pub.domain && g2.domain = c.domain &&
               g1.prefix = p && g2.prefix = p && (g1->g2) in
                  ^(co.revchain)))}
```

5 Reasoning About Failure and Mobility

In addition to the basic invariants (Sect. 4), there are other important aspects of formal analysis of networks that warrant a more quantitative analysis; among them are failure and mobility analysis. Failures and mobility of nodes can occur in a network, causing disruption and lack of content availability. To better compare how different network architectural components, *e.g.*, routing, impact the number of success and violation scenarios, we perform model counting [7]. While we can consider the probability for all instances as being equal, we can also calculate each instance's probability by additionally factoring in the real-world probability of individual elements causing failures and mobility, provided as external information (*e.g.*, the probability of a gateway failing when processing a content request, a route disconnecting while carrying a packet, *etc.*). Thus, we can provide a more realistic probabilistic analysis for the effect of failures and mobility using weighted model counting methods [5].

While the Alloy Analyzer (v4.20) [2] allows for a limited, graphical iteration over instances, it does not enable an explicit counting of instances in an efficient manner. To perform model counting, we wrote an application [1] that counts all SAT solutions, using the SAT4J solver [13] (SAT4J can be replaced by any off-the-shelf SAT solver). We feed the Alloy model and properties, in Kodkod format [17], to our application. Predicates and assertions are used for counting instances that satisfy or violate (counterexamples) respectively. Through this counting, we can also look into the details (relations and values) within each instance, and gain insight such as possible cause of violations (in case of counterexamples) and calculate the probability of occurrence of each instance in real-world scenarios. While we do not focus on the performance aspects of model counting in this paper, optimizations of this procedure can be leveraged for enhancing the scalability of our approach in case of very large problem sizes. At a minimum, our approach can provide a rough estimate of failure probabilities. Even if the model counting provided by the SAT solver is through "approximate" model counting (*e.g.*, using repetitive halving procedures) [4] rather than an "exact" one, it still gives us a good enough assessment of the degree of success and violation of properties.

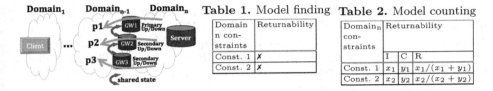

Fig. 4. Gateway failure scenario

Table 1. Model finding

Domain n con-straints	Returnability
Const. 1	✗
Const. 2	✗

Table 2. Model counting

Domain_n con-straints	Returnability		
	I	C	R
Const. 1	x_1	y_1	$x_1/(x_1 + y_1)$
Const. 2	x_2	y_2	$x_2/(x_2 + y_2)$

5.1   Failure

Our interoperability framework depends on gateways that retain state information. What would happen to a response packet if that state is lost at the gateway for any reason? For reliability, we consider state sharing between redundant gateways that have the same domains on either side. Figure 4 depicts an example for this. Consider the gateway that received the request and created the state as the *primary* gateway for the request ($GW1$ in the Fig.), and the replicas that have the shared state as the *secondary* gateways ($GW2$ and $GW3$). Formally, we add an extra condition to our reachability and returnability properties such that, for two routes to connect, the gateway attaching them must be up and running at the time the packet is received. Additionally, for returnability, the state information must be present at the gateway. If any gateway goes down, the corresponding potential path going through it ($p1$–3) back for the content cannot be leveraged. If the gateway is neighboring an NDN domain (*e.g.*, in $Domain_n$ or $Domain_{n-1}$), then the gateway has to be the primary only, for correct operation with the NDN reverse-path-forwarding (RPF) policy [20]. For other domain types, a secondary gateway that is active and has the shared state information is adequate to forward the response data back. We model the conditions representing this in Alloy as shown in Listing 5.1.

Listing 5.1. Failure scenario constraints: impact of gateway status on route connectivity

```
all r1,r2: Route, c:Connections | -- forward routes (request) condition
    (r1->r2) in c.connected <=> r1.acceptor = r2.initiator &&
    r1.initiator.status1 in Up && r2.initiator.status1 in Up
all r1,r2: RevRoute, c:Connections | -- reverse routes (response) condition
    (r1->r2) in c.connectedR <=> r1.acceptor = r2.initiator &&
    r1.initiator.status1 in Up && r2.initiator.status1 in Up &&
    ((r1.domain in NDNdomain || r2.domain in NDNdomain) =>
        r1.acceptor.type in Primary) -- NDNdomain enforces RPF policy
```

Gateways can go down due to various reasons such as completely failing or just losing state information due to a software failure. Our method can be used to reason about various scenarios and measure failure probability given an input configuration space, *i.e.*, a set of Alloy facts that set constraints on some objects or variables while relaxing others. As Table 1 shows, a simple model finding analysis does not provide a helpful comparison between different such constraints:it will say that both cases lead to counterexamples raised (*e.g.*, for the case that all gateways go down). To gain a better assessment of which constraint does better, we resort to model counting (Table 2). Using model counting, we can count (satisfying) instances (I) and counterexamples (C), and calculate (even if approximately [7]) the probability of reliability ($R = I/(I+C)$). This reliability indicates to what degree interoperability is impacted in presence of failure, given certain conditions (*i.e.*, choice of domain policies, *etc.*).

5.2   Mobility

To model and analyze mobility (Fig. 5), we add the notion of "time" to our model. In particular, we associate timeout values to state entries at gateways and *birthTime* and *deathTime* to routes (and similarly for reverse routes). We assume gateways are stationary, but other nodes can move, causing the "death" of their route (*route*1) to/from their closest gateway. A new route to the gateway is "born" (*route*2) after some time, assuming the existence of a domain-specific method to handle mobility. Temporal conditions must be incorporated into reachability/returnability properties. The most critical case is when a mobility event occurs while the packet is in-flight [21]. At high-level, the sum total latency formulated as *firstDeliveryAttempt+recovery+secondDeliveryAttempt*, must be below a certain *expiration* threshold (at every gateway and consumer). *firstDeliveryAttempt* is the incomplete partial delivery latency via *route*1 and *secondDeliveryAttempt* is the delivery via *route*2 (continuation in MF, and complete retransmission in IP and NDN). The *recovery* delay is the time it takes for the packet to be transmitted back on the new path again; it includes re-registration (MF and IP), FIB re-population (IP and NDN in case of provider mobility) and/or PIT re-population (for NDN in the case of consumer mobility) delays [16,20,21]. Using this formal method, we check properties in the presence of mobility, find appropriate values for a timeout threshold on gateways and investigate the effect of domain-specific mobility handling methods on interoperability. Listing 5.2 generally specifies how the reachability property (to deliver a named request) depends on the condition of mobility (stationary or mobile) and the domain policy on handling mobility (early binding or late binding). Returnability is similarly specified (for content). Predicates *stationary*, *mobileEarlyBinding*, and *mobileLateBinding* specify timing conditions for successful delivery assuming their corresponding conditions (details of the three properties are omitted here due to space but are in [1]). As shown in Fig. 5, we only consider intra-domain mobility here, *i.e.*, the mobile node changes its location and point of attachment, but stays within its domain.

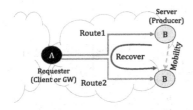

Fig. 5. Mobility scenario example: Route 2 established after B moves and changes its point of attachment

Table 3. Verif. scopes for properties of ICI services

Property	Client	GW	Repos/ Server/ Publisher	Domain	ContentID/ Keyword	Prefix	P'Tree	Content	Connections	Route	RevRoute	NodeID	Port	NDNreqID	MFreqID	Group	GroupID	Verif. Result
1.1	1	2	1	3	1			1	1	12	12	6						✓
1.2	1	2	1	3	1			1	1	12	12	6						✓
2.1	1	2	1	3	1			1	1	12	12	6	6	3	3			✓
2.2	1	2	1	3	1			1	1	12	12	6	6	3	3			✓
2.3.a	1	1	1	2	2			2	1	8	8	4	4	4	4			✓
2.3.b	2	1	1	2	1			1	1	8	8	5	4	4	4			✓
3.1	1	2	1	3	3	3	1	3	1	12	12					9	9	✓
3.2	1	2	1	3	3	3	1	3	1	12	12					9	9	✓
3.3	1	2	1	3	3	3	1	3	1	12	12					9	9	✓

Listing 5.2. Reachability in presence of mobility

```
pred reach[c:Client, p:Producer, cid:ContentID]{ -- a client and content producer
    (stationary[c,p,cid] && p.mobility in Stationary) -- producer p stationary
    || (mobileLateBinding[c,p,cid] && p.mobility in Mobility
        && Domain.binding in LateBinding) -- p mobile, domain does late binding
    || (mobileEarlyBinding[c,p,cid] && p.mobility in Mobility
        && Domain.binding in EarlyBinding)} -- p mobile, domain does early binding
```

6   Implementation and Results

We implemented the ICI framework discussed in our model in Sect. 3, with gateways for interoperation among IP, NDN, and MF (Fig. 1 as an example) in a software testbed (implementation details in [11]). This section provides the description and results of our analysis of the ICI framework (our Alloy source code is approximately 800 lines of code in total [1]).

To check for correctness, we performed verification (supported by Alloy Analyzer's model finding engine) of our ICI framework model, against the information-centric services properties (as specified in Sect. 4). In order to reach convincing proofs (as advised in [18]), we pick the scopes for verification in Alloy that are large enough to contain all necessary cases (*i.e.*, minimum number of actors and objects for each service), and small enough so that we do not encounter model explosion. The scopes, *i.e.*, upper bounds on the number of key objects, are provided in Table 3. For most properties, we consider 1 *Client*, 1 *Server*, 1 *Content*, and 1 *ContentID*. That is, different <client, request> pairs are considered independent of each other. However, for Properties 2.3.a/b, such a dependency matters, and we want to show lack of conflicts. For Property 2.3.a, we set 1 *Client* and 2 *Contents* (to generate scenarios where *one* client makes *two* separate request for *two* different contents), and for Property 2.3.b, we set 2 *Clients* and 1 *Content* (to look for conflicts between request for *one* content but by *two* clients). We use 3 *Domains* for most properties, as it contains all cases with 1, 2, or 3 domains of any type, *i.e.*, IP, NDN, or MF. Also, with upper

Table 4. Failure analysis results

Cases	Reachability			Returnability		
	I	C	R	I	C	R
No domain constraints	290	0	1.00	56	210	0.21
One NDN domain	176	0	1.00	8	168	0.04

Table 5. Mobility analysis results

Cases	Stationary			Mobile					
				Late binding			Early binding		
DL range	I	C	R	I	C	R	I	C	R
[0, 20]	100	8	0.92	72	24	0.75	92	64	0.58
[0, 18]	96	0	1.00	72	8	0.90	92	48	0.65
[0, 15]	84	0	1.00	64	0	1.00	92	24	0.79
[0, 10]	64	0	1.00	44	0	1.00	84	0	1.00

bound n on the total number of *Nodes*, *i.e.*, sum of *Clients*, *Servers*, and *GWs*, we specify the upper bound on the number of *Routes* (as well as *RevRoutes*) to be $n(n-1)$, enabling the existence of any possible (uni-directional) route. For pub/sub services (*i.e.*, Properties 3.1–3), we set 3 *Prefixes*, *ContentIDs*, and *Contents*, to capture inter-relationship of content IDs in a large enough namespace. Additionally, with the upper bound on *Domains* and *ContentIDs* both set at 3, we set the upper bound on total number of *Groups* (and *GroupIDs*) to be $3 \times 3 = 9$, so as to contain cases with one group per content ID per domain. The blank cells in Table 3 indicate either "N/A" or "no particular upper bound set", in which case Alloy picks a default value. Within this scope, our verification passes successfully for each property, showing that the stated properties are *invariants* of our ICI framework. In other words, the framework design ensures that *any sequence of interconnected IP, NDN, and MF domains are information-centrically interoperable*.

We use our proposed model counting approach to analyze scenarios with the failure of one or multiple gateways. The most important factor affecting returnability in scenarios with the possibility of failure, is domain-specific routing policies, in particular, whether or not it allows for a secondary (backup) gateway to relay the returning response content. Different domains have different policies; MF and IP decouple the forward (request) and return (response) paths, and they can be delivered through different gateways, while NDN strictly requires the two paths to be the same, due to RPF policy. To investigate the impact of that policy, we considered a scenario of two domains, with two gateways between them (one primary and one secondary), sharing state. Both gateways are *Up* (working) when the request is forwarded, and either *may* go *Down* (failing) when the response is one its way back. Table 4 shows different scenarios for reachability and returnability, with different domain constraints (with different routing policies). In particular, the two domain constraints we consider are the following: 1) no constraint on what any of the domains are; and 2) one domain is definitely NDN. The table shows the values of I (instances), C (counterexample), and R (reliability) for each scenario, as defined in Sect. 5. Our results for R in Table 4 prove that having an NDN domain on one side dramatically reduces the returnability reliability ratio, since basic NDN forwarding strictly forbids data coming back on a different path than the original path taken by the request.

When a content producer (server) moves while a content request is in-flight (Fig. 5), the domain's handling of mobility recovery determines the reachability probability. NDN and IP use early binding with retransmissions, while MF supports late binding with rerouting. We compare the impact of these mechanisms and techniques using our model counting method, with results shown in Table 5. Our modeled scenario consists of two nodes in a domain, one requester (client or gateway) and one server (producer) with a route established among them. The 'Stationary' columns in the table show reachability results in the stationary server case. With 'Mobile', the route dies due to a server mobility event (at time $t = 10$), leading to the birth of the second route. We set the re-registration and re-population delays to 1 each. Also, a retransmission is initiated 1 time unit after the mobility event. Different binding techniques for mobility, $i.e.$, late and early binding, are also shown in Table 5. We compare cases with different ranges for $Delivery$ $Latency$ (DL), which is time approximately needed for a packet to travel from requester to server. For a delivery latency range of $[0, 20]$, we see a higher R for stationary $vs.$ mobility cases. The reason is that when the server does not move, the original route stays active, thus providing a higher chance for requests to reach the server. Comparing the two binding techniques, late binding leads to higher chance of reachability compared to early binding, as it allows for packets to be re-routed on the newly-born route, rather than retransmitting from the original requester. These results serve as proof that under similar scenarios, late binding outperforms early binding in ICI. Also, changing the delivery latency ranges, we can find out at what points, reachability is an invariant (if ever) under mobility conditions. As the table shows, with ranges within $[0, 18]$, $[0, 15]$, and $[0, 10]$ (rows in Table 5 labeled in first column accordingly), reachability becomes an invariant in cases of Stationary, Late Binding, and Early Binding, respectively; as zero counterexamples are raised. With a small enough delivery latency ranges, namely $[0, 10]$, reachability becomes an invariant, no matter the mobility conditions or binding techniques. Our approach can be used to find such points of invariance, comparing different techniques, and prove them.

7 Conclusion

This paper presented an Alloy-based formal analysis model for information-centric interoperability (ICI) for Future Internet environments. We showed how model finding can be used to analyze basic (reachability and returnability) properties of ICI. Additionally, our proposed model counting approach analyzes failure and mobility scenarios, which we used to prove the negative impact of certain routing policies (particularly, reverse path forwarding), and the helpfulness of certain mobility-handling mechanisms (particularly, late binding), providing necessary confidence and guidelines for Future Internet interoperability.

Acknowledgements. This work was supported by the US Department of Commerce, National Institute of Standards and Technology (award 70NANB17H188) and US National Science Foundation grants CNS-1455815 and CNS-1818971.

References

1. https://www.cs.ucr.edu/~mjaha001/ICI.zip
2. Alloy: A Language and Tool for Relational Models. http://alloy.mit.edu/alloy/
3. Carofiglio, G., et al.: Enabling ICN in the internet protocol: analysis and evaluation of the hybrid-ICN architecture. In: ACM ICN (2019). https://doi.org/10.1145/3357150.3357394
4. Chakraborty, S., Meel, K.S., Vardi, M.Y.: A scalable approximate model counter. In: Schulte, C. (ed.) CP 2013. LNCS, vol. 8124, pp. 200–216. Springer, Heidelberg (2013). https://doi.org/10.1007/978-3-642-40627-0_18
5. Chavira, M., Darwiche, A.: On probabilistic inference by weighted model counting. AI **172**(6–7), 772–799 (2008). https://doi.org/10.1016/j.artint.2007.11.002
6. Chen, J., et al.: COPSS: an efficient content oriented publish/subscribe system. In: ACM/IEEE ANCS (2011). https://doi.org/10.1109/ANCS.2011.27
7. Gomes, C.P., Sabharwal, A., Selman, B.: Model counting: a new strategy for obtaining good bounds. In: AAAI (2006)
8. Jackson, D.: Alloy: a lightweight object modelling notation. TOSEM **11**(2), 256–290 (2002)
9. Jacobson, V., et al.: Networking named content. In: CONEXT (2009)
10. Jahanian, M., Ramakrishnan, K.K.: Name space analysis: verification of named data network data planes. In: ACM ICN (2019). https://doi.org/10.1145/3357150.3357406
11. Jahanian, M., et al.: Managing the evolution to future internet architectures and seamless interoperation. In: Proceedings of the 29th International Conference on Computer Communication and Networks (ICCCN) (2020)
12. Jahanian, M., et al.: Graph-based namespaces and load sharing for efficient information dissemination in disasters. In: ICNP (2019). https://doi.org/10.1109/ICNP.2019.8888047
13. Le Berre, D., Parrain, A.: The SAT4J library, release 2.2, system description. J. Satisf. Boolean Model. Comput. **7**, 59–64 (2010)
14. Li, Y., et al.: A survey on network verification and testing with formal methods: approaches and challenges. IEEE Commun. Surv. Tutor. **21**(1), 940–969 (2019)
15. McCauley, J., et al.: Enabling a permanent revolution in internet architecture. In: ACM SIGCOMM (2019). https://doi.org/10.1145/3341302.3342075
16. Raychaudhuri, D., et al.: MobilityFirst: a robust and trustworthy mobility-centric architecture for the future internet. ACM SIGMOBILE MCCR **16**(3), 2–13 (2012)
17. Torlak, E., Jackson, D.: Kodkod: a relational model finder. In: Grumberg, O., Huth, M. (eds.) TACAS 2007. LNCS, vol. 4424, pp. 632–647. Springer, Heidelberg (2007). https://doi.org/10.1007/978-3-540-71209-1_49
18. Zave, P.: A practical comparison of alloy and spin. Formal Aspects Comput. **27**(2), 239–253 (2015). https://doi.org/10.1007/s00165-014-0302-2
19. Zave, P.: A formal model of addressing for interoperating networks. In: Fitzgerald, J., Hayes, I.J., Tarlecki, A. (eds.) FM 2005. LNCS, vol. 3582, pp. 318–333. Springer, Heidelberg (2005). https://doi.org/10.1007/11526841_22
20. Zhang, L., et al.: Named data networking. ACM SIGCOMM CCR **44**(3), 66–73 (2014)
21. Zhang, Y., et al.: KITE: producer mobility support in named data networking. In: ACM ICN (2018). https://doi.org/10.1145/3267955.3267959

Experiences on Teaching Alloy
with an Automated Assessment Platform

Nuno Macedo[1,2], Alcino Cunha[1,2(✉)], José Pereira[2], Renato Carvalho[1,2],
Ricardo Silva[2], Ana C. R. Paiva[1,3], Miguel Sozinho Ramalho[1,3],
and Daniel Silva[3]

[1] INESC TEC, Porto, Portugal
[2] University of Minho, Braga, Portugal
`alcino@di.uminho.pt`
[3] University of Porto, Porto, Portugal

Abstract. This paper presents Alloy4Fun, a web application that
enables online editing and sharing of Alloy models and instances (including dynamic ones developed with the Electrum extension), to be used
mainly in an educational context. By introducing secret paragraphs and
commands in the models, Alloy4Fun allows the distribution and automated assessment of simple specification challenges, a mechanism that
enables students to learn the language at their own pace. Alloy4Fun
stores all versions of shared and analyzed models, as well as derivation
trees that depict how they evolved over time: this wealth of information
can be mined by researchers or tutors to identify, for example, learning breakdowns in the class or typical mistakes made by Alloy users.
Alloy4Fun has been used in formal methods graduate courses for two
years and for the latest edition we present results regarding its adoption by the students, as well as preliminary insights regarding the most
common bottlenecks when learning Alloy (and Electrum).

Keywords: Teaching formal methods · Alloy · Automated assessment

1 Introduction

Alloy [6] is a popular formal specification language, accompanied by a toolkit, to
describe and reason about software design. It is taught in several undergraduate
and graduate courses in formal methods, including graduate courses taught by
some of the authors at University of Minho (UM) and University of Porto (UP),
in Portugal. One of the reasons for this popularity is the support for automated
analysis provided by the Alloy Analyzer, an easy to download and install self-
contained executable written in Java. The Analyzer also allows instances (either
witness scenarios or counter-examples) to be graphically depicted using user-
customized themes, a popular feature both for experienced users and students.
Alloy is very effective in the specification and analysis of the static structures
that pervade software design, but requires the employment of well-established

A. Raschke et al. (Eds.): ABZ 2020, LNCS 12071, pp. 61–77, 2020.
https://doi.org/10.1007/978-3-030-48077-6_5

idioms, that introduce an explicit notion of state or time, if mutability is to be considered and temporal properties analyzed. To avoid this cumbersome and error-prone process, several extensions to Alloy have been proposed, including one by authors of this paper – Electrum [7] – which extends the Alloy language with variable structures and linear temporal logic (including past operators), also adding bounded and unbounded model checking engines to the Analyzer.

Despite such streamlined toolkit, over the many years we taught and researched with Alloy we identified some missing features and functionalities that could further ease its adoption and its usage in an educational context. The first is the lack of a straightforward mechanism to *share* simple Alloy models, instances[1] and associated themes. This would be particularly useful for students trying to get feedback from the tutors about specific counter-examples, or to submit exercise resolutions for evaluation. The second is the absence of some *automated assessment* functionality or online judge system for students to independently check the correctness of their exercise resolutions. Due to some limitations of the visualizer packaged with the Analyzer, we also felt the need for a more decoupled infrastructure to test alternative instance *visualization* features.

To address these limitations we developed Alloy4Fun, a web application that enables online editing and sharing of Alloy and Electrum models[2] and instances, including simple specification challenges in the form of duels where students attempt to discover a secret specified by the tutors. Such online platform also provided us the opportunity to collect information regarding Alloy usage patterns from an extended user base: one of the features of Alloy4Fun is thus the ability to record every interaction with the (anonymous) user, information that is made available to the creator of the challenges for subsequent analysis. Over the last two years, Alloy4Fun has been used in 3 editions of graduate courses on formal methods and a tutorial at an international venue, which has allowed us to quickly obtain insight on how students use the language, namely identify typical mistakes or learning breakdowns in the class.

This paper presents Alloy4Fun and reports on its application in teaching Alloy, starting with an overview of (and rationale for) its current features in Sect. 2. Section 3 reports on its deployment in a formal methods graduate course (Sect. 3.1), including our experience on defining exercises, results regarding usage and adoption of the platform (Sect. 3.2), and some preliminary insights on Alloy usage patterns and learning pitfalls (Sect. 3.3). Finally, Sect. 4 concludes the paper and presents some ideas for future work. Knowledge of Alloy is not required to understand the paper, but can help better appreciate some of the features of Alloy4Fun.

---

[1] In Alloy literature, specifications are usually referred to as models, and the results of animation/verification commands as model instances.

[2] Electrum is retro-compatible with Alloy: models without temporal features are valid Alloy, apart from protected keywords. For readability we will simply refer to Alloy throughout the paper, unless some Electrum-specific feature is being discussed.

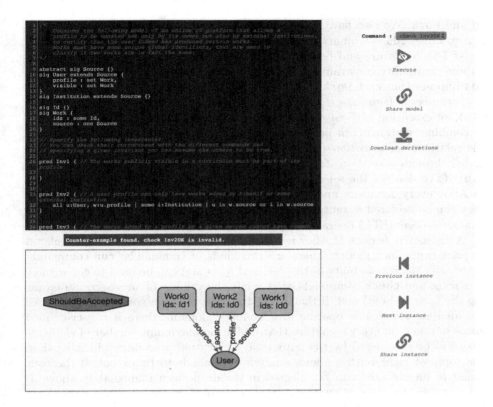

Fig. 1. A failed attempt to solve a challenge in the CV exercise.

2   Alloy4Fun Overview

The core of Alloy4Fun mimics in a web application the main features of the standalone Alloy Analyzer. After accessing alloy4fun.inesctec.pt (the URL where Alloy4Fun is currently deployed) the user gets an empty online editor (with syntax highlighting) where Alloy models can be written. An Alloy model consists of a sequence of paragraphs: each paragraph is either a *signature* (and the respective *fields*) declaration, a *fact* with a constraint that is assumed to hold, an *assertion* with a constraint to be checked, or an auxiliary *predicate* or *function* definition. Signatures introduce sets of elements (known in Alloy as *atoms*) and fields establish relations of arbitrary arity between those sets. Disjoint subset signatures can be declared by *extension*, and the parent signature can be marked as *abstract*, if it should only contain atoms present in its extensions. For example, the Alloy4Fun screen capture shown in Fig. 1 shows a model of an online *Curriculum Vitae* (CV) platform, an example that was used as an exercise in classes. This model declares a signature Source that is partitioned in two subsets, User and Institution. Two more signatures are declared in this example:

Id and Work. We also have several fields that relate atoms of these signatures. For example, ids is a binary relation that associates each atom of Work with its set of Ids. Signature and field declarations can have *multiplicities* attached to impose cardinality constraints. For example, the **some** in the declaration of field ids imposes that each Work should have at least one Id.

Formulas in facts, assertions, and predicates, are written in *Relational Logic* (RL), an extension of *First-Order Logic* (FOL) with operators that can be used to combine relations (*aka* predicates in FOL). The most frequently used one is the relational *composition* (written as .), an operator that allows us to "navigate" through a relation: for example, in predicate Inv2 of Fig. 1, expression u.profile denotes the set of atoms of signature Work associated with User u. In Alloy every signature and field is immutable. With the Electrum extension they can be declared as mutable, and formulas can also be specified with *Linear Temporal Logic* (LTL) operators.

A distinctive feature of Alloy is that analysis commands can also be declared as paragraphs in a model. There are two kinds of commands: **run** commands, that verify the satisfiability of the declared facts and can be used to get witness scenarios; and **check** commands, that verify the validity of an assertion (assuming the facts to hold) and, if that is not the case, return a counter-example. All the analysis commands operate in a bounded domain: there is a user-defined *scope* imposed on every signature that limits the maximum number of elements that will be considered by the automatic verification procedures. In Alloy4Fun the topmost right button allows analysis commands to be executed: the command to be executed can be selected in the drop-down immediately above. If witnesses (in the case of **run**) or counter-examples (in the case of a **check**) are found, they are depicted below the editor as graphs that, likewise in Analyzer, can be customised with user-defined themes.

Besides these core functionalities, Alloy4Fun has some new features (and some improvements to existing ones) when compared to the Analyzer, as described in the sequel. Currently, it also has some limitations, most notably the inability to choose the underlying SAT solver used to perform a given analysis, not being able to display an unsatisfiable core, and lack of support for Alloy's module system (except for the standard modules distributed with Alloy, which can be used). In the specific case of Electrum, Alloy4Fun lacks the more sophisticated trace exploration options available in the Electrum Analyzer [3], as described next.

Instance Visualization and Navigation. When compared to the Analyzer, Alloy4Fun follows a more lightweight approach to the user interface, allowing the most common theme customizations (like changing the color of the atoms of a given signature) to be performed quickly through a right-click menu on atoms or edges. We also stripped down a bit theme features to a subset that we identified as those more commonly used. Alloy4Fun themes allow color, shape, stroke, and visibility parametrization for signatures and fields, signature projection, and the display of fields as attributes inside atoms. Among the unsupported features we have, for example, the customization of the atom labels for each signature or the ability to hide only unconnected atoms of a particular signature. A new feature is the ability to select different layout algorithms to automatically organize nodes, which the user can then manually move. Unlike in the Analyzer, atom positions are preserved between the frames of projected instances, and when navigating the different states of a trace in the case of an Electrum (mutable) instance. In Fig. 1 a counter-example of a **check** command named Inv20K is being depicted with a user-defined theme. Unlike in the Analyzer, besides navigating to the next instance the user can also re-visit previously presented instances. In the case of Electrum, Alloy4Fun only allows one state of an instance trace to be visualised at a time (the Electrum Analyzer depicts two states side by side), and it is only possible to ask for a different next trace (the Electrum Analyzer has more sophisticated trace exploration options, for example it is possible to ask for trace with the same prefix up to the displayed state, but a different next state).

Sharing Models and Instances. The standard Alloy Analyzer provides limited support for model and instance sharing: they can be saved in separate files, which can then be shared using external tools (email, online repositories, etc), to be again opened at the destination for inspection or editing. When a visualization theme has been developed to ease the interpretation of instances, it must also be shared in an additional file. This sharing by saving/opening files rapidly becomes tedious and time consuming in some contexts, in particular for tutors of large classes that interact frequently with students (typically by email) to clarify doubts. Alloy4Fun provides the ability to easily share models and instances. After pressing the "share model" button a *permalink* is generated, that can later be used to access the model. Any theme defined by the user is also preserved when sharing, thus allowing instances of shared models to be depicted as intended by their creators. Concrete instances can also be shared via *permalinks*. The theme and positions of the depicted atoms and relations at the time of sharing are also preserved. This is a very handy feature since, likewise in the Analyzer, the positioning of atoms by the automatic layout mechanism is often not ideal, requiring manually rearrangement for better comprehension. For instance, the instance presented in Fig. 1 can be shared as depicted[3].

Anonymous Interaction. In Alloy4Fun there are no user accounts nor means to recover the *permalinks* of previously shared models and instances. The user is responsible for keeping track of relevant *permalinks* using some external mechanism (Alloy4Fun provides a "copy to clipboard" button to ease this task). The anonymity, namely the absence of user accounts, was a design choice made in order to keep the interaction with the web application as simple as possible, to maximize user exposure, and also to avoid dealing with privacy and security issues, namely the hassle of storing and managing user credentials and of implementing mandatory regulations concerning data protection.

Automatic Assessment. Although the Alloy specification language has very neat and simple syntax and semantics, many students struggle with its declarative

---

[3] http://alloy4fun.inesctec.pt/8Q4Sbjqj4KzHuvuNC.

nature, in particular those used to procedural programming [2]. One way to overcome this difficulty is by independently solving exercises proposed by tutors, but, even with automated analysis and visual feedback, it is often difficult for students to assess whether they reached the correct answer, and tutors are required to inspect and interpret the solutions (something not scalable for large classes). These problems could be mitigated with automatic assessment functionalities, allowing students to solve exercises at their own pace and without the constant need for face-to-face time with tutors. In recent years, auto-graders and online judges have become widely popular for learning how to program [10], and we believe this success could be replicated in the learning of formal methods in general, and Alloy in particular.

With this in mind, the user in Alloy4Fun has the ability to mark any paragraph of a model as *secret*, by adding the special comment *//SECRET* immediately before. When sharing a model with secret paragraphs two *permalinks* are generated: a private one that, when accessed, reveals the full model, including secrets; and a public one that, when accessed, only shows public paragraphs, but internally still considers the secret in analyses and still allows the execution of secret commands (whose names are public). Using a comment instead of a new keyword to mark secret paragraphs ensures compatibility with Alloy's default syntax, allowing users to copy and paste models from Alloy4Fun to the standalone Analyzer, and vice versa. Section 3.1 will describe how this feature can be used to create simple specification exercises in the form of duels, where the user/student tries to reach a secret specification. The instance shown in Fig. 1 was obtained precisely by accessing the public *permalink* of an exercise, and failing to solve a challenge, for which a counter-example was returned.

Mining Derivation Trees. A possible way to gain insight about the students' learning process is to have access to their attempts at solving the proposed exercises, and tool support to mine this corpus for useful data [8]. Again, such feature would also be useful for research, and was one of the reasons that led Microsoft to develop the www.rise4fun.com web service, that allows researchers to easily deploy their tools on the web and collect human-tool interactions for posterior mining [1] (besides other advantages of web tools, like increased exposure, since the need for downloading and installing is eliminated, and promoting reliability given the large amount of test cases that can be collected). One of the most popular examples available via Rise4Fun, and the inspiration for developing Alloy4Fun, is www.pex4fun.com, a web-based educational gaming environment for learning programming, where students can engage in coding duels where they attempt to write code equivalent to a tutor's secret implementation [12]. Pex [11], an advanced white box test-generation tool, is used on the background to find inputs that show discrepancies between the student's code and the secret implementation. However, the interaction with the outcome of the tools has limitations in Rise4Fun, which would prevent the implementation of key Alloy features such as instance iteration and customization. This has led us to implement our own solution rather than integrate Alloy in this service.

Every shared model and instance is stored by Alloy4Fun in its database. However, to enable the proponents of challenges to mine the submissions for useful information, every model for which a command was executed is also stored, along with the respective result (e.g., whether satisfiable or not, or whether errors were thrown). Moreover, for each model, the identifier of the model from which it derives and a time-stamp are also stored. This means that all the models that are developed after accessing a shared *permalink* end up forming a *derivation tree*. In the case of a *permalink* with secrets/challenges, a branch in this tree typically corresponds to an interactive session where one user/student is trying to solve the different challenges defined inside, and can be analyzed to determine, for example, how many challenges were solved or how many attempts were needed to solve each one. Every fork in branch represents a point where a user generated a new *permalink* for a model which was subsequently accessed multiple times. Alloy4Fun allows anyone in possession of the secret *permalink* of a model to download the respective derivation tree in an easy to process JSON format.

Implementation. Alloy4Fun was developed [9] with Meteor, a full-stack isomorphic JavaScript framework for developing web applications based on Node.js. The client uses CodeMirror as text editor and the Cytoscape.js graph visualization library to depict instances. Models and instances are stored in a MongoDB document-oriented database at the server. To execute commands, we encapsulated the Alloy Analyzer in a RESTful web service implemented in Java. Seamless deployment of both the application and the service in a server is performed using Docker. All the Alloy4Fun code is open-source and available at github.com/haslab/Alloy4Fun.

3   Experiences on Teaching with Alloy4Fun

In the first semester of the 2018/19 academic year we did a preliminary evaluation of Alloy4Fun in two graduate formal methods courses at UM and UP. The former taught Alloy for 6 weeks and had 22 students enrolled, and the latter for 4 weeks and had 156 students enrolled. Both courses had one weekly lecture and one weekly lab session. This experiment – which recorded almost 5000 interactions – allowed us to test a beta version of the application in a medium-sized audience to detect and fix bugs and identify possible design improvements. One major identified design improvement regarded a special "lock" comment available in the beta version to prevent the accidental editing of certain paragraphs that could render the challenges unsolvable (or trivially solvable). However, we noticed students rarely tried to change the model outside of the challenge predicates, and opted to remove this feature for simplicity and efficiency[4]. These first experiences also allowed us to identify which classes of exercises are better suited to be explored in Alloy4Fun, as well as how the visualization features can be explored to provide more intuitive feedback to the students.

---

[4] Note that Alloy4Fun was only used for self-study and not for student grading.

From this process resulted the first official release of Alloy4Fun, which has been used in the 2019/20 academic year in the UM graduate course and on an Alloy/Electrum tutorial at the World Congress on Formal Methods[5], with a refined set of specification exercises with challenges. The remainder of this section reports on the usage of the platform by the students during this latest instance of the UM graduate course, including preliminary results regarding the most common mistakes and difficulties when learning Alloy.

3.1 Alloy4Fun Exercises

The model secrets supported by Alloy4Fun can be used to create simple specification challenges in the form of duels, where the user/student tries to reach a secret specification. Such models – which we refer to as *exercises* – can have a public predicate that the student must fill-in, together with a secret **check** command that asserts (for a given scope) that such predicate is equivalent to the desired specification (typically in a separate secret predicate). Although useful for practicing the usage of logic (either relational or temporal) in the specification of properties, there are certain classes of problems for which the approach based on secret specifications is not well-suited, namely modeling exercises where the student is expected to freely declare signatures and fields.

The model shown in Fig. 1 was obtained precisely by accessing the public *permalink* of the CV exercise, which contains 4 challenges (in this case, simple problems where a natural language description of a desired property of the model is given for each of them). After filling the empty predicate (e.g., Inv2), the student can check whether it is a valid solution (e.g., by running secret command Inv2OK, for the case of Inv2), which will either return a "no counter-example found" message, meaning the challenge is solved, or a counter-example otherwise (as is the case in Fig. 1, showing that the specification of Inv2 is still not correct).

Figure 2 shows the secret implementation of challenge Inv2: predicate Inv2o specifies a correct solution for the challenge and command Inv2OK checks the equivalence between both. In exercises such as CV where several desired (and natural) properties of the model are solved in different challenges, we opted to check this equivalence assuming that the remaining properties hold: if that was not the case the student would get many counter-examples where it would not be clear why their specification failed, since they would be "polluted" with distracting problems corresponding to failures of other properties. This conditional check is the reason to include Inv1o, Inv3o, and Inv4o as assumptions in the equivalence check Inv2OK. Notice that in the preamble to the exercise the students are warned that they can assume the properties in the remaining challenges to be true when solving a particular challenge.

During the course we also noticed that the students found it hard to distinguish whether the provided counter-example represents a scenario where their solution was over-specified or under-specified. For this reason, in the challenges

---

[5] http://haslab.github.io/TRUST/tutorial.html.

```
//SECRET
abstract one sig RejectedBy {}
//SECRET
sig ShouldBeRejected, ShouldBeAccepted extends RejectedBy {}
...
pred Inv2 { // A user profile can only have works added by himself or some external institution

}
//SECRET
pred Inv2o { all u : User | u.profile.source in Institution+u }
//SECRET
check Inv2OK {
    (Inv1o and Inv3o and Inv4o and (some ShouldBeRejected iff (Inv2 and not Inv2o))) implies
    (Inv2 iff Inv2o) }
```

Fig. 2. The secret for the challenge `Inv2` of `CV` from Fig. 1.

used later in the course we opted to include two special atoms in the counter-example instance that signal whether an instance that should have been rejected or accepted by a correct specification, meaning their solution is under- or over-specified, respectively. As seen in Fig. 2 this can be achieved by introducing a singleton signature whose possible values are either `ShouldBeRejected` or `ShouldBeAccepted` and through a simple trick in the equivalence check, namely making the verification conditional to the existence of the `ShouldBeRejected` atom when the student solution incorrectly holds (or vice-versa).

The challenges used in this course were based on 6 different problems:

- `Trash`, a model of a file system trash bin.
- `Classroom`, a model a classroom management system.
- `Graph`, a specification of several standard properties of unlabeled graphs.
- `LTL`, a specification of several standard properties of labeled transition systems.
- `Production`, a model of an automated production line in a factory.
- `CV`, the *Curriculum Vitae* model used as running example in this paper.

For some of these problems we developed more than one variant (or *exercise*) focusing on different features of the language. Each variant was provided as a shared model to students and contained multiple challenges, as summarized in Table 1. The table lists the *permalink* and total number of challenges of each exercise (the columns F1 to F9 will be discussed in Sect. 3.3).

Challenges in these exercises range from trivial (e.g., asking to enforce simple inclusion dependencies or multiplicities), to more complex ones requiring the use of nested quantifiers or closures. As expected, the introduction of the Alloy (and Electrum) language and underlying logics in classes was gradual: FOL constructs were first presented, followed by the full set of RL operators, and finally the LTL operators specific to Electrum. To try to understand the impact of using relational operators, we introduced two variants of the first two problems: one where challenges were to be solved using only the FOL subset of Alloy, and another, introduced when students already had knowledge of RL, where they

Table 1. Alloy4Fun exercises shared for the 2019/20 year.

Id	Exercise	Permalink	Chall.	F1	F2	F3	F4	F5	F6	F7	F8	F9
1	Trash FOL	zA2MMSGy6iW8Mihep	10	0	1	2	0	0	0	0	0	0
2	Classroom FOL	Pdvipvrpr5hg7JKbs	15	6	6	9	4	0	0	0	0	0
3	Trash RL	WJdLnDL78m7mM7W4J	10	0	1	2	0	0	0	0	0	0
4	Classroom RL	i5u2pjKJt6Bz227QT	15	5	6	9	4	1	0	0	0	0
5	Graphs	28fwdmjL79X4SQ9EP	8	1	0	0	0	2	1	0	0	0
6	LTS	gqS3qTTn4B62NYmJX	7	4	2	6	6	0	2	0	0	0
7	Production	PKy7chamCieZyCix5	4	1	1	3	0	1	0	1	0	0
8	CV	X72J6js9fA3CKYQWX	4	3	0	3	0	0	1	0	0	0
9	Trash LTL	irRLJn7qbQq3xMFGp	20	0	0	1	0	0	0	0	5	14

could use all the standard Alloy operators to solve the challenges. For the Trash problem we also created a mutable variant, where challenges required the usage of the LTL operators of Electrum to be solved. Hence the total of 9 exercises described in Table 1. As an example, exercise CV (containing 4 challenges) is the one shown in Fig. 1.

3.2   Student Usage and Adoption

In the 2019/20 edition 17 students attended the UM course. Alloy was taught for 5 weeks and, for the first time in this course, Electrum was also taught for 4 additional weeks. In each week, a 1 h lecture was followed by a 2 h lab session. Alloy4Fun was used in the lab sessions that followed the lectures that introduced FOL, RL, and LTL, mainly as a way to practice the usage of these logics to specify natural language requirements.

In the lab sessions that addressed other aspects of the Alloy language and analysis not amenable for automated assessment, such as solving problems that required the development of a full model from scratch, students were expected to still use the Alloy Analyzer and locally manage their models. In principle, they could also have used Alloy4Fun to develop most of the problems addressed in those sessions, but we also wanted students to gain some experience in using the standard Analyzer, particularly since the current limitations of Alloy4Fun (presented in the beginning of Sect. 2, such as the lack of module support or the lack of sophisticated trace exploration options in the case of Electrum) might prove problematic for some more realistic problems. Thus, Alloy4Fun was only used in 4 lab sessions, each introducing a particular set of exercises – 1 session with Trash FOL and Classroom FOL after the FOL lecture, 2 sessions with Trash RL, Classroom RL and Graphs after the RL lecture, and 1 session with Trash LTL after the LTL lecture. Extra exercises (namely LTS, Production, and CV) were made available in the course website for the students to freely explore. Moreover, all exercises were kept available throughout the semester so that students could independently practice outside of the classes. During the course there were 3 evaluation points involving Alloy: a medium-size modeling project (developed with the standard Analyzer outside of the classes in groups of two students), an individual written exam, and finally a supplementary exam for students failing the first attempt.

After concluding the course, the main question we tried to answer was whether students found Alloy4Fun useful as an automated assessment platform while learning Alloy. More specifically: 1) have the students used Alloy4Fun regularly outside classes? 2) in particular, have they used it when studying for the exams? 3) have they found the sharing feature useful? 4) were the counter-examples useful to reach the correct solution? To answer these question we used two methods: an anonymous questionnaire and analysis of the data collected by Alloy4Fun. The questionnaire was answered by 13 of the 17 students, and, over the duration of the course, we collected almost 11000 interactions with the exercises, most of them resulting from the execution of commands (checking the correctness of challenges) and a small portion from sharing of models[6].

Concerning the first question, of the 13 students that answered the questionnaire, 9 said they used Alloy4Fun frequently outside classes, 3 only used it rarely, and 1 never used it. To the second question all of the 12 students that used it outside classes answered that they used it to study for the exam. Of these, 9 mentioned that when studying for the exam they actually repeated some of the exercises they had already solved before. The data collected throughout the semester, shown in Fig. 3, seems to corroborate these answers. Figure 3a depicts the usage of the platform over time, highlighting the classes where Alloy4Fun was mandatory and the evaluation points (first the project deadline, and later in the semester the two exams). Each entry in the dataset is either a *correct* (unsatisfiable) check, a *wrong* (satisfiable) check, an analysis that threw an *error* (e.g., parsing) or a model stored for *sharing*. Despite the peak of usage during the Alloy4Fun classes, we can see that the students have indeed relied on Alloy4Fun outside the classes, and in particular when studying for the written exam.

Figures 3b and 3c present statistics per exercise (below each exercise number we recall the number of challenges inside). Figure 3b presents the same execution information as Fig. 3a (except shares), with the addition of the number of successful analyses (i.e., without error) that threw a *warning*. This information is normalised taking into account the number of challenges in each exercise (i.e., the graph shows the average number of executions per challenge). This chart provides some evidence that most of the students attempted to solve all exercises, including some of those not used in class. For example, averaging the executions per challenge and per student, we have a maximum of around 10 for exercise 1 and a minimum of around 3 for exercise 7, and an overall average of around 6 attempts per challenge per student. Even taking into account failed attempts and repeated attempts to solve exercises already previously solved, it is relatively safe to infer that such numbers can only have resulted from having most of the class attempting to solve all exercises.

Figure 3c presents information regarding solving "sessions". Recall that a session is a branch in the derivation tree, typically recording the interaction of a

---

[6] This dataset is freely available in the Alloy4Fun GitHub repository.

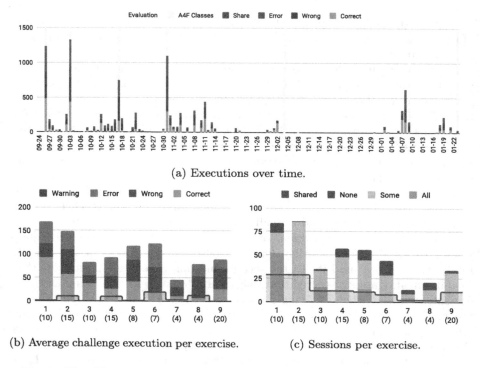

(a) Executions over time.

(b) Average challenge execution per exercise.   (c) Sessions per exercise.

Fig. 3. Alloy4Fun usage statistics by 17 students over a semester for 9 exercises.

student with Alloy4Fun while solving the challenges inside an exercise. For each exercise we depict how many session solved all its challenges, some of its challenges, or none. Of course, some students might have multiple sessions recorded for each attempt to solve an exercise, since they might not solve all the challenges in a single continuous session and access the original shared *permalink* several times, instead of generating a new *permalink* of a partial resolution for later resuming the work. Overall we identified 430 sessions, with an average of 48 sessions per exercise. Even with all the uncertainty, it is safe to say that indeed most students should have used Alloy4Fun frequently outside the classes (from our observation, during classes students mainly used a single session per exercise), including repeated attempts to solve exercises already previously solved (as reported in the questionnaire): for example, for Trash FOL around 50 sessions were recorded where all the challenges were solved, a strong indicator that each student should have solved it at least twice.

Concerning *permalinks*, 7 students mentioned that they generated them frequently to store their own solutions for later access, 3 did it rarely, and, somehow surprising, 3 never did it. Generating *permalinks* for the purpose of sharing with colleagues and tutors was even less common: only 5 students did it frequently, 4 rarely and 4 never. Figure 3c also depicts how many session had at least one *permalink* generated, and indeed we can see that, for most of the exercises, the number of *permalinked* sessions is clearly less than the number of students. Surprisingly, the share instance feature has not been used: there were only 2 generated *permalinks* for instances. These results seem to suggest that one of our main goals for Alloy4Fun – to simplify the sharing of models and instances – may actually not be that popular in an educational setting, but of course a more comprehensive study must be conducted to clarify that.

Concerning the last question, 10 students mentioned that counter-examples were frequently useful to help find the correct answer, but of these 4 only found them useful if they had the atoms that signal whether the shown counter-example should have been rejected or accepted by a correct specification. Unfortunately we have no data to corroborate this, but in principle Alloy4Fun could be used to check whether those atoms are indeed helpful or not, for example by giving two different versions of an exercise to different sets of students and then analyzing the results. This is one of the studies we intend to conduct in the near future.

Finally we also asked the students the overall question of whether they found Alloy4Fun useful for learning Alloy and Electrum: all of them agreed that was the case, with 8 of the 13 strongly agreeing.

3.3   Insights on Learning Alloy

Taking advantage of the collected data, we also tried to get some insights about how students learn Alloy, and in particular determine which features of the language pose more difficulties and should thus be addressed more carefully in lectures. To this end, we started by classifying a normalized version[7] of each challenge according to a set of required concepts, namely whether it requires:

F1 using more than 10 logic or relational operators
F2 a simple restriction of the multiplicity of a relation
F3 nested quantifications (ignoring multi-variable quantifications)
F4 manipulating ternary relations
F5 transitive closure over fields
F6 transitive closure over expressions (either relational expressions or relations by comprehension)
F7 reasoning about total orders (i.e, using the `ordering` module)
F8 a single temporal operator
F9 nested temporal operators

For each exercise, Table 1 presents the number of challenges that fall into each of these (non-exclusive) categories. Figure 4 compares the results of challenge execution classified under each category (also listing the total number of challenges for each). For each of the 9 categories, the number of correct (green) and wrong (red) executions are presented. Additionally, entry F0 collects the results of challenges that require none of the above concepts, and category All the results for all challenges. Of all the 7689 executions without errors, 3682 were correct (48%), meaning that in average each challenge required two attempts to be solved (after solving possible errors).

---

[7] Normalized specifications were expanded into almost pure FOL (or FO-LTL when temporal logic was required), using no relational operators except for closures.

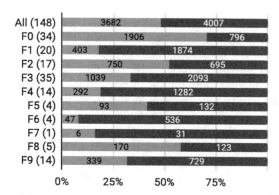

Fig. 4. Executions per class of challenge. (Color figure online)

As expected, challenges requiring none of the listed concepts (F0) were simpler (71% success rate), and those requiring more than 10 operators (F1) were notoriously more difficult (18% success rate). Contrary to our expectations, given that Alloy has special syntax for that purpose, challenges that required restricting the multiplicity of relations (F2) were only slightly easier than average (52%). As expected, the need to use nested quantifiers (F3) increases the difficulty of challenges (33% success rate). Concerning closures, usage of a closure operator over a relation (F5) was not very problematic (41% success rate), but challenges that required applying a closure operator to a relational expression (F6) were the most difficult to solve (8% success rate). We had some anecdotal evidence that closures were difficult for students, but this discrepancy between the two cases was rather surprising, meaning that special attention should be given to the later case in lectures. Other problematic concepts were the manipulation of ternary relations (F4) (19% success rate), and usage of the standard `ordering` module (F7) (16% success rate), both frequently used in Alloy specifications. The first result is aligned with our anecdotal evidence, and we already had special care with higher arity relations in lectures. The second is a bit more surprising, meaning that, likewise to closures of relational expressions, we should invest more lecture time in explaining how to use this module. Concerning Electrum, students seem to understand well the usage of a single temporal operator (F8) (58% success rate), but, as expected and likewise quantifiers, specifications requiring nesting of several temporal operators (F9) were more difficult (32% success rate).

We also collected statistics about typical errors and warnings, with Tables 2 and 3 presenting the 10 most commonly found error and warning messages, respectively. Concerning errors, as expected, the most frequent are basic parsing errors (corresponding to messages 1, 2, and 8, and including, for example, parenthesis problems or misspelled identifiers), totaling around 44% of the errors. Of the remaining, the most frequent are incorrectly applying logic operators to relational expressions and vice-versa (messages 3, 5, and 7), in total 28% of

Table 2. Most common error messages.

	Message	#
1	There are ... possible tokens that can appear here.	747
2	The name ... cannot be found.	444
3	This must be a formula expression.	432
4	**in** can be used only between 2 expressions of the same arity.	277
5	This must be a set or relation.	277
6	This cannot be a legal relational join.	220
7	This expression failed to be typechecked.	117
8	The "**all** x" construct is no longer supported.	85
9	~ can be used only with a binary relation.	58
10	This must be a unary set.	50

Table 3. Most common warning messages.

	Message	#
1	The join operation here always yields an empty set.	213
2	Subset operator is redundant, because the left & right subexpressions are always disjoint.	123
3	This variable is unused.	121
4	^ is redundant since its domain and range are disjoint.	25
5	= is redundant, because the left & right expressions always have the same value.	11
6	<: is irrelevant because the result is always empty.	10
7	& is irrelevant because the two subexpressions are always disjoint.	8
8	= is redundant, because the left & right expressions are always disjoint.	8
9	The value of this expression does not contribute to the value of the parent.	6
10	Subset operator is redundant, because the right subexpression is always empty.	3

the errors, and simple typing errors related to arity (messages 4, 6, 9, and 10), in total 26% of the errors. The reader unacquainted with Alloy could find the frequency of the former rather surprising, but this is a rather frequent error due to the syntactic similarity between some logical and relational operators (for example, **not** for negation vs. **no** for emptiness check or && for conjunction vs. & for intersection). Fortunately, Alloy has alternative syntax for many logic operators (for example, **and** for conjunction) and maybe instructors should recommend using that alternative instead. Concerning warnings, all but the third most common message (unused variables, 23% of the total warnings) are warnings about potentially irrelevant expressions – formulas that are trivially true or false or expressions that always denote an empty set – a testimony to the usefulness of Alloy's sophisticated type system [5].

4   Concluding Remarks and Future Work

We briefly presented Alloy4Fun, a web application for online editing and sharing of Alloy models and instances, that also allows the automated assessment of simple specification challenges. Its main intended use is in an educational context, and our preliminary evaluation in a graduate formal methods course provided evidence that students found the automated assessment feature useful for learning Alloy and Electrum (and the sharing feature less so). We also collected evidence that some features of the Alloy language are particularly problematic for students, and should be addressed with particular care by tutors.

We intend to continue using Alloy4Fun in our formal methods courses in the upcoming years, collecting more data to support more detailed and informed analyses about the language usage. Concerning the application itself, we intend to develop tools to simplify the mining of useful data from the derivation trees, possibly to be run server-side at the click of a button (with results visualized in the browser), to enable the timely identification of learning breakdowns. We also intend to incorporate in Alloy4Fun an alternative instance visualizer more amenable for dynamic systems [4].

Acknowledgements. We would like to thank Daniel Jackson for the helpful comments and suggestions about the design of Alloy4Fun. This work is financed by National Funds through the Portuguese funding agency, FCT - Fundação para a Ciência e a Tecnologia, within project UIDB/50014/2020. The third and forth authors were financed by the ERDF – European Regional Development Fund through the Operational Programme for Competitiveness and Internationalisation - COMPETE 2020 Programme and by National Funds through the Portuguese funding agency, FCT - Fundação para a Ciência e a Tecnologia, within project POCI-01-0145-FEDER-016826. The second author was also supported by the FCT sabbatical grant with reference SFRH/B-SAB/143106/2018.

References

1. Ball, T., de Halleux, P., Swamy, N., Leijen, D.: Increasing human-tool interaction via the web. In: Proceedings of the 11th ACM SIGPLAN/SIGSOFT Workshop on Program Analysis for Software Tools and Engineering, pp. 49–52. ACM (2013)
2. Boyatt, R., Sinclair, J.: Experiences of teaching a lightweight formal method. In: Proceedings of the 1st Workshop on Formal Methods in Computer Science Education, pp. 71–80 (2008)
3. Brunel, J., Chemouil, D., Cunha, A., Macedo, N.: Simulation under arbitrary temporal logic constraints. In: Proceedings of the 5th Workshop on Formal Integrated Development Environment, EPTCS, vol. 310, pp. 63–69 (2019)
4. Couto, R., Campos, J.C., Macedo, N., Cunha, A.: Improving the visualization of Alloy instances. In: Proceedings 4th Workshop on Formal Integrated Development Environment, EPTCS, vol. 284, pp. 37–52 (2018)
5. Edwards, J., Jackson, D., Torlak, E.: A type system for object models. In: Proceedings of the 12th ACM SIGSOFT International Symposium on Foundations of Software Engineering, pp. 189–199. ACM (2004)
6. Jackson, D.: Software Abstractions: Logic, Language, and Analysis, 2nd edn. The MIT Press, Cambridge (2012)
7. Macedo, N., Brunel, J., Chemouil, D., Cunha, A., Kuperberg, D.: Lightweight specification and analysis of dynamic systems with rich configurations. In: Proceedings of the 24th ACM SIGSOFT International Symposium on Foundations of Software Engineering, pp. 373–383. ACM (2016)
8. Mangaroska, K., Giannakos, M.N.: Learning analytics for learning design: a systematic literature review of analytics-driven design to enhance learning. IEEE Trans. Learn. Technol. **12**(4), 516–534 (2019)
9. Pereira, J.: A web-based social environment for Alloy. Master's thesis, Universidade do Minho, Escola de Engenharia (2016)

10. Sioson, A.A.: Experiences on the use of an automatic C++ solution grader system. In: Proceedings of the 4th International Conference on Information, Intelligence, Systems and Applications, pp. 1–6. IEEE (2013)

11. Tillmann, N., de Halleux, J.: Pex–white box test generation for.NET. In: Beckert, B., Hähnle, R. (eds.) TAP 2008. LNCS, vol. 4966, pp. 134–153. Springer, Heidelberg (2008). https://doi.org/10.1007/978-3-540-79124-9_10

12. Tillmann, N., de Halleux, J., Xie, T., Bishop, J.: Pex4Fun: a web-based environment for educational gaming via automated test generation. In: Proceedings of the 28th IEEE/ACM International Conference on Automated Software Engineering, pp. 730–733. IEEE (2013)

A Characterization of Distributed ASMs
with Partial-Order Runs

Egon Börger[1] and Klaus-Dieter Schewe[2(✉)]

[1] Dipartimento di Informatica, Università di Pisa, Pisa, Italy
boerger@di.unipi.it
[2] UIUC Institute, Zhejiang University, Haining, China
kdschewe@acm.org

Abstract. To overcome the practical limitations of partial-order runs of 'distributed ASMs' (Abstract State Machines) proposed by Gurevich, we have defined a concept of concurrent runs of multi-agent ASMs and could show that concurrent ASMs capture a natural language-independent axiomatic definition of concurrent algorithms, thus generalising Gurevich's seminal 'Sequential ASM Thesis' from sequential to concurrent algorithms. However, we remained intrigued by the fact that Blass and Gurevich used partial-order runs of distributed ASMs to explain runs of sequential recursive algorithms. We discovered that also the inverse simulation holds: for every distributed ASM with partial order runs, these runs can be described by runs of a sequential recursive algorithm. This surprising result clarifies the difference in expressivity between partial-order and concurrent runs.

1 Introduction

In [8, Sect. 2–3] the concept of sequential Abstract State Machines (*seq-ASMs*) has been defined for which the 'Sequential ASM Thesis' [7]—to capture the intuitive notion of sequential algorithm—could be proved from three natural postulates, see [9]. In [8, Sect. 6] the concept of sequential ASM runs is extended by *partial-order runs* of a specific class of multi-agent ASMs called *distributed ASMs*. However, contrary to the great variety of successful applications of sequential ASMs, the use of distributed ASMs with partial-order runs turned out to be impractical to adequately model concurrent systems. It has been replaced in [4] by a language-independent axiomatic characterization of concurrent runs, adding a fourth postulate (on the intuitive meaning of concurrency), together with a definition of concurrent ASMs, based upon which the Sequential ASM Thesis and its proof could be generalized to a Concurrent ASM Thesis—to capture the proposed intuitive notion of concurrent algorithms.

In reaction to some scepticism expressed in [13], whether recursive algorithms can be adequately defined by ASMs, partial-order runs of *distributed ASMs* have been used in [1] to simulate the computations of recursive algorithms.[1] For a long

---

[1] Already the definition of recursive ASMs in [10] uses a special case of this translation of recursive into *distributed* computations.

© Springer Nature Switzerland AG 2020
A. Raschke et al. (Eds.): ABZ 2020, LNCS 12071, pp. 78–92, 2020.
https://doi.org/10.1007/978-3-030-48077-6_6

time we have been intrigued by this proposal, since on the one side, a simple sequential extension of ASMs suffices for the specification of recursive algorithms (see for example [2]), on the other side partial-order runs of *distributed ASMs* turned out to be impractical for modeling truly concurrent systems (see [4]).

In Sect. 3 we review Gurevich's description of *distributed ASMs* with partial-order runs and analyse the proof that the runs of recursive algorithms can be defined as partial-order runs of distributed ASMs. The analysis reveals that the *distributed ASMs* used to define recursive runs by partial-order runs are finitely composed concurrent ASMs with non-deterministic sequential (nd-seq) components (see the definition in Sect. 3). In Sect. 4 we show the surprising discovery that also the inverse relation holds, namely: for every finitely composed concurrent algorithm with nd-seq components, if its concurrent runs are definable by partial-order runs, then the algorithm can be simulated by a recursive algorithm. This establishes the main result of this paper.

Theorem 1.1 (Main Theorem). *Recursive algorithms are behaviourally equivalent to finitely composed concurrent algorithms C with nd-seq components such that all concurrent C-runs are definable by partial-order runs.*[2]

The equivalence of runs of recursive ASMs and of partial-order runs of distributed ASMs makes it explicit in which sense concurrent ASM runs as characterized in [4] are more expressive than the 'partial-order runs of distributed ASMs' proposed in [8, Sect. 6].

We will also show that if the concurrent runs are restricted further to partial-order runs of a concurrent algorithm with a fixed finite number of agents and fixed non-deterministic sequential (nd-seq) programs, one can simulate them even by a non-deterministic sequential algorithm. An interesting example of this special case are partial-order runs of Petri nets and more generally of Mayr's Process Rewrite Systems [12].

For the proofs we use an axiomatic characterization of recursive algorithms as sequential algorithms enriched by call steps,[3] such that the parent-child relationship between caller and callee defines well-defined shared locations representing input and return parameters. This characterization is reviewed in Sect. 2 and is taken from [5] where it appears as Recursion Postulate and is added to Gurevich's three postulates for sequential ASMs [9] as basis for the proof of an ASM thesis for recursive ASMs.

We assume the knowledge of [8,9] and [4] and use without further explanations standard textbook notations for ASMs, including ambient ASMs [3, Ch. 4.1].

---

[2] We call $\mathcal{R}$ behaviourally equivalent to C if each $r \in \mathcal{R}$ can be simulated by a $c \in C$ and vice versa.

[3] To emphasize the sequential nature of recursive algorithms we sometimes use the term 'sequential recursive algorithm'. See [5] for the technical reason for this naming policy.

2   The Recursion Postulate

We start with a characteristic example to illustrate the intuitive idea of recursion which guided the formulation of the recursion postulate below.[4] Take the *mergesort* algorithm, which consists of a main algorithm *sort* and an auxiliary algorithm *merge*. Every call to (a copy, we also say an instance of) *sort* and every call to (an instance of) the *merge* algorithm could give rise to a new agent. However, these agents only interact by passing input parameters and return values, but otherwise operate on disjoint sets of locations. In addition, a calling agent always waits to receive return values, which implies that only one or (in case of parallel calls) a finite number of agents are active in any state.

If one considers mutual recursion, then this becomes slightly more general, as there is a finite family of algorithms calling (instances of) each other. Furthermore, there may be several simultaneous calls. E.g. in *mergesort*, *sort* calls two copies of itself, each sorting one half of the list of given elements. Such simultaneously called copies may run sequentially in one order or the other, in parallel or even asynchronously. This give rise to non-deterministic execution of multiple sequential algorithms.

Therefore, for a characterization of recursive algorithms and their computations we can rely on the capture of non-deterministic sequential algorithms by non-deterministic sequential ASMs.[5] Thus, to axiomatically define recursive algorithms and their runs it suffices to add to the three postulates for nd-seq algorithms a Call Step Postulate and a Recursive Run Postulate defined below, which together form the **Recursion Postulate**.

To characterize the input/output relation between the input provided by the caller in a call step and the output computed by the callee for this input we use the ASM function classification from [6] to distinguish between *input, output* and *local* (also called controlled) function symbols in the signature, the union of pairwise disjoint sets Σ_{in}, Σ_{out} and Σ_{loc} respectively. We call any nd-seq algorithm which comes with such a signature and also satisfies the Call Step Postulate below an *algorithm with input and output* (for short: *i/o-algorithm*). We can then define (sequential) recursive algorithms syntactically as collections of i/o-algorithms.

Definition 2.1. A *recursive algorithm* $\mathcal{R}$ is a finite set of i/o-algorithms with one distinguished *main* algorithm. The elements of $\mathcal{R}$ are called components of $\mathcal{R}$.

The independency condition for (possibly parallel) computations of different instances of the given algorithms requires that for different calls, in particular for different calls of the same algorithm, the state spaces of the triggered subcomputations are separated from each other. This encapsulation of subcomputations can be made precise by the concept of ambient algorithms where each *instance of*

---

[4] For a detailed analysis see [5].

[5] The proof for the Sequential ASM Thesis is easily extended from deterministic to non-deterministic algorithms, see [9, Sect. 9.2].

an algorithm has a unique context parameter for its functions, e.g. its executing agent (see [3, Ch. 4.1]), and is started in an initial state that only depends on its input locations.[6]

Now we are ready to formulate the postulate for call steps. In Sect. 3.4 we formalize this postulate by an ASM $\text{CALL}(t_0 \leftarrow N(t_1, \ldots, t_n))$ (see Definition 3.4 and its refinement in Sect. 4).

Postulate 1 (Call Step Postulate). When an i/o-algorithm p—the caller, viewed as parent algorithm—calls a finite number of i/o-algorithms $c_1, \ldots, c_n$— the callees, viewed as child algorithms $CalledBy(p)$—a *call relationship* (denoted as $CalledBy(p)$) holds between the caller and each callee. The caller activates a fresh instance of each callee c_i so that they can start their computations. These computations are independent of each other and the caller remains waiting— i.e. performs no step—until every callee has terminated its computation (read: has reached a final state). For each callee, the initial state of its computation is determined only by the input passed by the caller; the only other interaction of the callee with the caller is to return in its final state an output to p.

Definition 2.2. A *call relationship* holds for (instances of) two i/o-algorithms $\mathcal{A}^p$ (parent) and $\mathcal{A}^c$ (child) if and only if they satisfy the following conditions on their function classification:

- $\Sigma_{in}^{\mathcal{A}^c} \subseteq \Sigma^{\mathcal{A}^p}$ so that the parent algorithm is able to update input locations of the child algorithm. Furthermore, $\mathcal{A}^p$ never reads the input locations of $\mathcal{A}^c$.
- $\Sigma_{out}^{\mathcal{A}^c} \subseteq \Sigma^{\mathcal{A}^p}$ so that the parent algorithm can read the output locations of the child algorithm. Furthermore, $\mathcal{A}^p$ never updates output locations of $\mathcal{A}^c$.
- $\Sigma_{loc}^{\mathcal{A}^c} \cap \Sigma^{\mathcal{A}^p} = \emptyset$ (no other common locations).

Differently from runs of a nd-seq algorithm, where in each state at most one step of the nd-seq algorithm is performed, in a recursive run a sequential recursive algorithm $\mathcal{R}$ can perform in one step simultaneously one step of each of finitely many not terminated and not waiting called instances of its i/o-algorithms. This is expressed by the Recursive Run Postulate. In this postulate we refer to *Active* and not *Waiting* instances of components, which are defined as follows:

Definition 2.3. To be *Active* resp. *Waiting* in a state S is defined as follows:

$Active(q)$ **iff** $q \in Called$ **and not** $Terminated(q)$
$Waiting(p)$ **iff** **forsome** $c \in CalledBy(p)$ $Active(c)$
$Called = \{main\} \cup \bigcup_p CalledBy(p)$

---

[6] More precisely, one can define an instance of an algorithm $\mathcal{A}$ by adding a parameter a, say for an agent executing the instance $\mathcal{A}_a = (a, \mathcal{A})$ of $\mathcal{A}$. a can be used as environment parameter for the evaluation $val_S(t, a)$ of a term t in state S with the given environment. This yields for different agents a, a' different functions $f_a, f_{a'}$ as interpretation of the same function symbol f, so that the run-time interpretations of a common signature element f can be made to differ for different agents, due to different inputs which determine their initial states.

Called collects the instances of algorithms that are called during the run. The subset of *Called* which contains all the children called by p is denoted by $CalledBy(p)$. $Called = \{main\}$ and $CalledBy(p) = \emptyset$ are true in the initial state S_0, for each i/o-algorithm $p \in \mathcal{R}$. In particular, in S_0 the original component *main* is considered to not be $CalledBy(p)$, for any p.

Postulate 2 (Recursive Run Postulate). For a sequential recursive algorithm $\mathcal{R}$ with main component *main* a recursive run is a sequence $S_0, S_1, S_2, \ldots$ of states[7] together with a sequence $C_0, C_1, C_2, \ldots$ of sets of instances of components of $\mathcal{R}$ which satisfy the following constraints:

Recursive run constraint

- C_0 is the singleton set $C_0 = \{main\}$, i.e. every run starts with *main*,
- every C_i is a finite set of instances of components of $\mathcal{R}$ which are *Active* and not *Waiting* in state S_i,
- every S_{i+1} is obtained in one $\mathcal{R}$-step by performing in S_i simultaneously one step of each i/o-algorithm in C_i. Such an $\mathcal{R}$-step is also called a *recursive step* of $\mathcal{R}$.

Bounded call tree branching. There is a fixed natural number $m > 0$, depending only on $\mathcal{R}$, which in every $\mathcal{R}$-run bounds the number of callees which can be called by a call step.

Remark (on Call Trees). If in a recursive $\mathcal{R}$-run the main algorithm calls some i/o-algorithms, this call creates a finitely branched call tree whose nodes are labeled by the instances of the i/o-algorithms involved, with active and not waiting algorithms labeling the leaves and with the main (the parent) algorithm labeling the root of the tree and becoming waiting. When the algorithm at a leaf makes a call, this extends the tree correspondingly. When the algorithm at a child of a node has terminated its computation, we delete the child from the tree. The leaves of this (dynamic) call tree are labeled by the active not waiting algorithms in the run. When the main algorithm terminates, the call tree is reduced again to the root labeled by the initially called main algorithm.

Usually, it is expected that for recursive $\mathcal{R}$-runs each called i/o-algorithm reaches a final state, but in general it is not excluded that this is not the case.

In [5] the reader can find a definition of recursive ASMs together with a proof that they capture (are equivalent to) recursive algorithms as characterized by the Recursion Postulate. Here we use the postulate as a basis for the proof that recursive algorithms are captured by 'distributed ASMs with partial-order runs', as defined in [8].

---

[7] For the sake of simplicity we take a state as union of the states of the component instances in the run, in other words as state over the union of the individual signatures.

3 Recursive ASMs Are Distributed ASMs with Partial-Order Runs

Syntactically, a multi-agent (also called concurrent) algorithm $\mathcal{C}$ is defined as a family of algorithms $alg(a)$, each associated with ('indexed by') an agent $a \in Agent$ that executes the algorithm in a run. Each $(a, alg(a))$ resp. $alg(a)$ is called a component resp. (component) program of $\mathcal{C}$. This applies to distributed ASMs [8] as well as to recursive or concurrent algorithms and ASMs [4,5].

To investigate the simulation of recursive runs by partial-order runs of distributed ASMs (Sect. 3.4) we must explain what are finitely composed concurrent (Gurevich's 'distributed') algorithms (Sect. 3.1) and partial-order resp. concurrent runs (Sect. 3.2 resp. 3.3).

3.1 Finitely Composed Concurrent Algorithms

For recursive algorithms various restrictions on the syntactical definition of multi-agent algorithms have to be made most of which appear also for distributed ASMs in [8, Sect. 6].

First of all, although the components $alg(a)$ of concurrent algorithms are not necessarily sequential algorithms, to simulate specific concurrent algorithms by recursive ones, which are defined as families of nd-seq algorithms, we must restrict our attention to concurrent algorithms with sequential (though possibly non-deterministic) components.[8]

Second, for distributed ASMs it is stipulated in [8, p. 31] that the agents are equipped with instances of programs which are taken from 'a finite indexed set of single-agent programs'. This leads to what we call finitely composed concurrent algorithms or ASMs $\mathcal{C}$ where the components can only be copies (read: instances) of finitely many different nd-seq algorithms or ASMs, which we will call the program base of $\mathcal{C}$.

Third, for distributed ASMs it is stipulated in [8, 6.2, p. 31] that in initial states there are only finitely many agents, each equipped with a program. We reflect this by the (simplifying but equivalent) condition that the runs of a finitely composed concurrent algorithm or ASM must be started by executing a distinguished main component.

Fourth, for distributed ASMs it is stipulated in [8, p. 32] that 'An agent a can make a move at S by firing Prog(a) ... and change S accordingly. As part of the move, a may create new agents', which then may contribute by their moves to the run in which they were created. For this purpose we use the **new** function.

We summarize these constraints for distributed ASMs by the notion of *finitely composed* concurrent algorithms (read: concurrent ASMs).

Definition 3.1. A concurrent algorithm $\mathcal{C}$ is *finitely composed* iff (i)–(iii) hold:

---

[8] In fact it is shown in [5] that permitting the unbounded **forall** and **choose** constructs results in algorithms far more powerful than the recursive ones.

(i) There exists a finite set $\mathcal{B}$ of nd-seq algorithms such that each $\mathcal{C}$-program is of form **amb** a **in** r for some program $r \in \mathcal{B}$—call $\mathcal{B}$ the *program base* of $\mathcal{C}$.

(ii) There exists a distinguished agent a_0 which is the only one *Active* in any initial state. Formally this means that in every initial state of a $\mathcal{C}$-run, $Agent = \{a_0\}$ holds. We denote by *main* the component in $\mathcal{B}$ of which a_0 executes an instance. For partial-order runs of $\mathcal{C}$ defined below this implies that they start with a minimal move which consists in executing the program $\mathrm{asm}(a_0) = $ **amb** a_0 **in** *main*.

(iii) Each program in $\mathcal{B}$ may contain rules of form **let** $a = $ **new** (*Agent*) **in** r. Together with (ii) this implies that every agent, except the distinguished a_0, before making a move in a run must have been created in the run.

$\mathcal{C}$ is called *finite* iff *Agent* is finite.

3.2   Partial-Order Runs

In [8] Gurevich defined (for distributed algorithms) the notion of *partial-order run* by a partial order on the set of single moves of the agents which execute the component algorithms. For a nd-seq algorithm $\mathcal{A}$, to make one *move* means to perform one step in a state S.

Definition 3.2. Let $\mathcal{C} = \{(a, \mathrm{alg}(a))\}_{a \in Agent}$ be a concurrent algorithm, in which each $\mathrm{alg}(a)$ is an nd-seq algorithm. A *partial-order run* for $\mathcal{C}$ is defined by a set M of moves of instances of the algorithms $\mathrm{alg}(a)$ ($a \in Agent$), a function $\mathrm{ag} : M \rightarrow Agent$ assigning to each move the agent performing the move, a partial order $\leq$ on M, and an initial segment function σ such that the following conditions are satisfied:

finite history. For each move $m \in M$ its history $\{m' \mid m' \leq m\}$ is finite.

sequentiality of agents. The moves of each agent are ordered, i.e. for any two moves m and m' of one agent $\mathrm{ag}(m) = \mathrm{ag}(m')$ we either have $m \leq m'$ or $m' \leq m$.

coherence. For each finite initial segment $M' \subseteq M$ (i.e. such that for $m \in M'$ and $m' \leq m$ we also have $m' \in M'$) there exists a state $\sigma(M')$ over the combined signatures of the algorithms $(a, \mathrm{alg}(a))$ such that for each maximum element $m \in M'$ the state $\sigma(M')$ is the result of applying m to $\sigma(M' - \{m\})$.

3.3   Concurrent Runs

In a concurrent run as defined in [4], multiple agents with different clocks may contribute by their single moves to define the successor state of a state. Therefore, when a successor state S_{i+1} of a state S_i is obtained by applying to S_i multiple update sets U_a with agents a in a finite set $Agent_i \subseteq Agent$, each U_a is required to have been computed by $a \in Agent_i$ in a preceding state S_j, i.e. with $j \leq i$. It is possible that $j < i$ holds so that for different agents different $\mathrm{alg}(a)$-execution speeds (and purely local subruns to compute U_a) can be taken into account.

This can be considered as resulting from a separation of a step of an nd-seq algorithm $\text{alg}(a)$ into a *read step*—which reads location values in a state S_j—followed by a *write step* which applies the update set U_a computed on the basis of the values read in S_j to a later state S_i $(i \geq j)$. We say that a contributes to updating the state S_i to its successor state S_{i+1}, and that a *move* starts in S_j and contributes to updating S_i (i.e. it finishes in S_{i+1}). This is formally expressed by the following definition of concurrent ASMs and their runs.

Definition 3.3. Let $\mathcal{C}$ be a concurrent algorithm of component algorithms $pgm(a)$ (read: ASM rules) with associated agents $a \in Agent$. A *concurrent run* of $\mathcal{C}$ is defined as a sequence $S_0, S_1, \ldots$ of states together with a sequence $A_0, A_1, \ldots$ of finite subsets of $Agent$, such that S_0 is an initial state and each S_{i+1} is obtained from S_i by applying to it the updates computed by the agents in A_i, where each $a \in A_i$ computes its update set U_a on the basis of the location values (including the input and shared locations) read in some preceding state S_j (i.e. with $j \leq i$) depending on a.

Remark. In this definition we deliberately permit the set of $Agents$ to be infinite or dynamic and potentially infinite, growing or shrinking in a run. In Definition 3.2 above, the set of $Agents$ is fixed by the set M of moves.

3.4   Simulation of Recursive by Partial-Order Runs

We are now ready to specify recursive algorithms by *distributed ASMs*, following the thought proposed in [1]. For the sake of precision and simplicity we formulate the construction in terms of ASMs; due to the characterization theorems in [5] and [4] this implies no loss of generality.

Theorem 3.1. *Every recursive ASM $\mathcal{R}$ can be simulated by a finitely composed concurrent ASM $\mathcal{C}_{\mathcal{R}}$ with nd-seq ASM components for which every concurrent run of $\mathcal{C}_{\mathcal{R}}$ is definable by a partial-order run.*

Proof. Let $\mathcal{R}$ be a recursive ASM given with distinguished program *main*. We define a finitely composed concurrent ASM $\mathcal{C}_{\mathcal{R}}$ with program base $\{r^* \mid r \in \mathcal{R}\}$, where r^* is defined as

$$r^* = \textbf{if } Active(r) \textbf{ and not } Waiting(r) \textbf{ then } r.$$

In doing so, for each call rule $r = t_0 \leftarrow N(t_1, \ldots, t_n)$ in $\mathcal{R}$ we use for its translation the following ASM $\text{CALL}(t_0 \leftarrow N(t_1, \ldots, t_n))$, which rigorously defines the behavioral interpretation of the call rule r (for details see [5]):

Definition 3.4. $\text{CALL}(t_0 \leftarrow N(t_1, \ldots, t_n)) =$
  let $N(x_1, \ldots, x_n) = q$ // declaration of N
  let $v_1 = t_1, \ldots, v_n = t_n$ // input evaluation $val_S(t_i, \textbf{self})$ by caller
  let $t_0 = f(t_1', \ldots, t_k')$
  let $v_1' = t_1', \ldots, v_k' = t_k'$
  let $c = \textbf{new }(Agent)$

$pgm(c) := \mathbf{amb}\ c\ \mathbf{in}\ q\ //$ equip callee with its program instance
INSERT(c, $CalledBy(\mathbf{self})$)
INITIALIZE($q_c, v_1/x_1, \ldots, v_n/x_n, f(v'_1, \ldots, v'_k)/x_o$)
$CalledBy(c) := \emptyset$

Note that the call is a call-by-value and that $(f, (v'_1, \ldots, v'_k))$ denotes the output location whose value the caller expects to be updated by the callee with the return value.

By definition, r^* can only contribute a non-empty update set to form a state S_{i+1} in a concurrent run, if r is *Active* and not *Waiting*; this reflects that by the recursive run postulate, in every step of a recursive run of $\mathcal{R}$ only *Active* and not *Waiting* rules are executed.

The definition of r^* obviously guarantees that $\mathcal{C_R}$ simulates $\mathcal{R}$ step by step: in each run step the same *Active* and not *Waiting* rules r respectively r^* and their agents are selected for their simultaneous execution and their rules perform the same state change.

Note that by Definition 3.4 of CALL(i/o-*rule*), each agent operates in its own state space so that the view of an agent's step as read-step followed by a write-step is equivalent to the atomic view of this step. Note also that in a concurrent run of $\mathcal{C_R}$ the *Agent* set is dynamic, in fact it grows with each execution of a call rule, together with the number of instances of $\mathcal{R}$-components executed during a recursive run of $\mathcal{R}$.

It remains to define every concurrent run $(S_0, A_0), (S_1, A_1), \ldots$ of $\mathcal{C_R}$ by a partial-order run. For this we define an order on the set M of moves made during a concurrent run, showing that it satisfies the constraints on finite history and the sequentiality of agents, and then relate each state S_i of the run to the state computed by the set M_i of moves performed to compute S_i (from S_0), showing that M_i is a finite initial segment of M and that the associated state $\sigma(M_i)$ equals S_i and satisfies the coherence condition.

Each successor state S_{i+1} in a concurrent run of $\mathcal{C_R}$ is the result of applying to S_i the write steps of finitely many moves of agents in A_i. This defines the function ag, which associates agents with moves, and the finite set M_i of all moves finished in a state belonging to the initial run segment $[S_0, \ldots, S_i]$. Let $M = \cup_i M_i$. The partial order $\leq$ on M is defined by $m < m'$ iff move m contributes to update some state S_i (read: finishes in S_i) and move m' starts reading in a later state S_j with $i + 1 \leq j$. Thus, by definition, M_i is an initial segment of M.

To prove the finite history condition, consider any $m' \in M$ and let S_j be the state in which it is started. There are only finitely many earlier states $S_0, \ldots, S_{j-1}$, and in each of them only finitely many moves m can be finished, contributing to update S_{j-1} or an earlier state.

The condition on the sequentiality of the agents follows directly from the definition of the order relation $\leq$ and from the fact that in a concurrent run, for every move $m = (read_m, write_m)$ executed by an agent, this agent performs no other move between the $read_m$-step and the corresponding $write_m$-step in the run.

This leaves us to define the function σ for finite initial segments $M' \subseteq M$ and to show the coherence property. We define $\sigma(M')$ as result of the application of the moves in M' in any total order extending the partial order $\leq$. For the initial state S_0 we have $\sigma(\emptyset) = S_0$. This implies the definability claim $S_i = \sigma(M_i)$.

The definition of σ is consistent for the following reason. Whenever two moves $m \neq m'$ are incomparable, then either they both start in the same state or say m starts earlier than m'. But m' also starts earlier than m finishes. This is only possible for agents $ag(m) = a$ and $ag(m') = a'$ whose programs $pgm(a), pgm(a')$ are not in an ancestor relationship in the call tree. Therefore these programs have disjoint signatures, so that the moves m and m' could be applied in any order with the same resulting state change.

To prove the coherence property let M' be a finite initial segment, and let $M'' = M' \backslash M'_{\max}$, where $M'_{\max}$ is the set of all maximal elements of M'. Then $\sigma(M')$ is the result of applying simultaneously all moves $m \in M'_{\max}$ to $\sigma(M'')$, and the order in which the maximum moves are applied is irrelevant. This implies in particular the desired coherence property.   □

The key argument in the proof exploits the Recursion Postulate whereby for recursive runs of $\mathcal{R}$, the runs of different agents are initiated by calls and concern different state spaces with pairwise disjoint signatures, due to the function parameterization by agents, unless $pgm(a')$ is a child (or a descendant) of $pgm(a)$, in which case the relationship between the signatures is defined by the call relationship. Independent moves can be guaranteed in full generality only for algorithms with disjoint signatures.

4 Distributed ASMs with Partial-Order Runs Are Recursive ASMs

While Theorem 3.1 is not surprising, we will now show its less obvious inverse.

Theorem 4.1. *For each finitely composed concurrent ASM $\mathcal{C}$ with program base $\{r_i \mid i \in I\}$ of nd-seq ASMs such that all its concurrent runs are definable by partial-order runs, one can construct a recursive ASM $\mathcal{R}_{\mathcal{C}}$ such that each concurrent run of $\mathcal{C}$ can be simulated by a recursive run of $\mathcal{R}_{\mathcal{C}}$.*[9]

Proof. Let a concurrent $\mathcal{C}$-run $(S_0, A_0), (S_1, A_1), \ldots$ be given. If it is definable by a partial-order run $(M, \leq, ag, pgm, \sigma)$, the transition from $S_i = \sigma(M_i)$ to S_{i+1} is performed in one concurrent step by parallel independent moves $m \in M_{i+1} \backslash M_i$, where M_i is the set of moves which contributed to transform S_0 into S_i. Let $m \in M_{i+1} \backslash M_i$ be a move performed by an agent $a = ag(m)$ with rule $pgm(a) = \mathbf{amb}\ a\ \mathbf{in}\ r$, an instance of a rule r in the program base of $\mathcal{C}$. To execute the concurrent step by means of steps of a recursive ASM $\mathcal{R}_{\mathcal{C}}$, we simulate each

---

[9] One obtains even the behavioral equivalence via an inverse simulation of every recursive $\mathcal{R}_{\mathcal{C}}$-run by a concurrent $\mathcal{C}$-run if the delegates of $\mathcal{C}$-agents, called in the recursive run to perform the step of their *caller* in the concurrent run, act in an 'eager' way. See the remark at the end of the proof.

of its moves m by letting agent a act in the $\mathcal{R}_C$-run as *caller* of a named rule $out_r \leftarrow \text{ONESTEP}_r(in_r)$. The callee agent c acts as delegate for one step of a: it executes **amb** $a \in r$ and makes its program immediately *Terminated*.

To achieve this, we refine the CALL machine defined in Definition 3.4 such that upon calling $out_r \leftarrow \text{ONESTEP}_r(in_r)$, the delegate c created by the call becomes *Active* so that it can make a step to execute **amb** c **in** ONESTEP_r. It suffices to add to the component INITIALIZE the update *Terminated*(**amb** c **in** q) := *false*, which makes c *Active*. ONESTEP_r is defined to perform **amb** *caller*(c) **in** r and to terminate immediately (by setting *Terminated* to true). For ease of exposition we add to Definition 3.4 also the update *caller*(c) :=**self**, to distinguish agents in the concurrent run—the *callers* of ONESTEP_r-machines— from the delegates each of which simulates one step of its *caller* and immediately terminates its life cycle.

It remains to determine the input and output for calling ONESTEP_r. For the input we exploit the existence of a bounded exploration witness W_r for r. All updates produced in a single step are determined by the values of W_r in the state, in which the call is launched. So W_r defines the input terms of the called rule ONESTEP_r, combined in in_r. Analogously, a single step of r provides updates to finitely many locations that are determined by terms appearing in the rule, which defines out_r.

We summarize the explanations by the following definition:

$\mathcal{R}_C = \{ out_r \leftarrow \text{ONESTEP}_r(in_r) \mid r \in \text{ program base of } \mathcal{C} \}$
$\text{ONESTEP}_r =$
    amb *caller*(**self**) **in** r // the delegate executes the step of its *caller*
    Terminated(pgm(**self**)) := *true* // ... and immediately stops

Note that by the refined Definition 3.4, $out_r \leftarrow \text{ONESTEP}_r(in_r)$ triggers the execution of the delegate program **amb** c **in** ONESTEP_r. Let $a = caller(c)$. By definition, **amb** c **in** ONESTEP_r triggers **amb** c **in amb** a **in** r. Furthermore, since the innermost ambient binding counts, this machine is equivalent to the simulated machine **amb** a **in** r, as was to be shown.

Thus the recursive $\mathcal{R}_C$-run which simulates $(S_0, A_0), (S_1, A_1), \ldots$ starts by Definition 3.1 in S_0 with program **amb** a_0 **in** $in_{main} \leftarrow \text{ONESTEP}_{main}(out_{main})$. For the sake of notational simplicity we disregard the auxiliary locations of $\mathcal{R}_C$. Let

$A_i = \{ a_{i_1}, \ldots, a_{i_k} \} \subseteq Agent$ for some i_j and k depending on i
where forall $1 \leq j \leq k$
    $a_{i_j} = ag(m_{i_j}) \in M_{i+1} \backslash M_i$ **and** $pgm(a_{i_j}) = $ **amb** a_{i_j} **in** r_{i_j}

We use the same agents a_{i_j} for A_i in the $\mathcal{R}_C$-run, but with program $out_{r_{i_j}} \leftarrow \text{ONESTEP}_{r_{i_j}}(in_{r_{i_j}})$. Their step in the recursive run leads to a state S_i' where all callers a_{i_j} are *Waiting* and the newly created delegates c_{i_j} are *Active* and not *Waiting*. So we can choose them for the set A_i' of agents which perform the next $\mathcal{R}_C$ step, whereby

■ all rules r_{i_j} are performed simultaneously (as in the given concurrent run step), in the ambient of $caller(c_{i_j}) = a_{i_j}$ thus leading as desired to the state S_{i+1},

■ the delegates make their program *Terminated*, whereby their callers a_{i_j} become again not *Waiting* and thereby ready to take part in the next step of the concurrent run. We assume for this that whenever in the $\mathcal{C}$-run (not in the $\mathcal{R}_{\mathcal{C}}$ run) a new agent a is created, it is made not *Waiting* (by initializing $CalledBy(a) := \emptyset$).

□

Remark. Consider an $\mathcal{R}_{\mathcal{C}}$-run where each recursive step of the concurrent caller agents in A_i, which call each some ONESTEP program, alternates with a recursive step of all—the just called—delegates whose program is not yet *Terminated*. Then this run is equivalent to a corresponding concurrent $\mathcal{C}$-run.

Note that Theorem 4.1 heavily depends on the prerequisite that $\mathcal{C}$ only has partial-order runs.[10] With general concurrent runs as defined in [4] the construction would not be possible.

4.1 Partial Order Runs of Petri Nets

The semantics of Petri nets actually defines a rather special case of partial-order runs, namely runs one can describe even by a nd-seq ASM, as we show in this section.

A Petri net comes with a finite number of transition rules, each of which can be described by a nd-seq ASM (see [6, p. 297]). The special character of the computational Petri net model is due to the fact that during the runs, only exactly these rules are used. In other words there is a fixed association of each rule with an executing agent; there is no rule instantiation with new agents which could be created during a run. Therefore the states are the global markings of the net. The functions $\sigma(I)$ associated with the po-runs of the net yield for every finite initial segment I as value the global marking obtained by firing the rules in I.

For this particular kind of concurrent ASMs with partial-order runs one can define the concurrent runs by nd-seq ASMs, as we are going to show in this section.

Theorem 4.2. *For each finite concurrent ASM $\mathcal{C} = \{(a_i, r_i) \mid 1 \leq i \leq n\}$ with nd-seq ASMs r_i such that all its concurrent runs are definable by partial-order runs one can construct a nd-seq ASM $\mathcal{M}_{\mathcal{C}}$ such that the concurrent runs of $\mathcal{C}$ and the runs of $\mathcal{M}_{\mathcal{C}}$ are equivalent.*

---

[10] The other prerequisites in Theorem 4.1 appear to be rather natural. Unbounded runs can only result, if in a single step arbitrarily many new agents are created. Also, infinitely many different rules associated with the agents are only possible, if new agents are created and added during a concurrent run. Though this is captured in the general theory of concurrency in [4], it was not intended in Gurevich's definition of partial-order runs.

Corollary 4.1. *Partial-order Petri net runs can be simulated by runs of a non-deterministic sequential ASM.*[11]

Proof. We relate the states S_i of a given concurrent run of $\mathcal{C}$ to the states $\sigma(M_i)$ associated with initial segments M_i of a given corresponding partial order run $(M, \leq, ag, pgm, \sigma)$, where each step leading from S_i to S_{i+1} consists of pairwise incomparable moves in $M_{i+1} \backslash M_i$. We call such a sequence $S_0, S_1, \ldots$ of states a *linearised run* of $\mathcal{C}$. For $i > 0$ the initial segments M_i are non empty.

The linearized runs of $\mathcal{C}$ can be characterized as runs of a nd-seq ASM $\mathcal{M_C}$: in each step this machine chooses one of finitely many non-empty subsets of rules in $\mathcal{C}$ to execute them in parallel. Formally:

$\mathcal{M_C} = $ **choose** $\text{ALLRULESOF}(I_1) \mid \cdots \mid \text{ALLRULESOF}(I_n)$
where
$\quad \text{ALLRULESOF}(\{i_1, \ldots, i_k\}) =$
$\qquad r_{i_1}$
$\qquad \ldots$
$\qquad r_{i_k}$
$\quad \{I_1, \ldots, I_n\} = \{I' \neq \emptyset \mid I' \subseteq I\}$ // the non-empty subsets of I
$\quad n = 2^{|I|} - 1$

To complete the proof it suffices to show the following lemma. $\square$

Lemma 4.1. *The linearised runs of $\mathcal{C}$ are exactly the runs of $\mathcal{M_C}$.*

Proof. To show that each run $S_0, S_1, \ldots$ of $\mathcal{M_C}$ is a linearised run of $\mathcal{C}$ we proceed by induction to construct the partial-order run $(M, \leq)$ with its finite initial segments M_i. For the initial state $S_0 = \sigma(\emptyset)$ there is nothing to show, so let S_{i+1} result from S_i by applying an update set produced by $\text{ALLRULESOF}(J)$ for some non-empty $J \subseteq I$. By induction we have $S_i = \sigma(M_i)$ for some initial segment of a partial-order run $(M, \leq)$. As $\text{ALLRULESOF}(J)$ is a parallel composition, S_{i+1} results from applying the union of update sets $\Delta_{i_j} \in \Delta_{r_{i_j}}$ for $j = 1, \ldots, |J|$ to S_i. Each Δ_{i_j} defines a move m_{i_j} of some $ag(m_{i_j}) = a_{i_j}$, move which finishes in state S_i. We now have two cases:

(i) The moves m_{i_j} with $j \in J$ are pairwise independent, i.e. their application in any order produces the same new state. Then $(M, \leq)$ can be extended with these moves such that $M_{i+1} = M_i \cup \{m_{i_j} \mid j \in J\}$ becomes an initial segment and $S_{i+1} = \sigma(M_i)$ holds.

(ii) If the moves m_{i_j} with $j \in J$ are not pairwise independent, the union of the corresponding update sets is inconsistent, hence the run terminates in state S_i.

---

[11] We thank Wolf Zimmermann for pointing out that the argument applies more generally to Mayr's Process Rewrite Systems [12]. They have been used in [11] to verify protocols for services which may rise exceptions.

To show the converse we proceed analogously. If we have $S_i = \sigma(M_i)$ for all $i \geq 1$, then S_{i+1} results from S_i by applying in parallel all moves in $M_{i+1} - M_i$. Applying a move m means to apply an update set produced by some rule $r_j \in \mathcal{C}$ (namely the rule $pgm(ag(m))$) in state S_i, and applying several update sets in parallel means to apply their union Δ, which then must be consistent. So we have $S_{i+1} = S_i + \Delta$ with $\Delta = \bigcup_{j \in J} \Delta_{i_j}$ for some J, where each Δ_{i_j} is an update set produced by r_{i_j}, i.e. Δ is an update set produced by ALLRULESOF(J), which implies that the linearised run $S_0, S_1, \ldots$ is a run of $\mathcal{M}_\mathcal{C}$.   □

For the corollary it suffices to note that each Petri net transition can be described by a nd-seq ASM (see [6, p. 297]). The functions $\sigma(I)$ associated with the po-runs yield the global marking obtained by firing the rules in I.

5   Conclusions

While Gurevich's Sequential ASM Thesis [9] provides an elegant and satisfactory mathematical definition of the notion of sequential algorithm plus a proof that sequential algorithms are captured by sequential ASMs, this theory does not capture recursive algorithms. It lacks an appropriate call concept. In fact, in an attempt to solve this problem Blass and Gurevich in [1] invoked the notion of partial-order runs of 'distributed ASMs', which has been proposed in [8] as a concurrency concept for ASMs. We showed in this paper that these 'distributed ASMs' are finitely composed ASMs whose partial-order runs characterize (are equivalent to) recursive runs. Thus, partial-order runs of distributed ASMs do not capture the concept of concurrent algorithms (but see [4]).

References

1. Blass, A., Gurevich, Y.: Algorithms vs. machines. Bull. EATCS **77**, 96–119 (2002)
2. Börger, E., Bolognesi, T.: Remarks on turbo ASMs for functional equations and recursion schemes. In: Börger, E., Gargantini, A., Riccobene, E. (eds.) ASM 2003. LNCS, vol. 2589, pp. 218–228. Springer, Heidelberg (2003). https://doi.org/10.1007/3-540-36498-6_12
3. Börger, E., Raschke, A.: Modeling Companion for Software Practitioners. Springer, Heidelberg (2018). https://doi.org/10.1007/978-3-662-56641-1
4. Börger, E., Schewe, K.-D.: Concurrent abstract state machines. Acta Inform. **53**(5), 469–492 (2016). https://doi.org/10.1007/s00236-015-0249-7
5. Börger, E., Schewe, K.-D.: A behavioural theory of recursive algorithms (2020). http://arxiv.org/abs/2001.01862
6. Börger, E., Stärk, R.F.: Abstract State Machines: A Method for High-Level System Design and Analysis. Springer, Heidelberg (2003). https://doi.org/10.1007/978-3-642-18216-7
7. Gurevich, Y.: A new thesis. In: Abstracts, vol. 6, no. 4, p. 317. American Mathematical Society (1985)
8. Gurevich, Y.: Evolving algebras 1993: lipari guide. In: Börger, E. (ed.) Specification and Validation Methods, pp. 9–36. Oxford University Press (1995)

9. Gurevich, Y.: Sequential abstract-state machines capture sequential algorithms. ACM Trans. Comput. Logic **1**(1), 77–111 (2000)
10. Gurevich, Y., Spielmann, M.: Recursive abstract state machines. J. UCS **3**(4), 233–246 (1997)
11. Heike, C., Zimmermann, W., Both, A.: On expanding protocol conformance checking to exception handling. SOCA **8**(4), 299–322 (2013). https://doi.org/10.1007/s11761-013-0146-2
12. Mayr, R.: Process rewrite systems. Inf. Comput. **156**, 264–286 (1999)
13. Moschovakis, Y.N.: What is an algorithm? In: Engquist, B., Schmid, W. (eds.) Mathematics Unlimited - 2001 and Beyond, pp. 919–936. Springer, Heidelberg (2001). https://doi.org/10.1007/978-3-642-56478-9_46

A Logic for Reflective ASMs

Klaus-Dieter Schewe[1] and Flavio Ferrarotti[2(✉)]

[1] Zhejiang University, UIUC Institute, Haining, China
`kd.schewe@intl.zju.edu.cn`, `kdschewe@acm.org`
[2] Software Competence Center Hagenberg, Hagenberg, Austria
`flavio.ferrarotti@scch.at`

Abstract. Reflective algorithms are algorithms that can modify their own behaviour. Recently a behavioural theory of reflective algorithms has been developed, which shows that they are captured by reflective abstract state machines (rASMs). Reflective ASMs exploit extended states that include an updatable representation of the ASM signature and rules to be executed by the machine in that state. Updates to the representation of ASM signatures and rules are realised by means of a sophisticated tree algebra defined in the background of the rASM. In this paper the theory is taken further by an extension of the logic of ASMs to capture inferences on rASMs. The key is the introduction of terms that are interpreted by ASM rules stored in some location. We show that fragments of the logic with a fixed bound on the number of steps preserve completeness, whereas the full run-logic for rASMs becomes incomplete.

Keywords: Abstract state machine · Reflection · Logic · Tree algebra

1 Introduction

Reflection refers to the ability of an algorithm or program to modify its own behaviour. The concept is as old as computer science; it already appears in LISP [16], where programs and data are both represented uniformly as lists. General run-time and compile-time linguistic reflection in programming and database research have been investigated in general by Stemple, Van den Bussche and others in [18,19]. Recently, adaptivity and thus reflection has become a key aspect of (cyber-physical) systems [7]. Nonetheless, it is still not well understood and contains great challenges and risks. As it is hard to oversee how a system behaves after many adaptations, any uncontrolled application of reflection bears the risk of unpredictable and undesired outcomes. Thus, the challenge for rigorous methods is to enable static reasoning and verification of desired properties of reflective algorithms and systems, which requires to control an unbounded family of specifications.

The research reported in this paper has been partly funded by BMVIT, BMDW, and the Province of Upper Austria in the frame of the COMET Programme managed by FFG.

A. Raschke et al. (Eds.): ABZ 2020, LNCS 12071, pp. 93–106, 2020.
https://doi.org/10.1007/978-3-030-48077-6_7

Concerning the foundations of reflection we developed a behavioural theory of reflective sequential algorithms (RSAs) in [12] (see arXiv version in [9]), which extends and cleanses our previous sketch in [2]. The theory provides an axiomatic, language-independent definition of RSAs, defines an extension of sequential ASMs to reflective sequential ASMs (rsASMs), by means of which RSAs can be specified, and provides a proof that RSAs are captured by rsASMs. That is, rsASMs satisfy the postulates of the axiomatisation, and any RSA as stipulated by the axiomatisation can be defined by a behaviourally equivalent rsASM. The notion of *behavioural equivalence* is slightly weaker than the corresponding notion for sequential or parallel algorithms, as there is no need to require that changes to the represented algorithm are exactly the same, as long as the application of the algorithm to the core part of the structure yields the same results.

In [13] we sketched how to generalise the theory to reflective parallel algorithms [11], which requires an integration of the behavioural theory of synchronous parallel algorithms [3]. Leaving this general aspect aside the generalisation of just the reflective sequential ASMs to reflective ASMs is rather straightforward. For deterministic ASMs this was done in [10]. In a nutshell, in each step of a reflective ASM (rASM) the rule is taken from a dedicated location *self*, which uses a tree structure to represent the signature and rule, and a sophisticated tree algebra to manipulate tree values [14]. We also exploit partial updates in the form of [15] to minimise clashes that may otherwise result from simultaneously updating *self* by several parallel branches.

In this paper we address the fundamental question how desired properties of a reflective algorithm can be verified. As rASMs capture reflective algorithms, this requires extending the logic of ASMs [4,5,17]. We observe that in these logics the rules defining an ASM only enter as extra-logical constants r that are expanded in atomic formulae $[r]\varphi$ (the application of r to the current state leads to a state satisfying the formula φ), $\text{upd}(r, X)$ (the rule r yields an update set X in the current state), and $\text{upm}(r, \ddot{X})$ (the rule r yields an update multiset $\ddot{X}$ in the current state). In an rASM, however, the rule to be applied in the current state is stored itself in the state in a sublocation of a location *self*. We therefore explore the idea to treat r in formulae as variables that are interpreted by a rule stored in the current state. Furthermore, as reasoning about reflective algorithms only makes sense for multiple steps, we also extend the one-step ASM logic to a multiple-step logic. The precise definition of such a logic and the completeness proof for a fragment of the logic are the key contributions of this paper.

In Sect. 2 we present rASMs as extensions of ASMs. Section 3 is dedicated to the introduction of the logic of ASMs, which follows our previous work in [4]. The core of the paper is Sect. 4, where we formally develop the extension of the logic dealing with reflection and investigate completeness. We conclude with a brief summary and outlook in Sect. 5.

2   Reflective Abstract State Machines

We assume general familiarity with ASMs as defined in [1]. The extension to reflective ASMs requires to define a background structure that covers trees and operations on them, a dedicated variable *self* that takes as its value a tree representation of an ASM signature and rule, and the extension of rules by partial updates. Due to space limitations our presentation must be terse—nevertheless the details are given in [9,10,12]. Note that the omitted details include the sophisticated tree algebra defined for the representation of rules and the access to them. We use some of its operators, but they can be correctly understood from the context.

Let Σ be an ASM signature, i.e. a set of function symbols. Partial assignments are defined as follows: Whenever $f \in \Sigma$ has arity n and *op* is an operator of arity $m + 1$, t_i $(i = 1, \ldots, n)$ and t'_i $(i = 1, \ldots, m)$ are terms over Σ, then $f(t_1, \ldots, t_n) \Leftarrow^{op} t'_1, \ldots, t'_m$ is a rule. The informal meaning is that we evaluate the terms as well as $f(t_1, \ldots, t_n)$ in the current state S, then apply *op* to $\mathrm{val}_S(f(t_1, \ldots, t_n)), \mathrm{val}_S(t'_1), \ldots, \mathrm{val}_S(t'_m)$ and assign the resulting value v to the location $(f, (\mathrm{val}_S(t_1), \ldots, \mathrm{val}_S(t_n)))$. Conditions for compatibility and the collapse of an update multiset into an update set have been elaborated in detail in [15].

For the dedicated location storing the self-representation of an ASM it is sufficient to use a single function symbol *self* of arity 0. Then in every state S the value $\mathrm{val}_S(self)$ is a complex tree comprising two subtrees for the representation of the signature and the rule, respectively. That is, in the tree structure we have a root node o labelled by **self** with exactly two successor nodes, say o_0 and o_1, labelled by **signature** and **rule**, respectively. So we have $o \prec_c o_0$, $o_0 \prec_s o_1$ and $o \prec_c o_1$, where $\prec_c$ and $\prec_s$ denote, respectively, the child and sibling relationships. The subtree rooted at o_0 has as many children $o_{00}, \ldots, o_{0k}$ as there are function symbols in the signature, each labelled by **func**. Each of the subtrees rooted at o_{0i} takes the form $\mathbf{func}\langle\mathbf{name}\langle f\rangle\mathbf{arity}\langle n\rangle\rangle$ with a function name f and a natural number n. The subtree rooted at o_1 represents the rule of a sequential ASM as a tree.

The inductive definition of trees representing rules is rather straightforward. For instance, an assignment rule $f(t_1, \ldots, t_n) := t_0$ is represented by a tree of the form $\mathbf{update}\langle\mathbf{func}\langle f\rangle\mathbf{term}\langle t_1 \ldots t_n\rangle\mathbf{term}\langle t_0\rangle\rangle$, and a partial assignment rule $f(t_1, \ldots, t_n) \Leftarrow^{op} t'_1, \ldots, t'_m$ is represented by a tree of the form $\mathbf{partial}\langle\mathbf{func}\langle f\rangle\mathbf{func}\langle op\rangle\mathbf{term}\langle t_1 \ldots t_n\rangle\mathbf{term}\langle t'_1 \ldots t'_m\rangle\rangle$.

The *background of an rASM* is defined by a background class $\mathcal{K}$ over a background signature $V_\mathcal{K}$. It must contain an infinite set *reserve* of reserve values and an infinite set Σ_{res} of reserve function symbols, the equality predicate, the undefinedness value *undef*, and a set L of labels **self**, **signature**, **rule**, **func**, **name**, **arity**, **update**, **term**, **if**, **bool**, **par**, **let**, **partial**. The background class must further define truth values and their connectives, tuples and projection operations on them, natural numbers and operations on them, trees in T_L and tree operations, and the function $\mathbf{I}$, where $\mathbf{I}x.\varphi$ denotes the unique x satisfying condition φ.

If B is a base set, then an *extended base set* is the smallest set B_{ext} containing B that is closed under adding function symbols in the reserve Σ_{res}, natural numbers, the terms $\mathbb{T}$ with respect to B and Σ_{res}, and terms of the tree algebra defined over Σ_{res} with labels in L as defined above. Furthermore, we use $\hat{\mathbb{T}}_{ext}$ to denote the union of the set $\mathbb{T}_{ext}$ of terms with $\hat{\Sigma}_{ext}$ and the set of rules.

The background must further provide functions: $drop : \hat{\mathbb{T}}_{ext} \rightarrow B_{ext}$ and $raise : B_{ext} \rightarrow \hat{\mathbb{T}}_{ext}$ for each base set B and extended base set B_{ext}, and a derived *extraction function* $\beta : \mathbb{T}_{ext} \rightarrow \bigcup_{n \in \mathbb{N}} \mathbb{T}^n$, which assigns to each term defined over the extended signature Σ_{ext} and the extended base set B_{ext} a tuple of terms in $\mathbb{T}$ defined over Σ and B.

A *reflective ASM* (rASM) $\mathcal{M}$ comprises an (initial) signature Σ containing a 0-ary function symbol *self*, a background as defined above, and a set $\mathcal{I}$ of initial states over Σ closed under isomorphisms such that any two states $I_1, I_2 \in \mathcal{I}$ coincide on *self*. Furthermore, $\mathcal{M}$ comprises a state transition function τ on states over extended signature Σ_S with $\tau(S) = S + \Delta_{r_S}(S)$, where the rule r_S is defined as $raise(rule(val_S(self)))$ over the signature $\Sigma_S = raise(signature(val_S(self)))$.

In this definition we use extraction functions *rule* and *signature* defined on the tree representation of a sequential ASM in *self*. These are simply defined as $signature(t) = subtree(\mathbf{I}o.root(t) \prec_c o \land label(o) = \texttt{signature})$ and $rule(t) = subtree(\mathbf{I}o.root(t) \prec_c o \land label(o) = \texttt{rule})$.

3   The Logic of Abstract State Machines

We now look briefly into a simplified version of the logic of non-deterministic ASMs as defined in [4]. The simplification concerns the distinction between db-terms and algorithmic terms that is necessary, if explicit meta-finite states are considered. Here we just consider a single uniform signature Σ, so terms are defined in the usual way. However, we have to keep in mind that rASMs have a rich set of operators in their background that are used to build terms. Furthermore, as we are dealing with non-determinism there is a need to consider also ρ-terms of the form $\rho_v(t \mid \varphi)$, where ρ is a multiset operator defined in the background, φ is a formula, t is a term, and v is a variable. A *pure term* is defined as a term that does not contain any sub-term which is a ρ-term.

In order to define formulae inductively we extend the set of first-order variables with a countable set of second-order (relation) variables of arity r for each $r \geq 1$.

1. If s and t are terms, then $s = t$ is a formula.
2. If $t_1, \ldots, t_r$ are terms and X is a second-order variable of arity r, then $X(t_1, \ldots, t_r)$ is a formula.
3. If r is a rule and X is a second-order variable of arity 3, then $\mathrm{upd}(r, X)$ is a formula.
4. If r is a rule and $\ddot{X}$ is a second-order variable of arity 4, then $\mathrm{upm}(r, \ddot{X})$ is a formula.
5. If φ and ψ are formulae and x is a first-order variable, then $\neg\varphi$, $\varphi \lor \psi$ and $\forall x(\varphi)$ are formulae.

6. If φ is a formula and X is a second-order variable, then $\forall X(\varphi)$ is a formula.
7. If φ is a formula and X is a second-order variable of arity 3, then $[X]\varphi$ is a formula.

Note that we use second-order variables of arity 3 and 4 to capture update sets and update multisets, respectively.

The semantics of the logic is defined by Henkin structures. A *Henkin prestructure* S is a state of signature Σ with base set B extended with a new universe D_n of n-ary relations for each $n \geq 1$, where $D_n \subseteq \mathcal{P}(B^n)$.

Variable assignments ζ into a Henkin prestructure S are defined as usual: $\zeta(x) \in B$ for each first-order variable x, and $\zeta(X) \in D_n$ for each second-order variable X of arity n.

Then the interpretation of a term in a Henkin prestructure S with a variable assignment ζ is defined as usual; for ρ-terms $t = \rho_v(t' \mid \varphi)$ we have $val_{S,\zeta}(t) = \rho(\{\!\!\{ val_{S,\zeta[v \mapsto a_i]}(t') \mid a_i \in B \text{ and } [\![\varphi]\!]_{S,\zeta[v \mapsto a_i]} = true \}\!\!\})$.

We extend this interpretation to formulae. For a second-order variable X of arity 3 we abuse the notation by writing $val_{S,\zeta}(X) \in \Delta(r, S, \zeta)$ meaning that there is a set $U \in \Delta(r, S, \zeta)$ such that $(f, a_0, a_1) \in U$ iff $(c_f^S, a_0, a_1) \in val_{S,\zeta}(X)$. Analogously, for a second-order variable $\ddot{X}$ of arity 4 we write $val_{S,\zeta}(\ddot{X}) \in \ddot{\Delta}(r, S, \zeta)$ meaning that there is a multiset $\ddot{U} \in \Delta(r, S, \zeta)$ such that $(f, a_0, a_1) \in \ddot{U}$ with multiplicity n iff there are exactly $b_1, \ldots, b_n$ pairwise different values such that $(c_f^S, a_0, a_1, b_i) \in val_{S,\zeta}(X)$ for every $1 \leq i \leq n$. If φ is a formula, then its truth value on S under ζ (denoted as $[\![\varphi]\!]_{S,\zeta}$) is defined by the following rules:

– If φ is of the form $s = t$, then $[\![\varphi]\!]_{S,\zeta} = \begin{cases} true & \text{if } val_{S,\zeta}(s) = val_{S,\zeta}(t) \\ false & \text{otherwise} \end{cases}$.

– If φ is of the form $X(t_1, \ldots, t_r)$, then

$$[\![\varphi]\!]_{S,\zeta} = \begin{cases} true & \text{if } (val_{S,\zeta}(t_1), \ldots, val_{S,\zeta}(t_n)) \in val_{S,\zeta}(X) \\ false & \text{otherwise} \end{cases}.$$

– If φ is of the form $upd(r, X)$, then

$$[\![\varphi]\!]_{S,\zeta} = \begin{cases} true & \text{if } val_{S,\zeta}(X) \in \Delta(r, S, \zeta) \\ false & \text{otherwise} \end{cases}.$$

– If φ is of the form $upm(r, \ddot{X})$, then

$$[\![\varphi]\!]_{S,\zeta} = \begin{cases} true & \text{if } val_{S,\zeta}(\ddot{X}) \in \ddot{\Delta}(r, S, \zeta) \\ false & \text{otherwise} \end{cases}.$$

– If φ is of the form $\neg\psi$, then $[\![\varphi]\!]_{S,\zeta} = \begin{cases} true & \text{if } [\![\psi]\!]_{S,\zeta} = false \\ false & \text{otherwise} \end{cases}$.

– If φ is of the form $\alpha \vee \psi$, then

$$[\![\varphi]\!]_{S,\zeta} = \begin{cases} true & \text{if } [\![\alpha]\!]_{S,\zeta} = true \text{ or } [\![\psi]\!]_{S,\zeta} = true \\ false & \text{otherwise} \end{cases}.$$

– If φ is of the form $\forall x(\psi)$, then

$$[\![\varphi]\!]_{S,\zeta} = \begin{cases} true & \text{if } [\![\psi]\!]_{S,\zeta[x \mapsto a]} = true \text{ for all } a \in B \\ false & \text{otherwise} \end{cases}.$$

– If φ is of the form $\forall X(\psi)$, where X is a second-order variable of arity n, then

$$[\![\varphi]\!]_{S,\zeta} = \begin{cases} true & \text{if } [\![\psi]\!]_{S,\zeta[X \mapsto R]} = true \text{ for all } R \in D_n \\ false & \text{otherwise} \end{cases}.$$

– If φ is of the form $([X]\psi)$, then

$$[\![\varphi]\!]_{S,\zeta} = \begin{cases} false & \text{if } \zeta(X) \text{ represents an update set } U \\ & \quad \text{such that } U \text{ is consistent and } [\![\psi]\!]_{S+U,\zeta} = false \\ true & \text{otherwise} \end{cases}.$$

For a sentence φ to be valid in the given Henkin semantics, it must be true in all Henkin prestructures. This is a stronger requirement than saying that φ is valid in the standard Tarski semantics. A sentence that is valid in Tarski semantics is true in those Henkin prestructures, for which each universe D_n is the set of all relations of arity n.

The universes D_n of the Henkin prestructures should not be arbitrary collections of n-ary relations. Thus, it is reasonable to restrict our attention to some collections of n-ary relations that we can define, i.e. we restrict our attention to Henkin structures.

A *Henkin structure* is a Henkin prestructure S that is closed under definability, i.e. for every formula φ, variable assignment ζ and arity $n \geq 1$, we have that $\{\bar{a} \in A^n \mid [\![\varphi]\!]_{S,\zeta[a_1 \mapsto x_1,\ldots,a_n \mapsto x_n]} = true\} \in D_n$.

The main result in [4] states that the logic for ASMs defined here is complete with respect to Henkin semantics.

4   Reasoning About Reflection

Let us now investigate the extension of the logic above to handle reflection. The main difference of rASMs to ordinary ASMs is that in each step a different rule r is applied, and this rule is part of the current state. In the one-step logic of ASMs described in the previous section a rule is treated as a fixed extra-logical constant appearing only in formulae of the form $\text{upd}(r, X)$ and $\text{upm}(r, \ddot{X})$, and the meaning of these formulae depends on the actual rule r.

4.1 Extension of the Logic of ASMs

In an rASM $val_S(self)$ is a tree value t and $rule(t)$ (defined at the end of Sect. 2) is the subtree representing the actual rule of the rASM in state S. Then $raise(rule(val_S(self)))$ is the rule r_S of the rASM in state S, or phrased differently, we obtain this rule by interpretation of the term

$$\mathbf{therule} \;=\; raise(subtree(\mathbf{I}o.root(self) \prec_c o \wedge label(o) = \mathtt{rule}).$$

That is, the only extension to the logic required to capture reflection is the treatment of the first argument of $\mathrm{upd}(r, X)$ and $\mathrm{upm}(r, \ddot{X})$ as a term that is then evaluated in the state S. If the result is not a rule, these formulae remain undefined.

However, for a single machine step this extension is rather irrelevant, as in an rASM the main rule does not change within a single step. Thus, we have to take multiple steps into account. For these we introduce two additional predicates r-upd and r-upm with the following informal meaning:

- r-upd(n, X) means that n steps of the reflective ASM yield the update set X, where in each step the actual value of *self* is used.
- r-upm$(n, \ddot{X})$ means that n steps of the reflective ASM yield the update multiset $\ddot{X}$.

To be more precise, X and $\ddot{X}$ in predicates r-upd(n, X) and r-upd$(n, \ddot{X})$ are the *union* of the n update sets and n updates multisets, respectively, yielded by the reflective ASM in n steps.

Clearly, we have r-upd$(1, X) \;\leftrightarrow\; \mathrm{upd}(\mathbf{therule}, X)$, and analogously, r-upm$(1, X) \leftrightarrow \mathrm{upm}(\mathbf{therule}, \ddot{X})$. For the generalisation to arbitrary values of n we exploit the definition of $\mathrm{upd}(r, X)$ and $\mathrm{upm}(r, \ddot{X})$ for sequence rules to inductively define axioms for r-upd and r-upm. We further need the definition of consistent update sets in the logic:

$$\mathrm{conUSet}(X) \;\equiv\; \bigwedge_{c_f \in \mathcal{F}_{dyn}} \forall xyz\Big((X(c_f, x, y) \wedge X(c_f, x, z)) \to y = z\Big)$$

for the set $\mathcal{F}_{dyn}$ of constants representing the dynamic function symbols in Σ. Then we can use $\mathrm{con}(r, X)$ to expresses that X represents one of the possible update sets generated by a rule r and that X is consistent:

$$\mathrm{con}(r, X) \equiv \mathrm{upd}(r, X) \wedge \mathrm{conUSet}(X).$$

We further define

$$\text{r-upd}(n + 1, X) \;\leftrightarrow\; \big(\text{r-upd}(1, X) \wedge \neg\mathrm{conUSet}(X)\big)\vee$$
$$\big(\exists Y_1 Y_2(\text{r-upd}(1, Y_1) \wedge \mathrm{conUSet}(Y_1) \wedge [Y_1]\text{r-upd}(n, Y_2)\wedge$$
$$\bigwedge_{c_f \in \mathcal{F}_{dyn}} \forall xy(X(c_f, x, y) \leftrightarrow ((Y_1(c_f, x, y) \wedge \forall z(\neg Y_2(c_f, x, z))) \vee Y_2(c_f, x, y)))))$$

as well as

$$\mathrm{upm}(n+1, \ddot{X}) \leftrightarrow \Big(\mathrm{r\text{-}upm}(1, \ddot{X}) \wedge$$

$$\forall X \Big(\bigwedge_{c_f \in \mathcal{F}_{dyn}} \forall x_1 x_2 (X(c_f, x_1, x_2) \leftrightarrow \exists \mathbf{x}_3 (\ddot{X}(c_f, x_1, x_2, \mathbf{x}_3))) \wedge$$

$$\neg \mathrm{conUSet}(X) \Big) \Big) \vee \Big(\exists \ddot{Y}_1 \ddot{Y}_2 \Big(\mathrm{r\text{-}upm}(1, \ddot{Y}_1) \wedge$$

$$\forall Y_1 \Big(\bigwedge_{c_f \in \mathcal{F}_{dyn}} \forall x_1 x_2 (Y_1(c_f, x_1, x_2) \leftrightarrow \exists \mathbf{x}_3 (\ddot{Y}_1(c_f, x_1, x_2, \mathbf{x}_3))) \wedge$$

$$\mathrm{conUSet}(Y_1) \wedge [Y_1] \mathrm{r\text{-}upm}(n, \ddot{Y}_2) \Big) \wedge$$

$$\bigwedge_{c_f \in \mathcal{F}_{dyn}} \forall x_1 x_2 \mathbf{x}_3 \big(\ddot{X}(c_f, x_1, x_2, \mathbf{x}_3) \leftrightarrow (\ddot{Y}_2(c_f, x_1, x_2, \mathbf{x}_3) \vee$$

$$(\ddot{Y}_1(c_f, x_1, x_2, \mathbf{x}_3) \wedge \forall y_2 y_3 (\neg \ddot{Y}_2(c_f, x_1, y_2, \mathbf{y}_3))))) \Big) \Big).$$

4.2 Completeness

Let $\mathcal{L}_{asm}^{(r)}$ denote the logic of rASMs resulting from these extensions using **therule** and predicates r-upd(n, X) and r-upm(n, X) for arbitrary n. Let $\mathcal{L}_{asm}^r$ denote the further extended logic of rASMs, in which in addition quantification over n is permitted. Let us call $\mathcal{L}_{asm}^{(r)}$ the *multi-step logic* of rASMs, and $\mathcal{L}_{asm}^r$ the *run logic* of rASMs.

Even without updating the rule in every step it is obvious that the run logic $\mathcal{L}_{asm}^r$ subsumes a full dynamic logic over runs of ASMs. As such it is impossible to achieve completeness.

Theorem 1. *The run logic $\mathcal{L}_{asm}^r$ of rASMs is incomplete.*

Concerning the multi-step logic $\mathcal{L}_{asm}^{(r)}$ of rASMs the situation is not so obvious. We may continue a sublogic $\mathcal{L}_{asm}^{(r,n)}$ using a fixed value of n and formulae of the form r-upd(m, X) and r-upm(m, X) with fixed $m \leq n$. For such a sublogic we can extend the completeness result of the logic of ASMs using similar arguments.

Theorem 2. *For each $n \in \mathbb{N}$ the bounded fraction $\mathcal{L}_{asm}^{(r,n)}$ of the multi-step logic $\mathcal{L}_{asm}^{(r)}$ of rASMs is complete.*

The remaining part of this section is dedicated to prove this key result.

First note that every subformulae of the form r-upd(m, X) and of the form r-upm(m, X) that occurs in a $\mathcal{L}_{asm}^{(r,n)}$-formulae can be replaced by their corresponding definitions above. This is possible, because we have only bounded finite values for $m = 1 \ldots n$ to consider.

Thus, the axioms and rules of the derivation system remain the same as for the logic of ASMs [4,5]. Starting point is the natural formalism L_2 as defined in [6] for the relational variant of second-order logic on which the logic is based.

L_2 uses the usual axioms and rules for first-order logic, with quantifier rules applying to second-order variables as well as first-order variables, and with the stipulation that the range of the second-order variables includes at least all the relations definable by the formulae of the language. A deductive calculus for L_2 is obtained by augmenting the axioms and inference rules of first-order logic as follows:

- $\exists X \forall v_1, \ldots, v_k(X(v_1, \ldots, v_k) \leftrightarrow \varphi)$, where $k \geq 1$, $v_1, \ldots, v_k$ are first-order variables, and X is a k-ary second-order variable that does not occur freely in the formula φ.
- $\forall X(\varphi) \to \varphi[Y/X]$, provided the arity of X and Y coincides.
- $\dfrac{\psi \to \varphi[Y/X]}{\psi \to \forall X(\varphi)}$, provided Y is not free in ψ.

In addition to these axioms and rules and standard axioms and rules for first-order logic with equality, the logic $\mathcal{L}_{asm}^{(r,n)}$ comprises the following:

- The axioms for $\mathrm{upd}(r, X)$ and $\mathrm{upm}(r, X)$. Since here we do not need to consider explicit meta-finite states, these axioms are a simplified version of Axioms U1–U7 and Axioms Ü1–Ü7 in Section 7.2 and 7.3 in [4], respectively. For instance, Axiom U1 which states that X represents an update set yielded by the assignment rule $f(t) := s$ iff it contains exactly one update and this update is $((f, t), s)$, can be written as:

  U1: $\mathrm{upd}(f(t) := s, X) \leftrightarrow X(c_f, t, s) \wedge$
  $$\forall zxy(X(z, x, y) \to z = c_f \wedge x = t \wedge y = s)$$

- The distribution axiom and the necessitation rule from the axiom system **K** of modal logic, and *modus ponens*, which allow us to derive all modal properties that are valid in Kripke frames.
- The axiom $\neg \mathrm{conUSet}(X) \to [X]\varphi$ asserting that if an update set X is not consistent, then there is no successor state obtained from applying X to the current state—thus $[X]\varphi$ is interpreted as true for any formula φ.
- The axiom $\neg[X]\varphi \to [X]\neg\varphi$ describing the deterministic accessibility relation in terms of $[X]$.
- The Barcan axiom $\forall v([X]\varphi) \to [X]\forall v(\varphi)$, where v is a first-order or second-order variable.
- Axioms $\varphi \wedge \mathrm{upd}(r, X) \to [X]\varphi$ and $\mathrm{con}(r, X) \wedge [X]\varphi \to \varphi$ for static and pure φ asserting that the interpretation of static and pure formulae is the same in all states.
- The frame axiom $\mathrm{conUSet}(X) \wedge \forall z(\neg X(c_f, x, z)) \wedge f(x) = y \to [X]f(x) = y$ and the update axiom $\mathrm{conUSet}(X) \wedge X(c_f, x, y) \to [X]f(x) = y$ asserting the effect of applying an update set.
- The axiom $\mathrm{upm}(r, X) \to \exists Y(\mathrm{upd}(r, Y))$ stating that if a rule r yields an update multiset, then it also yields an update set.
- The restricted axiom of universal instantiation $\forall v(\varphi(v)) \to \varphi[t/v]$, if φ is pure or t is static, t is a term free for v in $\varphi(v)$.

– The rule of universal generalisation $\dfrac{\psi \to \varphi[v'/v]}{\psi \to \forall v(\varphi)}$ if v' is not free in ψ.
– The axiom

$$\exists X(\text{upd}(\mathbf{seq}\ r_1\ r_2\ \mathbf{endseq}, X) \wedge [X]\varphi) \leftrightarrow$$
$$\exists X_1(\text{upd}(r_1, X_1) \wedge [X_1]\exists X_2(\text{upd}(r_2, X_2) \wedge [X_2]\varphi)).$$

from dynamic logic asserting that executing a sequence rule is equivalent to executing its sub-rules sequentially.
– The extensionality axiom

$$r_1 \equiv r_2 \to (\exists X_1.\text{upd}(r_1, X_1) \wedge [X_1]\varphi) \leftrightarrow \exists X_2.\text{upd}(r_2, X_2) \wedge [X_2]\varphi.$$

For the proof of completeness we proceed in the same way as for the corresponding completeness proof for the logic of ASMs in [4]. First for operators defined in the background, in particular the multiset functions used in ρ-terms, are treated as standard non-axiomatised functions. This allows us to assume without loss of generality that formulae do not contain ρ-terms.

Then we turn formulae into variants of formulae of first-order logic with types. For this we create a modified signature Σ^T, which contains the function symbols from Σ, a unary relation symbol T_n for each $n \geq 1$, an $(n+1)$-ary relation symbol E_n for each $n \geq 1$, and unary relation symbols T_0 and T_r. With these we proceed as follows:

1. Turn formulae $\text{upd}(r, X)$ and $\text{upm}(r, X)$ into formulae of the form $T_r(x) \to \text{upd}(x, X)$ and $T_r(x) \to \text{upm}(x, X)$, where $T_r(x)$ asserts that x is a tree term representing a rule.
2. Bring all remaining atomic formulae into the form $v_1 = v_2$, $f(v_2) = v_1$ or $X(v_1, \dots, v_n)$.
3. Eliminate all modal operators expressing them by means of the formula $\text{conUSet}(X)$.
4. Replace each atomic (second-order) formula of the form $X(t_1, \dots, t_n)$ by $E_n(t_1, \dots, t_n, X)$, and relativise quantifiers over individuals using T, and quantifiers over n-ary relations in D_n for some $n \geq 1$ to T_n.

The main difference to the similar reduction applied in [4] is that subformulae $\text{upd}(r, X)$ and $\text{upm}(r, X)$ cannot be completely eliminated. However, by using T_r and tree terms we turn these formulae into first-order formulae with types. Then the axioms for $\text{upd}(r, X)$ and $\text{upm}(r, X)$ have to be adapted to this modification as well. In the case of upd, we define a new axiom that replaces Axioms U1–U7 and has the form

$$\text{upd}(x, X) \leftrightarrow \varphi_{U1}(x, X) \vee \dots \vee \varphi_{U7}(x, X),$$

where $\varphi_{U1}, \dots, \varphi_{U7}$ are modified versions of the formulae in the right-hand side of Axioms U1–U7, respectively. In particular, φ_{U1} can be defined as follows:

$$\exists x_0 x_1 x_2 x_3 x_f x_t x_s \big((x_0 = \mathbf{I}o.root(x) \prec_c o \wedge label(o) = \mathbf{update})\wedge$$

$$label(x_1) = \texttt{func} \wedge label(x_2) = \texttt{term} \wedge label(x_3) = \texttt{term} \wedge$$

$$x_0 \prec_c x_1 \wedge x_0 \prec_c x_2 \wedge x_0 \prec_c x_3 \wedge x_2 \prec_s x_3 \wedge x_1 \prec_c x_f \wedge x_2 \prec_c x_t \wedge x_3 \prec_c x_s$$

$$\exists y_f y_t y_s (y_f = vals(raise(x_f)) \wedge y_t = vals(raise(x_t)) \wedge y_s = vals(raise(x_s)) \wedge$$

$$\wedge X(y_f, y_t, y_s) \wedge \forall z_f z_t z_s (X(z_f, z_f, z_s) \rightarrow z_f = y_f \wedge z_t = y_t \wedge z_s = y_s)))$$

Note that for simplicity we have assumed, w.l.o.g. (see [4] among others), that the arity of the functions in the update rules is 1.

Due to space limitations, we leave the definition of the remaining formulae $\varphi_{U2}, \ldots, \varphi_{U7}$ as a simple exercise to the reader. Likewise the definition of a new axiom that replaces Axioms Ü1–Ü7 is also left as an easy exercise to the reader.

Lemma 1. *A formula φ of $\mathcal{L}_{asm}^{(r,n)}$ is true in a Henkin prestructure S iff the transformed formula φ^* is true in a first-order structure S^* over Σ^T that is uniquely determined by S.*

The first direction of Lemma 1, i.e., if an $\mathcal{L}_{asm}^{(r,n)}$-formula φ is true in S, then φ^* is true in S^*, can be proven by structural induction. We only need to apply the transformation described above to each of the cases in the definition of the set of $\mathcal{L}_{asm}^{(r,n)}$-formulae and then check that the resulting first-order formulae is satisfied by the corresponding state S^*. Likewise, the second direction of Lemma 1 can be proven by structural induction on the definition of first-order formulae, in this case using the inverse of the transformation described above. We omit these proofs since both are quite long, but technically straightforward.

Thus, if φ^* is valid, then φ is true in all Henkin structures. Note that the converse does not always hold. For instance $\exists x(x = x)$ is true in all Henkin structures (since by definition the domain of S is not empty), but $\exists x(T_0(x) \wedge (x = x))$ is not valid. In general, *not every* Σ^T-structure is an S^* structure for some Henkin Σ-structure S. However, if a Σ^T-structure S^* satisfies the following properties, then it corresponds to a Henkin structure S (cf. [6]):

1. Σ-correctness:
   - $T_r(c)$ for nullary function symbols $self \in \Sigma$,
   - $T_0(c)$ for all nullary function symbols $c \in \Sigma$ other than $self$, and
   - $\bigwedge_{1 \le i \le n} T_0(x_i) \rightarrow T(f(x_1, \ldots, x_n))$ for every $f \in \Sigma$.
2. Non-emptiness: $\exists x(T_0(x) \vee T_r(x))$.
3. Disjointness:
   - $T_i(x) \rightarrow \neg T_j(x)$ for $i, j \ge 0$ with $i \ne j$,
   - $T_r(x) \rightarrow \neg T_i(x)$ and $T_i(x) \rightarrow \neg T_r(x)$ for $i \ge 0$.
4. Elementhood: $E_n(x_1, \ldots, x_n, y) \rightarrow T_n(y) \wedge (T_0(x_1) \vee T_r(x_1)) \wedge \cdots \wedge (T_0(x_n) \vee T_r(x_n))$ for $n \ge 1$.
5. Extensionality: $T_n(x) \wedge T_n(y) \wedge \forall \bar{z}(E_n(\bar{z}, x) \leftrightarrow E_n(\bar{z}, y)) \rightarrow x = y$ for $n \ge 1$.
6. Comprehension: $\exists y \forall \bar{x}(E_n(\bar{x}, y) \leftrightarrow \psi)$ for $n \ge 1$ and y non-free in ψ.

Lemma 2. *If A is a first-order structure of signature Σ^T which satisfies properties 1–6 above and $sub(A)$ is the sub-structure of A induced by the elements of $\bigcup_{n \geq 0}(T_n)^A \cup (T_r)^A$, then for some Henkin structure S of signature Σ, $sub(A)$ is the structure S^* determined by S.*

Proof. Given A with domain $dom(A)$, we define S as follows:

- $dom(S) = (T_0)^A \cup (T_r)^A$ is the base set (of individuals) of S.
- For each $n \geq 1$, the universe D_n of n-ary relations consists of the sets $\{\bar{a} \in (dom(S))^n \mid (E_n)^A(\bar{a}, s)\}$ for all $s \in (T_n)^A$.
- The interpretation of function symbols $f \in \Sigma$ is the same as in A but restricted to arguments from $dom(S)$.

By the Σ-correctness, non-emptiness and comprehension properties of A, we get that S is a Henkin structure.

We claim that $sub(A)$ is isomorphic to S^* via function $g : dom(S^*) \to dom(sub(A))$ where

$$
g(x) = \begin{cases} x & \text{if } x \in (T_0)^{S^*} \cup (T_r)^{S^*} \\ \{\bar{a} \in ((T_0)^{S^*} \cup (T_r)^{S^*})^n \mid (E_n)^{S^*}(\bar{a}, x)\} & \text{if } x \in (T_n)^{S^*} \text{ for } n \geq 1 \end{cases}
$$

First, we note that g is well defined by the disjointness property and by the fact that, by definition of S and S^*, every element x in $dom(S^*)$ is in $\bigcup_{n \geq 0}(T_n)^{S^*} \cup (T_r)^{S^*}$. That g is surjective follows from the definition of S^* from A and the fact that $dom(sub(A))$ is the restriction of $dom(A)$ to $dom(S^*)$. By the extensionality property, we get that g is injective. By definition we get that g preserves the function symbols in Σ as well as the relation symbols T_n for every $n \geq 0$. Finally, for every $n \geq 1$, we get that g preserves E_n by the elementhood property.   □

Let Ψ be the set of formulae listed under properties 1–6 above, we obtain the following Henkin style completeness theorem:

Theorem 3. *An $\mathcal{L}_{asm}^{(r,n)}$-formula φ is true in all Henkin structures iff φ^* is derivable in first-order logic from Ψ (i.e., iff $\Psi \vdash \varphi^*$).*

Proof. Assume that $\Psi \vdash \varphi^*$, and let S be a Henkin structure. Then $S^* \models \Psi$ and therefore $S^* \models \varphi^*$. By Lemma 1, we get that $S \models \varphi$.

Conversely, assume that φ is true in all Henkin structures. Towards showing $\Psi \models \varphi^*$, let us assume that $A \models \Psi$, and let $sub(A)$ be its substructure generated by the elements of $\bigcup_{n \geq 0}(T_n)^A \cup (T_r)^A$. Then by Lemma 2, $sub(A) = S^*$ for some first-order structure S^* determined by a Henkin structure S. Since by assumption we have that $S \models \varphi$, it follows from Lemma 1 that $S^* \models \varphi^*$ and therefore $sub(A) \models \varphi^*$. But each quantifier in φ^* is relativised to $(T_n)^A$ for some $n \geq 1$, and then we also have that $A \models \varphi^*$. We have shown that $\Psi \models \varphi^*$, and then, by the completeness theorem of first-order logic, we get that $\Psi \vdash \varphi^*$.   □

It is easy to see that the proof system that we have described earlier in this section is sound. Thus, if φ is a formula derivable in $\mathcal{L}_{asm}^{(r,n)}$, then φ is true in all Henkin structures. It is then immediate from Theorem 3 that φ^* is derivable in first-order logic from Ψ. On the other hand, via an easy but lengthy induction on the length of the derivations, we get the following.

Lemma 3. φ^* *is derivable in first-order from* Ψ *iff* φ *is derivable in* $\mathcal{L}_{asm}^{(r,n)}$.

Theorem 3 and Lemma 3 immediately imply that $\mathcal{L}_{asm}^{(r,n)}$ is complete.

5   Conclusion

We have shown before that reflective algorithms are captured by reflective abstract state machines (rASMs), which exploit extended states that include an updatable representation of the main ASM rule to be executed by the machine in that state. Updates to the representation of ASM signatures and rules are realised by means of a sophisticated tree algebra. This enables the rigorous specification of reflective algorithms and thus adaptive systems and is one step in the direction of controlling the risk associated with systems that can change their own behaviour.

In this paper we made another step in this direction by providing an extension of the logic of ASMs to rASMs. For this we replaced extra-logical constants representing rules by terms that are subject to interpretation in the current state. As reasoning about reflective algorithms only makes sense for multiple steps, we also extend the one-step ASM logic to a multiple-step logic, and prove that for a sublogic with the number of steps bound to a fixed constant we preserve the completeness of the logic, whereas the logic in general will be incomplete.

By providing such a logic we show that it is possible to reason *statically* over specifications that are highly *dynamic* and even *unbounded* in the sense that the behaviour of the system after a sequence of adaptations is not known at all at the time the system is specified. This is of tremendous importance for the application of rigorous methods to truly adaptive systems. Even more, by showing that fragments of the logic that deal with bounded sequences of steps are still complete we even enable tool support for such reasoning.

The use of the logic in an extension of proof obligations for the refinement of rASMs in the line of [8] will be the next step in our research.

References

1. Börger, E., Stärk, R.: Abstract State Machines. Springer, Heidelberg (2003). https://doi.org/10.1007/3-540-36498-6
2. Ferrarotti, F., Schewe, K.-D., Tec, L.: A behavioural theory for reflective sequential algorithms. In: Petrenko, A.K., Voronkov, A. (eds.) PSI 2017. LNCS, vol. 10742, pp. 117–131. Springer, Cham (2018). https://doi.org/10.1007/978-3-319-74313-4_10

3. Ferrarotti, F., Schewe, K.-D., Tec, L., Wang, Q.: A new thesis concerning synchronised parallel computing - simplified parallel ASM thesis. Theor. Comput. Sci. **649**, 25–53 (2016)

4. Ferrarotti, F., Schewe, K.-D., Tec, L., Wang, Q.: A complete logic for Database Abstract State Machines. Logic J. IGPL **25**(5), 700–740 (2017)

5. Ferrarotti, F., Schewe, K.-D., Tec, L., Wang, Q.: A unifying logic for non-deterministic, parallel and concurrent Abstract State Machines. Ann. Math. Artif. Intell. **83**(3–4), 321–349 (2018)

6. Leivant, D.: Higher order logic. In: Handbook of Logic in Artificial Intelligence and Logic Programming, Deduction Methodologies, vol. 2, pp. 229–322. Oxford University Press (1994)

7. Riccobene, E., Scandurra, P.: Towards ASM-based formal specification of self-adaptive systems. ABZ 2014. LNCS, vol. 8477, pp. 204–209. Springer, Heidelberg (2014). https://doi.org/10.1007/978-3-662-43652-3_17

8. Schellhorn, G.: Verification of ASM refinements using generalized forward simulation. J. UCS **7**(11), 952–979 (2001)

9. Schewe, K., Ferrarotti, F.: Behavioural theory of reflective algorithms I: reflective sequential algorithms. CoRR, abs/2001.01873 (2020)

10. Schewe, K.-D.: Concurrent reflective Abstract State Machines. In: 19th International Symposium on Symbolic and Numeric Algorithms for Scientific Computing, (SYNASC 2017), pp. 30–35. IEEE Computer Society (2017)

11. Schewe, K.-D.: Behavioural theory of reflective algorithms II: reflective parallel algorithms (2019, under review)

12. Schewe, K.-D., Ferrarotti, F.: Behavioural theory of reflective algorithms I: reflective sequential algorithms (2019, under review)

13. Schewe, K.-D., Ferrarotti, F., Tec, L., Wang, Q., An, W.: Evolving concurrent systems: behavioural theory and logic. In: Proceedings of the Australasian Computer Science Week Multiconference, (ACSW 2017), pp. 77:1–77:10. ACM (2017)

14. Schewe, K.-D., Wang, Q.: XML database transformations. J. UCS **16**(20), 3043–3072 (2010)

15. Schewe, K.-D., Wang, Q.: Partial updates in complex-value databases. In: Information and Knowledge Bases XXII, Frontiers in Artificial Intelligence and Applications, vol. 225, pp. 37–56. IOS Press (2011)

16. Smith, B.C.: Reflection and semantics in LISP. In: Proceedings of the 11th ACM SIGACT-SIGPLAN Symposium on Principles of Programming Languages, POPL 1984, pp. 23–35. ACM (1984)

17. Stärk, R., Nanchen, S.: A logic for abstract state machines. J. Univ. Comput. Sci. **7**(11), 952–979 (2001)

18. Stemple, D., et al.: Type-safe linguistic reflection: a generator technology. In: Fully Integrated Data Environments, Esprit Basic Research Series, pp. 158–188. Springer, Heidelberg (2000). https://doi.org/10.1007/978-3-642-59623-0_8

19. Van den Bussche, J., Van Gucht, D., Vossen, G.: Reflective programming in the relational algebra. J. Comput. Syst. Sci. **52**(3), 537–549 (1996)

Analysing PROB's Constraint Solving Backends
What Do They Know? Do They Know Things? Let's Find Out!

Jannik Dunkelau$^{(\boxtimes)}$, Joshua Schmidt , and Michael Leuschel

Institut für Informatik, Heinrich-Heine-Universität Düsseldorf, Universitätsstraße 1, 40225 Düsseldorf, Germany
{jannik.dunkelau,joshua.schmidt,michael.leuschel}@hhu.de

Abstract. We evaluate the strengths and weaknesses of different backends of the PROB constraint solver. For this, we train a random forest over a database of constraints to classify whether a backend is able to find a solution within a given amount of time or answers *unknown*. The forest is then analysed in regards of feature importances to determine subsets of the B language in which the respective backends excel or lack for performance. The results are compared to our initial assumptions over each backend's performance in these subsets based on personal experiences. While we do employ classifiers, we do not aim for a good predictor, but are rather interested in analysis of the classifier's learned knowledge over the utilised B constraints. The aim is to strengthen our knowledge of the different tools at hand by finding subsets of the B language in which a backend performs better than others.

Keywords: Constraint solving · Machine learning · Decision trees · Feature importances · Association rules · Automated tool selection

1 Introduction

Besides its native CLP(FD)-based backend, the validation tool PROB [30] offers various backends for solving constraints, e.g. encountered during symbolic verification. In previous work [18,19], we trained neural networks to decide for a given constraint which backend should be used. We compared two approaches: one based on feature vectors derived from domain knowledge, and one based on encoding constraints as images. While we achieved promising results with the image-based approach, it was not possible to extract a comprehensible explanation about how the predictions were made. In follow-up work [34] the experiment was replicated with decision trees [5] using the same feature sets as before. This was motivated by the fact that decision trees are a transparent machine learning algorithm allowing to extract and interpret the learned decision rules and thus the acquired knowledge.

© Springer Nature Switzerland AG 2020
A. Raschke et al. (Eds.): ABZ 2020, LNCS 12071, pp. 107–123, 2020.
https://doi.org/10.1007/978-3-030-48077-6_8

In this paper we will expand on the decision tree approach and further analyse the relative importances of the features used for deciding whether the different backends of PROB will be successful or not. Moreover, we will compare these results with our *a priori* assumptions about the subdomains in which each backend should work well. While we will display achievable classification performances for our predictors, we do not aim for a good performance, but instead for an analysis over the whole dataset. In particular, we are not interested in replacing the decision function in PROB with a predictor presented in this paper. The goal is to find subsets of the B language in which a backend performs better than others to strengthen our knowledge of the different tools at hand. With the gathered information we may be able to improve the PROB constraint solver and to obtain more suitable features sets for related machine learning tasks for B in the future.

2   Primer on PROB and its Backends

PROB [29,30] is an animator, model checker, and constraint solver for the formal specification language B [1]. The B language allows to specify, design, and code software systems as well as to perform formal proof of their properties. When using PROB, properties can be checked exhaustively on a state space using various model checking techniques. B is rooted in predicate logic with arithmetic and set theory. At the heart of PROB is a constraint solver for the B language. PROB's constraint solver is used for many tasks. During animation it has to find suitable parameters for the B operations and compute the effect of executing an operation, during disproving [26] it is used to find counter examples to proof obligations. The constraint solver is also used for test case generation, symbolic model checking or program synthesis.

PROB has actually not one but three constraint solving backends and each backend has a variety of options. In Sects. 2.1, 2.2 and 2.3 we will introduce each backend, outline their differences, and summarise our a priori assumptions about their performances on subdomains of the B language.

2.1   The Native CLP(FD) Backend

PROB's kernel [29] is implemented in SICStus Prolog [11] using features such as co-routines for delayed constraint propagation, or mutable variables for its constraint store. The CLP(FD) finite domain library [10] is used for integers and enumerated set elements. The library has a limited precision of 59 bits. PROB handles overflows by custom implementations and also supports unbounded domains as well as symbolic representations for infinite or large sets. Some specific features of the PROB constraint solver are that it computes all solutions to a constraint using backtracking. This is important as constraints are often used within set comprehensions. It is also important for model checking to ensure that the entire state space is constructed. PROB can deal with higher-order sets, relations and functions.

Subdomains in Which CLP(FD) Presumably Performs Better. First and foremost, the CLP(FD) backend of ProB is the only backend supporting all constructs available in B. It is thus the default backend. It performs best for constraints arising in animation, where usually a small number of variables (operation parameters) have to be enumerated. In this context, it can deal well with large data values.

Generally speaking, ProB performs well on constraints using enumerated sets, booleans and/or bounded integers as base types. It performs reasonably well on unbounded intervals if interval reasoning can be applied. While ProB is very good at model finding, it can only detect unsatisfiability by exhaustively enumerating all values remaining after deterministic propagation. In case of unbounded data structures, ProB cannot exhaustively enumerate all cases and is much less powerful. While CLP(FD) cannot natively handle unbounded domains or quantifiers, ProB's backend contains several custom extensions to do so. A key limitation of the CLP(FD) backend is that it has no features such as backjumping, conflict-driven clause learning, or random restarts. In consequence, the backend can get stuck in the search space repeatedly enumerating invalid values which SAT or SMT solvers would rule out by learning.

2.2 The Kodkod Backend

An alternative backend [35] for ProB makes use of Alloy's Kodkod library [38] to translate constraints to propositional logic, which are then solved by a SAT solver. For instance, sets are translated as bit vectors. In particular, a subset x of the interval 0..2 would be translated into three propositional logic variables x_0, x_1, x_2 where x_i is true if $i \in x$ holds. The constraint $\{1, 2\} \subseteq x$ can then be translated to the propositional logic formula $x_1 \wedge x_2$. As Kodkod does not allow higher-order values, any such constraint is not passed to Kodkod and is instead dealt with by ProB's default CLP(FD) backend after Kodkod has found a solution for the other constraints.

When using this backend, ProB will first perform an interval analysis and determine which variables have a finite scope and a first-order type. The constraint is then partitioned into a part sent to Kodkod and a part solved by ProB. During solving, the SAT solver is called first. For every solution obtained by the SAT solver, ProB's CLP(FD) backend solves the remaining constraints. By default, Kodkod's Sat4j [28] SAT solver is selected.

Subdomains in Which Kodkod Presumably Performs Better. The strengths and weaknesses of the backend based on Kodkod stem from its internal reliance on SAT solving. While modern SAT solvers are very fast when it comes to solving very large boolean formulae, encoding B into propositional logic underlies certain restrictions. SAT encodings can only be used for data types known to be finite. In particular, one has to assign an upper and lower bound for integers and set sizes. Thus, integer overflows might occur and it is hard to ensure soundness and completeness. Furthermore, arithmetic operations have to be encoded in propositional logic as well such as binary adders. This leads to additional over-head when generating a conjunctive normal form, especially for large bit widths. The designers of Alloy argue [25] that lack of integers is not disadvantageous in general, as integer constraints are often of secondary nature. In B models, this is not the case. In summary, this backend is not good for arithmetic, large relations, infinite domains, higher-order constraints, or data structures.

In contrast, SAT solving is ideal for problems involving relations as those can be expressed in a way suitable for Kodkod's backends [35]. Furthermore, given that Kodkod is originally used as a backend for analysing Alloy it has been tuned towards constraints involving operations on relations. For instance, the relational image or transitive closure operations of B are handled efficiently by the translation to SAT using Kodkod.

2.3   The Z3 Backend

The third backend of PROB translates B constraints to SMT-LIB formulae and targets the SMT solvers Z3 [13] and CVC4 [3]. Here we focus on the Z3 bind-ing [27] only. The translation works by rewriting the B constraints into a normal form using a core subset of the B operators which can be mapped to SMT-LIB. Additional variables, set comprehensions, and quantifiers are introduced for those operators which have no counterpart in SMT-LIB or Z3, e.g. cardinality, or mini-mum and maximum of an integer set. Functions and relations are translated to the Array theory of SMT-LIB. The DPLL(T) [21] algorithm underlying SMT solvers is fundamentally different from CLP(FD). Just like for the SAT translation, SMT solvers can perform backjumping and conflict-driven clause learning.

Subdomains in Which Z3 Presumably Performs Better. SMT solvers such as Z3 are very good at proof for B and Event-B (cf. [14,15]). Our experience in the context of model finding is that Z3 is good at detecting inconsistencies, in particular on infinite domains. For example, Z3 is able to detect that the constraint $x < y \land y < x$ is unsatisfiable. The other two backends are unable to detect this using their default settings. Note that PROB is able to detect this inconsistency if one enables an additional set of propagation rules based on CHR.

On the downside, Z3 often has difficulties to deal with quantifiers. Moreover, the translation from B to SMT-LIB does not yet support various operators such as general union or general sum nor does it support iteration and closure operators. Constraints using one of these operators are not translated to SMT-LIB at all and the backend returns *unknown*. In summary, the Z3 backend is good at detecting inconsistencies and reasoning over infinite domains, but for constraints involving quantifiers, larger data values or cardinality computations it often answers *unknown*.

3   Primer on Decision Trees and Random Forests

We utilise techniques of supervised machine learning to train a classifier for B constraints, which we will then further analyse in Sect. 6.

The notion of machine learning covers a family of algorithms which are able to improve their predictions using a dataset processed at a so called *training* time. For *supervised* machine learning, this dataset consists of tuples $(x, y) \in D \subseteq X \times Y$, where $x \in X$ represents the input data and $y \in Y$ is the corresponding *ground truth*, which is the correct class label to be predicted by the employed algorithm. For instance, for a binary classification task, the ground truth can be either 0 or 1. Usually, $X = \mathbb{R}^d$ corresponds to a d-dimensional feature space, where each problem instance to be classified is represented as a feature vector $x = \langle x_1, \ldots, x_d \rangle$. Each x_i hereby refers to a specific characterisation, i.e. *feature*, of the problem instance. During training time, the algorithm is supposed to learn the mapping $x \mapsto y$ for each $(x, y) \in D$ by generalising over recurring patterns in the input data X. It is important that this learned mapping is accurate yet as general as possible, so as to cover yet unseen problem instances. A classifier is said to *overfit* on the training data if its performance in classifying unseen data is significantly worse. To detect possible overfitting, the resulting classifier's performance is evaluated on a separate *test set*, i.e. a data set which was not experienced during training time.

In this article, we employed decision trees as the machine learning algorithm of choice. They correspond to a supervised learning method where the training data at the root of the decision tree is progressively split into smaller subsets using a feature-based splitting criterion. At the leaves of the decision tree only subsets with the same ground truth remain. Such subsets are referred to as pure subsets.

A variety of splitting criteria exist. For example, the CART algorithm [5] is based on the *Gini impurity* $i(t)$ [31] of a node t defined as

$$i(t) = 1 - \sum_{c \in C} p_c(t)^2$$

with C being the set of possible classes, and $p_c(t)$ is the relative frequency of the elements in t belonging to the class c. For a pure subset t' of a class c, $p_c(t')$ will be 0 for $c \neq c'$ and 1 for $c = c'$. Hence $i(t') = 0$. For an evenly distributed node t'' we have $i(t'') = 1 - \frac{1}{|C|}$, where $|C|$ is the cardinality of C.

The goal of the decision tree learning algorithm is to reach an impurity of 0 with as few splits as possible. For any split of t into two sub-nodes t_L and t_R, we thus measure the *impurity decrease* by

$$d(t_{(L,R)}) = i(t) - i(t_L)\frac{|t_L|}{|t|} - i(t_R)\frac{|t_R|}{|t|} .$$

The split which maximises the impurity decrease is finally chosen and the algorithm is called recursively on t_L and t_R respectively. A decision tree is shown in Fig. 1, where leaves represent actual classes.

3.1 Random Forests

Random forests [7] are a bagging approach [6] to decision trees, i.e. instead of only training a single decision tree, a set of k decision trees $(T_i)_{1 \leq i \leq k}$ is trained.

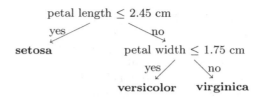

Fig. 1. Decision tree classifier over a set of iris flowers [20]. The species iris setosa, iris versicolor, or iris virginica is classified based on petal length and width.

Each tree is trained on a random subset of the training samples as well as a random subset of features. This randomisation ensures most trees in the set to be distinct from each other. For example, the impurity decrease of common features will vary between the training samples, leading to different choices of splitting. Due to bagging, the relatively unstable nature of decision trees is countered and the technique is less prone to overfitting.

A measure for the relative importances of each feature in a random forest is the *mean decrease importance* [7]. The mean decrease importance of a feature averages the impurity decrease per feature over each decision tree in the forest. Hence, it is a measure of the average impurity decrease the feature offers [2,37].

3.2   Rationale for Using Random Forests

While we had multiple classification algorithms to choose from, we finally settled on random forests. This choice was motivated by our need for a strong and interpretable classifier.

In previous work [19] we used convolutional neural networks, but we were unable to extract the knowledge accumulated by the classifciation due to the black box nature of the neural networks. Hence, we started to use decision trees, as one can easily extract classification rules after the training phase. These rules are comprehensible and can be interpreted by non-experts as well. Decision trees also offer insights about the relevancy of features: the closer to the root a split over a specific feature is done, the more impact it has for the decision process.

Alternate machine learning approaches are linear regression and clustering approaches. For linear regression the relevance of features could be extracted by examining the relative differences in their coefficients. However, this would not yield direct rules describing *why* a particular prediction was made. As we are particularly interested in extractable knowledge from trained classifiers and reasoning for the given predictions, we favoured decision trees over linear regression. On a similar note, we decided against clustering. However, a clustering approach for grouping similar constraints together presents an interesting alternative approach to be studied in future work.

In the end, we decided to utilise random forests for the present article. Although they are again blackbox algorithms, they consist of interpretable pieces, which can be analysed for more general rules [16,23].

4   Related Work

The related work in the field is split into two categories: machine learning powered algorithm portfolios for SMT solving, and knowledge extraction from tree ensemble learners such as random forests. To the best of our knowledge, no intersection of both categories exists yet in literature, as we do in this article.

Healy et al. [24] conducted a solver portfolio for the Why3 platform [4]. The solver selection was done via decision trees which predicted the anticipated runtime of a proof obligation for each solver, and choosing the fastest one. James P. Bridge [8] used support vector machines for automating the heuristic selection for the E theorem solver [36]. While he was able to improve the already implemented auto-mode in E, he also investigated picking a minimal feature set which ultimately consisted of only two to three features.

Yang et al. [39] analysed decision trees to extract minimal feature subsets which need to be flipped to achieve a more favourable outcome. Their application area was customer relationship management with focus on increasing the amount of loyal customers, i.e. detect what needs to be done to turn a regular customer into a loyal customer. Similarly, Cui et al. [12] proposed an integer linear program on random forests for finding the minimal subset of features to change for obtaining a different classification. Deng [16] proposed interpretable trees (inTrees) for interpreting tree ensembles. In their paper, they propose a set of metrics to extract learned knowledge from a tree ensemble such as a random forest. This includes the actual rules learned in an ensemble as well as frequent variable interactions. Narayanan et al. [32] extracted the most common patterns for failing solid state drives in datacenters using inTrees. In their work, they found that these extracted patterns match with previously made observations.

5   Experimental Setup

In this section, we briefly outline the training data and the feature set in use.

5.1   The Training Data

For acquiring the constraints for the training data, we extracted B predicates from the public PROB examples repository[1] and constructed more complex constraints inspired by PROB's enabling analysis [17] or discharging proof obligations [26]. Each backend was given a timeout of 25 s to decide whether the constraint has a solution or is a contradiction. Constraints for which a definite answer was found build up the positive class for a solver. The negative class is made up of the other outcomes: timeouts, errors, or the answer *unknown*. Overall, the class distribution was imbalanced, as for instance only about 35% of samples belonged to the negative class for the CLP(FD) backend. Yet, we do not deem this as a problem because the decision trees are trained with respect to a weighted training set.

---

[1] https://www3.hhu.de/stups/downloads/prob/source/ProB_public_examples.tgz.

The choice of the 25 s timeout was arbitrary. However, we evaluated how much more constraints are assigned to the positive class compared to using PROB's default timeout of 2.5 s. The CLP(FD) backend is able to solve 65.47% of the constraints using a timeout of 2.5 s, while the Kodkod and Z3 backends solve 64.65% and 21.52% respectively. When increasing the timeout by factor 10 to 25 s, these percentages increase to 65.48% for CLP(FD) (+0.01%), 64.67% for Kodkod (+0.02%), and 21.53% for Z3 (+0.01%). As the percentage of solvable constraints for each backend only increased by a rather insignificant amount, we deemed the unsolvable constraints as complex enough for our analysis approach. We did not test with higher timeouts.

For each backend's analysis we had around 170,000 unique samples.

5.2   The Feature Set

For training the decision trees, we created a manually selected set of 109 features (further referred to as F109) which mainly consists of characteristics such as the amount of arithmetic operations per top level conjunct, or the ratio of intersections of all used set operators. Further features consist of maximum and mean nesting depths for certain language constructs such as negations and powersets, or the amount of unique identifiers per top level conjunct and number of interactions between them. Additionally, identifiers are grouped into unbounded, semi-bounded (only upper or lower bound), and fully bounded (both, upper and lower bound) identifiers. This grouping is sensitive to whether the boundaries are explicitly set (e.g. $a < 5$) or only bounded by another identifier ($a < b$).

As we are interested in the knowledge gathered by the random forests over the whole corpus of B constraints at our disposal, we will not split the dataset into sets for training and testing for our final analysis as is common for classification tasks aiming for a good predictor. However, as a sanity check that the selected features are indeed discriminatory enough to actually learn weaknesses and strengths of each backend, we still analysed the predictive performances of a random forest for each backend on a classical split into datasets for training and testing. For measuring performance, we utilised the metrics *accuracy*, *balanced accuracy* [9], and the F_1-*score* [22].

Each prediction of a classifier can either be a true positive (tp), true negative (tn), false positive (fp) or false negative (fn), i.e. the prediction can be either correct or false corresponding to either the positive or negative classes 1 and 0. The utilised performance metrics are defined as follows:

$$accuracy = \frac{tp + tn}{tp + tn + fp + fn},$$

$$balanced\ acc. = \frac{1}{2}\left[\frac{tp}{tp + fn} + \frac{tn}{tn + fp}\right].$$

Accuracy describes the percentage of the test data which were classified correctly. Balanced accuracy is most suitable for an unbalanced dataset in which the distribution of classes is not equal. It averages the percentage of correctly

Table 1. Random Forest classification performances over the set of 109 features.

Backend	Dataset	Accuracy	Balanced acc.	F_1-score
CLP(FD)	F109	0.947	0.926	0.966
Kodkod	F109	0.926	0.906	0.950
Z3	F109	0.919	0.873	0.797

predicted samples per class. The F_1-score is defined as the harmonic mean over the notions *precision* and *recall* [22]:

$$precision = \frac{\text{tp}}{\text{tp} + \text{fp}}, \quad recall = \frac{\text{tp}}{\text{tp} + \text{fn}}$$

$$F_1 = 2 * \frac{precision * recall}{precision + recall}.$$

Precision describes the probability of a positive prediction to be correct. Recall describes the probability for samples of the positive class to be classified as such.

Table 1 shows the results of this sanity check. We used 80% of the data for training, whereas the performance measures were taken on the remaining 20%. Each classification task was concerned with whether a backend would return a definitive answer for a given constraint (satisfiable or unsatisfiable) or would yield *unknown*. As the performance scores are all higher than 0.9, we deem the feature set F109 to be suitable for our purposes.

6 Analysis and Results

For each backend, we trained a random forest with 50 trees using the Gini impurity decrease splitting criterion to predict whether the respective backend can find an answer or leads to *unknown*. As machine learning framework, we employed the Scikit-learn Python library [33].

For the following analysis, we utilised the whole dataset of 109 features (F109) as the training set and did not make use of a test set, as we are more interested in an analysis of random forests containing information of all the data.

Please note that this work only considered the default settings of each backend. It is relevant to mention that multiple settings exist which could influence the respective outcomes of the analysis. For instance, although the CLP(FD) backend has problems with detecting inconsistencies over unbounded domains such as $x < y \wedge y < x$, one can activate an additional CHR propagation which improves detection of inconsistencies in general as mentioned in Sect. 2.3.

6.1 Feature Importances

In order to gain a deeper insight in the feature set we compute the *Gini importance* which is the mean decrease importance of a feature within a random forest using the Gini impurity as the splitting criterion.

Table 2. Top ten features for each backend ranked by the Gini importance.

Backend	Most important features (descending)
CLP(FD)	*Function application*, *max conjunct depth*, *forward compositions*, *relational overrides*, *nested logic with conj.*, *nested logic with implications*, equalities, *function variables*, subset ratio, *identifier count*
Kodkod	*Function application*, *function vars*, *forward compositions*, set op., *nested logic with conj.*, *nested logic with disj.*, nested logic with impl., *avg. powerset nesting*, *identifier count*, *relational overrides*
Z3	Relational operators, domain ops, functions, *function vars*, *avg. powerset nesting*, domain restrictions, unbounded domains, *identifier count*, *max. conjunct depth*, *function application*

Table 2 shows the top ten features that are necessary to classify the data at hand for each backend. The common features of the three subsets are *highlighted*. These indicate a particularly high importance as they are used for each of the three backends' decisions.

The CLP(FD) backend and the Kodkod backend have the most features in common. The most important feature for both backends is the presence of function applications. Indeed, a function application is a complex operation for a constraint solver since it entails for example the well-definedness condition that the applied value is an element of the function's domain. Both backends' classifiers favour the presence of nested logic formulae with further possibly nested conjunctions, disjunctions, and implications, indicating more involved constraint as well. The initial assumption that the Kodkod backend is better suited to solve constraints over relations is strengthened by the high ranking of the ratio of relational compositions and overrides in the top 10 features. Of course, the overall higher similarity of the top ranked features for the CLP(FD) and Kodkod backend is influenced by the fact that constraints that cannot be translated to SAT are solved by PROB.

The gathered feature set for the classifier of the Z3 backend favours the presence of relational operations, in particular, the presence of domain operations. As initially expected, the feature representing the presence of unbounded domains has a high importance as well.

While this analysis allows for selection of features for the sole purpose of classification, it does not yet give us info as of why a feature ranks high. For instance, it remains uncertain whether the presence of relational operators correlates to Z3's positive or negative class. This will be analysed in Sect. 6.2.

Classifying on Reduced Feature Sets. While we are mostly interested in analysing which language subsets are hard for a backend to solve, we can evaluate the significance of the most relevant features (determined via Gini relevance as done above) by conducting a regular classification over only these relevant features. When using the ranked feature sets to find a minimal set of features, we have to consider that at least one feature exists for each B data type or group of operations, e.g. relational operators, as the dataset might be biased to specific operations. For instance, the fact that the presence of arithmetic operations is not ranked high does not mean there should be no such feature at all in general.

Table 3. Random forest classification performances for minimised feature sets.

Backend	Dataset	Accuracy	Balanced acc.	F_1-score
CLP(FD)	F10	0.914	0.875	0.944
	F50	0.947	0.929	0.966
Kodkod	F10	0.887	0.853	0.923
	F50	0.924	0.907	0.949
Z3	F10	0.875	0.804	0.747
	F50	0.916	0.870	0.795

We created two features sets containing the top 10 and 50 ranked features for each backend, referred to as F10 and F50 respectively. The results presented in Table 3 show that the minimised subsets of 50 features capture the problem domain as well as the larger set containing 109 features presented in Table 1. The smaller subset containing 10 features already shows good performance but does not perform as well as the one using 50 features, indicating that the problem at hand is complicated at least.

6.2   Association Rule Analysis

Our main goal is to determine how the backends perform on different subsets of the B language. For this we performed an association rule analysis using the inTrees framework [16], thereby identifying frequent feature interactions as well as determining those syntax elements which increase the chance of unsolvability for each backend. For the analysis, we interpret paths from the root to the leaves of each decision tree in the forest as a single rule. Each node in these trees corresponds to a feature along with a threshold value for deciding which path to follow. An example based on the decision tree from Fig. 1 is given in Fig. 2.

petal length (<) $\Rightarrow$ setosa
petal length (>) $\wedge$ petal width (<) $\Rightarrow$ versicolor
petal length (>) $\wedge$ petal width (>) $\Rightarrow$ virginica

Fig. 2. Association rules extracted from the decision tree in Fig. 1.

Different paths might be identical up to the respective threshold values. In our analysis, we discard the threshold values and only consider the tendency (below or above threshold) for each rule. This way we can compare rules without having to worry about mismatching threshold values while still accounting for the feature's tendency. Table 4 displays several rules that were collected from the random forest trained for each backend.

Deng [16] uses two metrics for the association rules, *support* and *confidence*. Given two rules $a = \{C_a \Rightarrow Y_a\}$ and $b = \{C_b \Rightarrow Y_b\}$ where C_a, C_b are the respective conditions and Y_a, Y_b the respective outcomes. Rule b is said to be in the support of rule a iff $C_a \subseteq C_b$. That is, each feature used in C_a is also used in C_b (with equal threshold tendency). Let $\sigma(a) = \{r \mid r$ is in the support of $a\}$ denote the support set of a. The confidence of an association rule a is then defined as $c(a) = |\{\{C_r \Rightarrow Y_r\} \in \sigma(a) \mid Y_r = Y_a\}|/|\sigma(a)|$, i.e. the ratio of rules in the support of a with the same outcome as a.

For a deeper analysis of the subproblems' performances for each backend, we calculated the support and confidence of the respectively 250,000 shortest rules of the corresponding random forests.

Table 4. Exemplary association rules with their corresponding support and confidence values (Supp. and Conf. respectively). The operators < and > indicate whether the feature value is above or below the learned threshold.

Backend	Rule	Supp.	Conf.
CLP(FD)	Function applications (<) ∧ conjunctions (<) ∧ quantifiers (>) ∧ logic operators (>) ∧ functions (>) ⟹ negative	853	0.69
Kodkod	Function applications (<) ∧ conjunctions (<) ∧ disjunctions (>) ∧ implications (<) ∧ powersets (<) ∧ inequality (>) ∧ quantifiers (<) ∧ lambda-expression ratio (<) ∧ relational inversions (<) ∧ sequences (<) ∧ ⟹ negative	2413	0.79
Z3	Relational operations (<) ∧ functions (<) ∧ unbounded variables (>) ∧ set inclusions (member, subset) (<) ∧ sequences (<) ∧ set operations (<) ⟹ positive	24,479	0.69

Analysis for CLP(FD). For PROB's native backend, most rules with high support only had a confidence of 50%, rendering them insignificant for our analysis. While higher confidence rules had less support such as the one presented in Table 4, they allowed for a look on certain subareas in the problem domain in which the backend struggles to find an answer for.

Main concern for the backend appears to be function applications because they are the most relevant feature for deciding whether the CLP(FD) backend is able to satisfy or reject a constraint according to the analysis in Sect. 6.1.

The implementation of function applications in PROB consists of many special cases such as different treatment for partial or total functions. Moreover, function applications entail a well-definedness condition leading to more involved constraints and possibly weaker propagation. In particular, the constraint solver has to deduce that the values applied to a function are part of its domain which increases complexity drastically if domains are (semi-)unbounded. The multitude of such cases might emphasise the overall complexity for constraint solving and be the reason for function applications leading to negative predictions. This finding suggests the need for a more involved statical analysis of constraints with function types by means of discarding well-definedness constraints early to allow for a more aggressive propagation of function applications. Thus the solver would not need to wait for verification of whether an element actually resides in a function's domain or not.

Further findings show that the use of implications, equivalences, nested powersets as well as operations on powersets contribute to the probability for the backend to answer *unknown* for a given constraint, as do operations concerning multiple variables representing functions and unbounded domains.

Comparing this to our initial presumptions made in Sect. 2.1, the particular difficulty associated with function application was mostly unexpected. Furthermore, while we did not anticipate implications or equivalences to have such significance, their role for unsolvability might be caused by a lot of backtracking inside the constraint solver for satisfying these constraints. The analysis did not bring up further results mismatching our assumptions from Sect. 2.1.

Analysis for Kodkod. The Kodkod backend struggles with arithmetic and powersets, which was to be expected. As already observed with the native backend, we also found an increase in logical operators to increase the constraint complexity significantly. An increase in logical operators naturally increases the nesting depth of the top-level conjuncts, leading to much more involved constraints. The use of functions only appears to be a problem for Kodkod if these are not manipulated by relational operators, rendering Kodkod as a more suitable choice over CLP(FD) in these cases. We generally found our expectations from Sect. 2.2 met regarding Kodkod's handling of relations.

Most positive rules favouring relational operators only showed a small support but had high confidence values and mostly differed in a single feature describing a different relational operator. If one was to generalise these rules into a singular one which is independent of the particular relational operator, these rules should be able to support each other while maintaining their high confidence. This suggests the use of relational operators for the Kodkod backend.

Note again that the Kodkod backend has a fallback to the CLP(FD) backend for non-translated structures, hence both backends perform similar overall.

Analysis for Z3. Contrary to the two backends presented above, the Z3 backend's association rule analysis delivered many high-support/high-confidence rules for the positive class. Table 4 shows one such rule with high support and confidence. Since the analysis did not provide rules with high support and confidence for the negative class, we compared absence of syntax elements in the positive rules to their existence in low-support negative rules for analysis of areas where Z3 does not perform well.

The results suggest that Z3 handles unbounded domains well and favours integer variables and inequality constraints. This is in line with our expectations from Sect. 2.3. However, we observed good performance for relational operators as well which goes against initial presumptions, although this is correlated to the amount of domain restrictions in use. Otherwise, Z3 lacks performance with quantifiers, set comprehensions, powersets, or set operations (as was expected).

The main issues for the Z3 backend are the non-translated operators as well as highly involved translations as outlined in Sect. 2.3. Revisiting these translations and comparing their implementations to those of well-performing syntax elements might allow to increase the backend's performance on further language subsets significantly. For instance, the translation of relational operators might inspire the translation of certain set operators.

7   Conclusion

In this article, we identified subproblems of the B language for which the individual PROB constraint solving backends performed better or worse respectively.

While our findings generally matched our expectations stated in Sects. 2.1, 2.2 and 2.3, we found certain results which we did not explicitly expect. For instance, our evidence suggests a difficulty for dealing with function applications as well as implications and equivalences. Involved constraints containing many nested conjunctions and disjunctions also increased the chance for the backends to return *unknown*. Surprisingly, the Z3 backend performed much better on relational operators as expected. As a consequence, our analysis identified the need for a more sophisticated handling of function application and nested logic operators.

As by-product of this work, we were also able to train well-performing classifiers for each backend, which can be used for automated backend selection.

The experimental data as well as corresponding Jupyter notebooks are available on GitHub:

https://github.com/jdnklau/prob-backend-analysis.

Acknowledgements. Computational support and infrastructure was provided by the "Centre for Information and Media Technology" (ZIM) at the University of Düsseldorf (Germany).

References

1. Abrial, J.R.: The B-Book: Assigning Programs to Meanings. Cambridge University Press, New York (1996)
2. Archer, K.J., Kimes, R.V.: Empirical characterization of random forest variable importance measures. Comput. Stat. Data Anal. **52**(4), 2249–2260 (2008). https://doi.org/10.1016/j.csda.2007.08.015
3. Barrett, C., et al.: CVC4. In: Gopalakrishnan, G., Qadeer, S. (eds.) CAV 2011. LNCS, vol. 6806, pp. 171–177. Springer, Heidelberg (2011). https://doi.org/10.1007/978-3-642-22110-1_14
4. Bobot, F., Filliâtre, J.C., Marché, C., Paskevich, A.: Why3: Shepherd your herd of provers. In: Boogie 2011: First International Workshop on Intermediate Verification Languages, Wrocław, Poland, pp. 53–64, August 2011
5. Breiman, L., Friedman, J., Olshen, R., Stone, C.: Classification and Regression Trees. Wadsworth and Brooks, Monterey (1984)
6. Breiman, L.: Bagging predictors. Mach. Learn. **24**(2), 123–140 (1996). https://doi.org/10.1007/BF00058655
7. Breiman, L.: Random forests. Mach. Learn. **45**(1), 5–32 (2001)
8. Bridge, J.P.: Machine learning and automated theorem proving. Technical report, University of Cambridge, Computer Laboratory (2010)
9. Brodersen, K.H., Ong, C.S., Stephan, K.E., Buhmann, J.M.: The balanced accuracy and its posterior distribution. In: 2010 International Conference on Pattern Recognition, pp. 3121–3124. IEEE, August 2010. https://doi.org/10.1109/ICPR.2010.764
10. Carlsson, M., Ottosson, G., Carlson, B.: An open-ended finite domain constraint solver. In: Glaser, H., Hartel, P., Kuchen, H. (eds.) PLILP 1997. LNCS, vol. 1292, pp. 191–206. Springer, Heidelberg (1997). https://doi.org/10.1007/BFb0033845
11. Carlsson, M., et al.: SICStus Prolog User's Manual, vol. 3. Swedish Institute of Computer Science Kista, Sweden (1988)
12. Cui, Z., Chen, W., He, Y., Chen, Y.: Optimal action extraction for random forests and boosted trees. In: International Conference on Knowledge Discovery and Data Mining KDD 2015, pp. 179–188. Association for Computing Machinery, New York (2015). https://doi.org/10.1145/2783258.2783281
13. de Moura, L., Bjørner, N.: Z3: An efficient SMT solver. In: Ramakrishnan, C.R., Rehof, J. (eds.) TACAS 2008. LNCS, vol. 4963, pp. 337–340. Springer, Heidelberg (2008). https://doi.org/10.1007/978-3-540-78800-3_24
14. Déharbe, D., Fontaine, P., Guyot, Y., Voisin, L.: SMT solvers for Rodin. In: Derrick, J., et al. (eds.) ABZ 2012. LNCS, vol. 7316, pp. 194–207. Springer, Heidelberg (2012). https://doi.org/10.1007/978-3-642-30885-7_14
15. Déharbe, D., Fontaine, P., Guyot, Y., Voisin, L.: Integrating SMT solvers in Rodin. Sci. Comput. Program. **94**, 130–143 (2014). https://doi.org/10.1016/j.scico.2014.04.012
16. Deng, H.: Interpreting tree ensembles with intrees. Int. J. Data Sci. Anal. **7**(4), 277–287 (2019). https://doi.org/10.1007/s41060-018-0144-8
17. Dobrikov, I., Leuschel, M.: Enabling analysis for Event-B. Sci. Comput. Program. **158**, 81–99 (2018). https://doi.org/10.1016/j.scico.2017.08.004
18. Dunkelau, J.: Machine learning and AI techniques for automated tool selection for formal methods. In: Proceedings of the PhD Symposium at iFM'18 on Formal Methods: Algorithms, Tools and Applications, University of Oslo, September 2018. https://doi.org/10.18154/RWTH-CONV-236485

19. Dunkelau, J., Krings, S., Schmidt, J.: Automated backend selection for PROB using deep learning. In: Badger, J.M., Rozier, K.Y. (eds.) NFM 2019. LNCS, vol. 11460, pp. 130–147. Springer, Cham (2019). https://doi.org/10.1007/978-3-030-20652-9_9

20. Fisher, R.A.: The use of multiple measurements in taxonomic problems. Ann. Eugen. **7**(2), 179–188 (1936)

21. Ganzinger, H., Hagen, G., Nieuwenhuis, R., Oliveras, A., Tinelli, C.: DPLL(T): fast decision procedures. In: Alur, R., Peled, D.A. (eds.) CAV 2004. LNCS, vol. 3114, pp. 175–188. Springer, Heidelberg (2004). https://doi.org/10.1007/978-3-540-27813-9_14

22. Goutte, C., Gaussier, E.: A probabilistic interpretation of precision, recall and F-score, with implication for evaluation. In: Losada, D.E., Fernández-Luna, J.M. (eds.) ECIR 2005. LNCS, vol. 3408, pp. 345–359. Springer, Heidelberg (2005). https://doi.org/10.1007/978-3-540-31865-1_25

23. Hara, S., Hayashi, K.: Making tree ensembles interpretable. In: ICML Workshop on Human Interpretability in Machine Learning (WHI 2016) (2016)

24. Healy, A., Monahan, R., Power, J.F.: Evaluating the use of a general-purpose benchmark suite for domain-specific SMT-solving. In: Symposium on Applied Computing SAC 2016, pp. 1558–1561. ACM (2016). https://doi.org/10.1145/2851613.2851975

25. Jackson, D.: Alloy: a lightweight object modelling notation. Trans. Softw. Eng. Methodol. **11**(2), 256–290 (2002)

26. Krings, S., Bendisposto, J., Leuschel, M.: From failure to proof: the PROB disprover for B and Event-B. In: Calinescu, R., Rumpe, B. (eds.) SEFM 2015. LNCS, vol. 9276, pp. 199–214. Springer, Cham (2015). https://doi.org/10.1007/978-3-319-22969-0_15

27. Krings, S., Leuschel, M.: SMT solvers for validation of B and Event-B models. In: Ábrahám, E., Huisman, M. (eds.) IFM 2016. LNCS, vol. 9681, pp. 361–375. Springer, Cham (2016). https://doi.org/10.1007/978-3-319-33693-0_23

28. Le Berre, D., Parrain, A.: The Sat4J library, release 2.2. J. Satisf. Boolean Model. Comput. **7**, 59–64 (2010). System description

29. Leuschel, M., Bendisposto, J., Dobrikov, I., Krings, S., Plagge, D.: From animation to data validation: the ProB constraint solver 10 years on. In: Boulanger, J.-L. (ed.) Formal Methods Applied to Complex Systems: Implementation of the B Method, pp. 427–446. Wiley, Hoboken (2014)

30. Leuschel, M., Butler, M.: ProB: a model checker for B. In: Araki, K., Gnesi, S., Mandrioli, D. (eds.) FME 2003. LNCS, vol. 2805, pp. 855–874. Springer, Heidelberg (2003). https://doi.org/10.1007/978-3-540-45236-2_46

31. Loh, W.: Classification and regression tree methods. In: Wiley StatsRef: Statistics Reference Online. American Cancer Society, September 2014. https://doi.org/10.1002/9781118445112.stat03886

32. Narayanan, I., et al.: SSD failures in datacenters: what? when? and why? In: Systems and Storage Conference, p. 7. ACM (2016)

33. Pedregosa, F., et al.: Scikit-learn: machine learning in Python. J. Mach. Learn. Res. **12**, 2825–2830 (2011)

34. Petrasch, J.: The decision does not fall far from the tree: automatic configuration of predicate solving. Master's thesis, Heinrich Heine Universität Düsseldorf, Universitätsstraße 1, 40225 Düsseldorf, April 2018

35. Plagge, D., Leuschel, M.: Validating B,Z and TLA$^+$ Using PROB and Kodkod. In: Giannakopoulou, D., Méry, D. (eds.) FM 2012. LNCS, vol. 7436, pp. 372–386. Springer, Heidelberg (2012). https://doi.org/10.1007/978-3-642-32759-9_31

36. Schulz, S.: E-a brainiac theorem prover. Ai Commun. **15**(2,3), 111–126 (2002)
37. Strobl, C., Boulesteix, A.L., Zeileis, A., Hothorn, T.: Bias in random forest variable importance measures: Illustrations, sources and a solution. BMC Bioinf. **8**(1), 25 (2007). https://doi.org/10.1186/1471-2105-8-25
38. Torlak, E., Jackson, D.: Kodkod: a relational model finder. In: Grumberg, O., Huth, M. (eds.) TACAS 2007. LNCS, vol. 4424, pp. 632–647. Springer, Heidelberg (2007). https://doi.org/10.1007/978-3-540-71209-1_49
39. Yang, Q., Yin, J., Ling, C.X., Chen, T.: Postprocessing decision trees to extract actionable knowledge. In: International Conference on Data Mining, pp. 685–688. IEEE, November 2003. https://doi.org/10.1109/ICDM.2003.1251008

Programming the CLEARSY Safety Platform with B

Thierry Lecomte[✉]

ClearSy, 320 Avenue Archimède, Aix en Provence, France
thierry.lecomte@clearsy.com

Abstract. The CLEARSY Safety Platform (CSSP) is aimed at easing the development and the deployment of safety critical applications, up to the safety integrity level 4 (SIL4). It relies on the smart integration of the B formal method, redundant code generation and compilation, and a hardware platform that ensures a safe execution of the software. This paper exposes the programming model of the CSSP used to develop control & command applications based on digital I/Os.

Keywords: B method · Safety critical · Programming model

1 Introduction

In many industrial standards, formal methods are highly recommended when developing safety critical software for the highest safety levels. However formal methods are highly recommended just like many other non-formal (combination of) techniques, as these recommendations are setup collectively and represent the industrial best practices. Convinced that formal methods could help to obtain better products [4,5,7,8], more easily certifiable, a generic, safe execution platform has been researched for years, combining safety electronics and defect-free proven software. The software model is proved to be defect-free - complying with its formal specification and without programming errors. The code generators and the compilers are not defect-free. They are not required to be defect-free as the defects are detected with divergent behaviour during execution. The CLEARSY Safety Platform was initially an in-house development project before being funded by the R&D collaborative project *LCHIP* (Low Cost High Integrity Platform) to obtain a generic version of the platform (i.e. not only aimed at railway systems). *LCHIP* [6] is aimed at allowing any engineer to develop a function by using its usual Domain Specific Language (DSL) and to obtain this function running safely on a hardware platform. With an automatic development process, the B formal method will remain "behind the curtain" in order to avoid expert transactions over several languages (domain specific language, B language, interactive proof). Indeed the programs developed with the CLEARSY Safety Platform are considerably simpler than metro automatic pilot, with few properties, simpler algorithms and hence with an expected excellent automatic proof ratio. The integration of third party provers/solvers is also

© Springer Nature Switzerland AG 2020
A. Raschke et al. (Eds.): ABZ 2020, LNCS 12071, pp. 124–138, 2020.
https://doi.org/10.1007/978-3-030-48077-6_9

expected to improve automatic proof. Based on our previous certification experience, the safety demonstration of a safety case does not require any specific feature for the input B model; it could be handwritten or the by-product of a translation process. Several DSLs are being connected (or planned to be) based on an Open API (Bxml).

This paper introduces the CLEARSY Safety Platform, presents and explains the evolution of the supported B0 modelling language. The shape of the programs developed for this platform are tightly linked with the specific mission of the platform: ensuring a safety (see Sect. 3.3) out of reach of the developer who cannot alter it.

This paper is structured in five parts. The Terminology is first introduced as some terms and concepts are quite specific. Then a description of the CLEARSY Safety Platform is provided with a focus on its safety features. Third the programming model is introduced; the simplification of the proof is also discussed. Exploitation and dissemination are then exposed. Finally conclusion and perspectives are discussed.

2   Terminology

This chapter clarifies a number of unusual terms and concepts used in this paper.

Atelier CSSP is Atelier B extended with diverse code generator toolchain, bootloader, and a new project type (CSSP project).

B0 is a subset of the B language [1] that must be used at implementation level. It contains deterministic substitutions and concrete types. B0 definition depends on the target hardware associated to a code generator [2]. Most railways product lines use their own own specific code generator.

Bxml is an XML interface to B models, supported by Atelier B.

CRC stands for cyclic redundancy check, is an error-detecting code commonly used in digital networks and storage devices to detect accidental changes to raw data.

CSSP abbreviation of CLEARSY Safety Platform. The CLEARSY Safety Platform is made up of a hardware execution platform, an IDE enabling the generation of diverse binaries from a single B model, and a certification kit describing its safety features as well as the safety constraints exported to the hosting system.

Diversity intentional differences between redundant components, to reduce the likelihood of common failures due to systematic causes that would reduce the benefit of redundancy [3].

Fault tolerance is the property that enables a system to continue operating properly in the event of the failure of some of its components. In our case, any electronic part including the processors.

HEX is a file format that conveys binary information in ASCII text form. It is commonly used for programming microcontrollers, EPROMs, and other types of programmable logic devices.

PLC stands for programmable logic controller, is an industrial digital computer which has been ruggedized and adapted for the control of any activity that requires high reliability control and ease of programming and process fault diagnosis.

Safety refers to the control of recognized hazards in order to achieve an acceptable level of risk.

SIL put for Safety Integrity Level, is a relative level of risk-reduction provided by a safety function. Its range is usually between 0 and 4, SIL4 being the most dependable and used for situations where people could die.

Reliability is the ability of a system to perform its required functions under stated conditions for a specified time.

3   The CLEARSY Safety Platform

3.1   Rationale

Developing a safety computer from scratch is not something you easily decide because of the effort required to obtain such a device. Two kinds of device are currently available on the market for safety critical applications: PLCs and SIL3/SIL4-ready boards. Large companies building trains have their own in-house devices but they are not publicly available. PLCs provide a strict, certified environment from which it is impossible to escape, requiring systems to be designed and programmed in specific ways. On the contrary, SIL3/SIL4-ready boards offer more freedom, come with hardware features not incompatible with the standards but where the safety principles have to be fully programmed by the developer in C or similar language.

To overcome this inconvenience, CLEARSY decided to develop its own solution based on the combination of redundant hardware and proven software developed with B. Producing its own hardware would reduce by an order of magnitude its cost compared to PLCs and SILx-ready boards while using Atelier B would allow more freedom and more control on the software development. The decision to go for B was easily taken as it is highly recommended by the industry standard for SIL4 software development. B is also the central formal technology we have been using during more than 20 years for most of safety critical software development. Finally the CLEARSY Safety Platform is aimed at easing the certification process, as the safety principles, embedded in the electronics design and the B software, are out of reach of the developer who cannot alter them.

3.2   Description

The CLEARSY Safety Platform (abbreviated as CSSP in the rest of the document) is a new technology, both hardware and software, combining a software development environment based on the B language and a secured execution hardware platform, to ease the development of safety critical applications.

It relies on a software factory that automatically transforms function into binary code that runs on redundant hardware. The starting point is a text-based, B formal model that specifies the function to implement. This model may contain static and dynamic properties that define the functional boundaries of the target software. The B project is automatically generated (Fig. 5), based on the inputs/outputs configuration (numbers, names). The project contains all the machines and implementation components required to program the CLEARSY Safety Platform. From the developer's point of view, only one function (name *user_logic*) has to be specified (machine *logic*) and implemented properly (implementation *logic_i*).

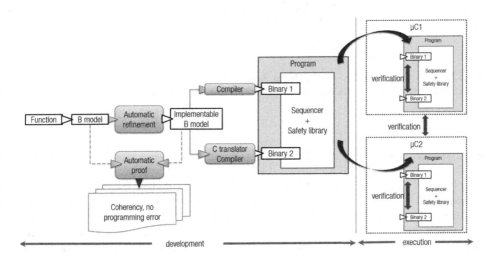

Fig. 1. The safe generation and execution of a function on the double processor.

The implementable model is then translated using two different chains:

- Translation into C ANSI code, with the C4B Atelier B code generator (instance I_1). This C code is then compiled into HEX[1] binary code with an off-the-shelf compiler (gcc).
- Translation into MIPS Assembly then to HEX binary code, with a specific compiler developed for this purpose (instance I_2). The translation in two steps

---

[1] A file format that conveys binary information in ASCII text form. It is commonly used for programming micro-controllers.

allows to better debug the translation process as a MIPS assembly instruction corresponds to a HEX line.

The software obtained is the uploaded on the execution platform to be executed by two micro-controllers (Fig. 2).

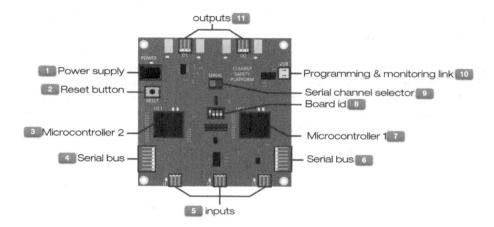

Fig. 2. The CLEARSY Safety Platform Starter Kit 0 (SK_0) – documentation available at https://github.com/CLEARSY/CSSP-Programming-Handbook

3.3 Safety

These two different instances I_1 and I_2 of the same function are then executed in sequence, one after the other, on two PIC32 micro-controllers. Each micro-controller hosts both I_1 and I_2, so at any time 4 instances of the function are being executed on the micro-controllers. The results obtained by I_1 and I_2 are first compared locally on each micro-controller then they are compared between micro-controllers by using messages. In case of a divergent behaviour (at least one of the four instances exhibits a different behaviour), the faulty micro-controller reboots. The sequencer and the safety functions are developed once for all in

```
initialisation

while(true) {
            execute I1
            execute I2
            perform safety verifications
    }
```

Fig. 3. The pseudo-code of the sequencer.

B by the IDE design team and come along as a library. This way, the safety functions are out of reach of the developers and cannot be altered. The safety is based on several features such as:

- the detection of a divergent behaviour,
- micro-controller liveness regularly checked by messages,
- the detection of the inability for a processor to execute an instruction properly[2],
- the ability to command outputs[3],
- memory areas (code, data for the two instances) are also checked (no overlap, no address outside memory range),
- each output needs the two micro-controllers to be alive and providing respectively power and command, to be active (permissive mode). In case of misbehaviour, the detecting micro-controller deactivate its outputs and enter an infinite loop doing nothing.

The code generators are different (code generation paths, specification, programming languages, development teams) and as such common failure modes are neglected. Some of the tools part of the tool-chain have been "certified by usage" since 1998 (B parser, B compiler, C code generator), but the newest tools of this tool-chain have no history to rely on for certification. It is not a problem for railway standards as the whole product is certified (with its environment, its development and verification processes, etc.), hence it is not required to have every tool certified. Instead the main feature used for the safety demonstration is the detection of a misbehaviour among the 4 instances of the function and the 2 microcontrollers. This way, similar bugs that could affect at the same time and with the same effects two independent tools are simply neglected. In its current shape, the CLEARSY Safety Platform provides an automatic way of transforming a proven B model into a program that safely executes on a redundant platform while the developer does not have to worry about the safety aspects.

3.4 Target Applications

The execution platform is based on two PIC32 micro-controllers[4]. The processing power available is sufficient to update 50k interlocking Boolean equations per second, compatible with light-rail signalling requirements. The execution platform can be redesigned seamlessly for any kind of mono-core processor if a higher level of performance is required.

---

[2] All instructions are tested regularly against an oracle.

[3] Outputs are read to check if commands are effective, a system not able to change the state of its outputs has to shutdown.

[4] PIC32MX795F512L providing 105 DMIPS at 80 MHz.

The IDE provides a restricted modelling framework for software where:

- No operating system is used.
- Software behaviour is cyclic (no parallelism).
- No interruption modifies the software state variables.
- Supported types are Boolean and integer types (and arrays of).
- Only bounded-complexity algorithms are supported (the price to pay to keep the proof process automatic).

4   Programming Model

Target CSSP applications are controllers. They execute the following infinite loop: read inputs, perform computation, then set outputs. If a failure happens, the board deactivates the outputs (they are all OFF – not powered) and enters an infinite loop doing nothing (Fig. 4). The only way to exit this loop is to reset the board. The program in Flash memory is copied into RAM and then its execution starts. If the failure is permanent, the board keeps restarting with the outputs deactivated – the board remains in a safe, restrictive state.

Fig. 4. A CSSP is either able to execute its software properly (transfer function F) (left) or is not able (right) and hence does nothing while its outputs are deactivated.

4.1   Development Process

A CSSP project (Fig. 5) is a B project generated from a CSSP board configuration where I/O are selected (some inputs/outputs pins may not be used) and named. This generated B project is made of:

- the interface with the safety library, containing the definition of all the types (and related constants) that may be used in a CSSP project, as well as specific operators (arithmetic, logic) and operations (access to current time, message to print on serial channel),
- the model of the function to program, that has:
  - a read-only access to the safety library, the digital inputs status (OFF, ON), the current time since the last rest/power-on, and
  - the ability to modify the digital outputs (OFF, ON).

Programming the CSSP consists in modifying the components *user_ctx* and *logic*, and to possibly add other components to be imported by *logic_i*.

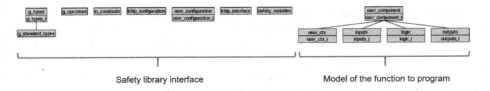

Fig. 5. A CSSP project.

4.2 Pragmas

A component cannot contain both constants (SETS, CONSTANTS) and variables. Constants are hosted by context machines (machines without variables, with possibly read-only operations). The compiler is made aware of this situation by the use of one and only one pragma in each implementation:

- CONSTANTS, to indicate a constants-only module
- SAFETY_VARS, to indicate a variables-only module

```
IMPLEMENTATION
    logic_i
REFINES
    logic
SEES
    g_types,
    g_operators,
    io_constants,
    lchip_interface,
    user_ctx,
    inputs

    // pragma SAFETY_VARS
```

```
IMPLEMENTATION
    user_ctx_i
REFINES
    user_ctx

    // pragma CONSTANTS
SEES
    g_types
VALUES
    DELTA_T = 1000 // 1000 ms == 1s
END
```

Fig. 6. Two examples of pragmas.

4.3 Types and Operators

The types available in implementation are:

- *uint8_t, uint16_t, uint32_t.* These types (unsigned integers coded on 8, 16 and 32 bits) are preferred to the generic type INT, to get a better control over variable memory size and overflow. Automatic casting is performed when for example a *uint16_t* variable is combined with a *uint8_t* value. The reverse situation generates a warning from the B32 compiler.
- *BOOL*

The values of the digital inputs and outputs (*IO_OFF*, *IO_ON*) are stored as *uint8_t* and not as Boolean. It is because a memory glitch could easily transform a 0 in 1 (or a 1 in 0) without being easily detected. Having these values coded with 8 bits (with a sufficient Hamming distance) make this undetected modification unlikely to occur. Moreover setting one output with a value different from *IO_OFF* and *IO_ON* is detected during execution by the CLEARSY Safety Platform which enters panic mode.

In order to automate as much as possible the proof process, the arithmetic operators able to overflow – +, −, x – are replaced by non-overflowing operators. These operators are modelled as modulo operators (Fig. 7), preventing an overflow to happen. These operators are defined for the 3 supported arithmetic types as lambda functions and implemented with native functions in the safety library. These operators avoid to generate overflow proof obligations and enable a better automation of the proof process. However well-definedness proof obligations remain and when using the integer division/, the denominator has to be proved different from 0.

```
add_uint32 = %(x1,x2).(x1 : uint32_t & x2 : uint32_t | (x1 + x2) mod (MAX_UINT32 + 1)) &
sub_uint32 = %(x1,x2).(x1 : uint32_t & x2 : uint32_t | (x1 - x2 + MAX_UINT32 + 1) mod (MAX_UINT32 + 1)) &
mul_uint32 = %(x1,x2).(x1 : uint32_t & x2 : uint32_t | (x1 * x2) mod (MAX_UINT32 + 1)) &
add_uint16 = %(y1,y2).(y1 : uint16_t & y2 : uint16_t | (y1 + y2) mod (MAX_UINT16 + 1)) &
sub_uint16 = %(y1,y2).(y1 : uint16_t & y2 : uint16_t | (y1 - y2 + MAX_UINT16 + 1) mod (MAX_UINT16 + 1)) &
mul_uint16 = %(y1,y2).(y1 : uint16_t & y2 : uint16_t | (y1 * y2) mod (MAX_UINT16 + 1)) &
add_uint8 = %(y1,y2).(y1 : uint8_t & y2 : uint8_t | (y1 + y2) mod (MAX_UINT8 + 1)) &
sub_uint8 = %(y1,y2).(y1 : uint8_t & y2 : uint8_t | (y1 - y2 + MAX_UINT8 + 1) mod (MAX_UINT8 + 1)) &
mul_uint8 = %(y1,y2).(y1 : uint8_t & y2 : uint8_t | (y1 * y2) mod (MAX_UINT8 + 1)) &
```

Fig. 7. Arithmetic operators redefined.

Bitwise operators (and, or, xor, not, shift left logical, shift right logical) have been added similarly (Fig. 8). They allow programs to operate more easily at bit level. They are defined for 8, 16, and 32 bit sizes.

```
bitwise_sll_uint32 : uint32_t*uint8_t --> uint32_t &
bitwise_srl_uint32 : uint32_t*uint8_t --> uint32_t &
bitwise_not_uint32 : uint32_t --> uint32_t &
bitwise_and_uint32 : uint32_t*uint32_t --> uint32_t &
bitwise_xor_uint32 : uint32_t*uint32_t --> uint32_t &
bitwise_or_uint32 : uint32_t*uint32_t --> uint32_t &
```

Fig. 8. Bitwise operators added.

4.4   Time

Time is defined as a *uint32_t* and represent a number of milliseconds. The operation *get_ms_tick* returns the number of milliseconds elapsed since the last reset or power on. Storing the current time and then checking its difference with a future current time allows one to program timers.

4.5   I/O

Inputs and outputs valid values are *IO_OFF* and *IO_ON*. To get the value of
an input, use the operation get_xxx where xxx is the name given to the input.
The operation returns a *uint8_t*. To set the value of an output, use the operation
set_xxx where xxx is the name you gave to the output.

4.6   Substitutions

The B0, implementation language, supported by the CLEARSY Safety Platform
is more strict than the one supported by the C code generator C4B. The main
reason for not providing as much freedom to the develop is to keep the B32
compiler simple in order to more easily convince the safety auditor during the
certification process. Several substitutions are constrained as follow:

- *IF THEN ELSE* supports only single condition. If a test is a disjunc-
  tion/conjunction of several expressions, the test will have to be nested into
  several levels. Testing operators are restricted to $<$, $\leq$ and $=$.
- assignments are restricted to two operands on the right hand term in order
  to avoid to manipulate the stack. The valuation with the addition of more
  than two operands will have to be decomposed in successive additions with
  two operands.
- variables declared in a VAR substitution have to be typed first with a substitution "becomes such that".

```
user_logic =                              res <-- triAND(v1, v2, v3) =
BEGIN                                     BEGIN
     VAR il_, i2_, i3_ IN                     res := IO_OFF;
         il_ :( il_ : uint8_t );              IF v1 = IO_ON THEN
         i2_ :( i2_ : uint8_t );                  IF v2 = IO_ON THEN
         i3_ :( i3_ : uint8_t );                      IF v3 = IO_ON THEN
                                                          res := IO_ON
         il_ <-- get_I1;                              END
         i2_ <-- get_I2;                          END
         i3_ <-- get_I3;                      END
                                          END;
         O1 <-- triAND(il_, i2_, i3_);
         O2 <-- negIO(O1)
     END
END;
```

Fig. 9. Local variables in *user_logic* are types before use. Tests in *triAND* are nested
because only single conditions are supported.

5   Ease to Prove Models

One of the objectives of the CLEARSY Safety Platform is to make to proof process fully automated. The use of the modulo arithmetic operators contributes

directly to this objective. The low complexity of the target lightweight applications (smaller and simpler than metro automatic pilots for example) is another reason to keep the proof effort low.

However given the modelling choices made for the arithmetic operators and the heavy use of lambda functions, we had to make sure that trivial arithmetic assignment with these operators would lead to proof obligations that are provable automatically. The analysis of the proof obligations initially that were not demonstrated automatically led to the addition of several proof elements (Fig. 10):

- two rules to handle properly any predicate containing 2**x. These rules appear in the PatchProver, a slot for mathematical rules to be applied for any project. PatchProverA means that these rules are applied after (A put for After) the one iteration of the main prover.
- several proof tactics in the User_Pass of several components.

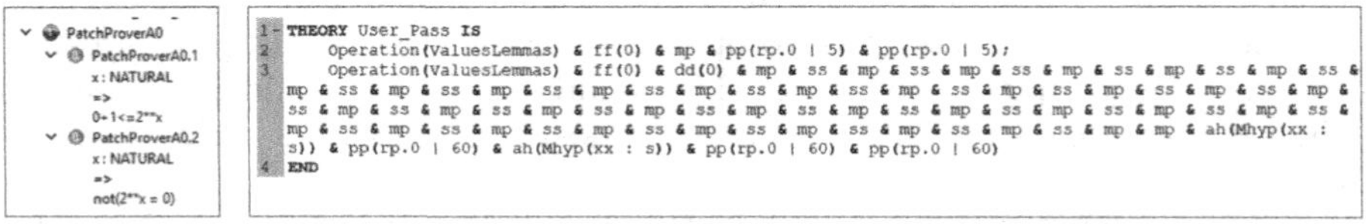

Fig. 10. Mathematical rules and User_Pass proof tactic defined to automate proof.

Finally the default CSSP project generated after creation is fully proved automatically with the following scenario: select all components, prove force 0, prove user pass. It also applies for the two examples provided with the Atelier CSSP: Clock and Combinatorial (Fig. 11).

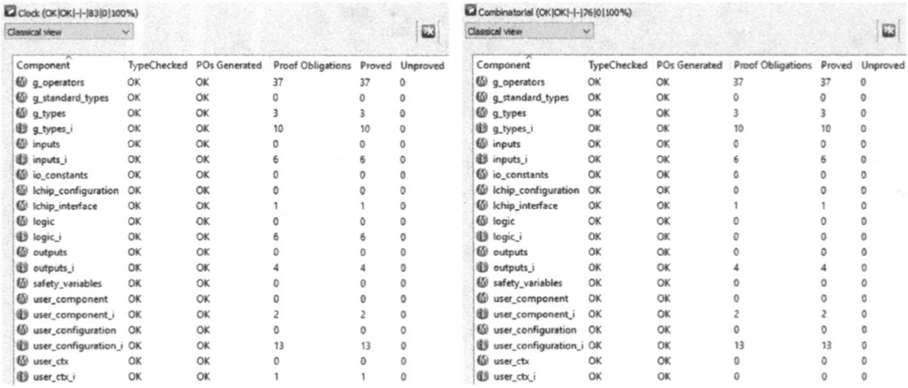

Clock (OK|OK|-|-|83|0|100%)

Classical view

Component	TypeChecked	POs Generated	Proof Obligations	Proved	Unproved
g_operators	OK	OK	37	37	0
g_standard_types	OK	OK	0	0	0
g_types	OK	OK	3	3	0
g_types_i	OK	OK	10	10	0
inputs	OK	OK	0	0	0
inputs_i	OK	OK	6	6	0
io_constants	OK	OK	0	0	0
lchip_configuration	OK	OK	0	0	0
lchip_interface	OK	OK	1	1	0
logic	OK	OK	0	0	0
logic_i	OK	OK	6	6	0
outputs	OK	OK	0	0	0
outputs_i	OK	OK	4	4	0
safety_variables	OK	OK	0	0	0
user_component	OK	OK	0	0	0
user_component_i	OK	OK	2	2	0
user_configuration	OK	OK	0	0	0
user_configuration_i	OK	OK	13	13	0
user_ctx	OK	OK	0	0	0
user_ctx_i	OK	OK	1	1	0

Combinatorial (OK|OK|-|-|76|0|100%)

Classical view

Component	TypeChecked	POs Generated	Proof Obligations	Proved	Unproved
g_operators	OK	OK	37	37	0
g_standard_types	OK	OK	0	0	0
g_types	OK	OK	3	3	0
g_types_i	OK	OK	10	10	0
inputs	OK	OK	0	0	0
inputs_i	OK	OK	6	6	0
io_constants	OK	OK	0	0	0
lchip_configuration	OK	OK	0	0	0
lchip_interface	OK	OK	1	1	0
logic	OK	OK	0	0	0
logic_i	OK	OK	0	0	0
outputs	OK	OK	0	0	0
outputs_i	OK	OK	4	4	0
safety_variables	OK	OK	0	0	0
user_component	OK	OK	0	0	0
user_component_i	OK	OK	2	2	0
user_configuration	OK	OK	0	0	0
user_configuration_i	OK	OK	13	13	0
user_ctx	OK	OK	0	0	0
user_ctx_i	OK	OK	0	0	0

Fig. 11. Both project Clock and Combinatorial, provided with Atelier CSSP are fully proved automatically with added rules and predefined tactics.

Of course, the added rules and tactics are not sufficient to automatically prove all the proof obligations generated for the CSSP but provide a basis for reuse and extension, together with existing mathematical rules (including packages s1 and b1, added after the end of the development of the automatic metro line 14 in Paris, and able to simplify arithmetic predicates and expressions).

6   Reaching the Limits

The CSSP is intrinsically different from an Arduino as its offers safety features. However if these safety features cannot be demonstrated, a CSSP is not distinguishable from an Arduino. The following situations allow to demonstrate some of the safety features:

- **2oo2 principle:** corrupting the memory is not easily performed as it requires generating perturbing electromagnetic field and some luck to indeed modify the memory. Instead the CSSP software interface provides two functions, *get_instance_id()* and *get_processor_id()*, which allow to program a behaviour dependent on the software instance and on the processor executing the software. In this case, a divergent behaviour could be obtained leading to the panic mode.
- **regular synchronisation between microcontrollers:** the two microcontrollers are expected to synchronise every 100ms maximum by checking the signature (CRC) of their memory spaces. Executing a loop with for example 100 millions steps would similarly trigger the panic mode.

It is also possible to reach the RAM limit by allocating large tables (containing 48700 *uint8_t* for example) or change the board id (jumper) during program execution to respectively prevent or stop its execution.

7   Dissemination

A first starter kit, SK_0, containing the IDE and the execution platform, was released by the end of 2017[5], presented and experimented at the occasion of several hands-on sessions organized at university sites in Europe, North and South America. Audience was diverse, ranging from automation to embedded systems, mechatronics, computer science and formal methods. Results obtained are very encouraging:

- Teaching formal methods is eased as students are able to see their model running in and interacting with the physical world. It was the occasion to demonstrate how formal methods could be used with embedded systems and IoT. Fruitful discussions took place about how to specify/guaranty performances, what can or cannot be proved with such systems, etc.

---

[5] https://www.clearsy.com/en/our-tools/clearsy-safety-platform/.

- Less theoretic student profiles (computer science, mechatronics, automation) may be introduced/educated to more abstract aspects of computation. *clock* and *combinatorial* exercises were a starting point for specification enrichment and the discovery of the formal proof. Of course, the pedagogical objective in term of formalization was lower than with more formal profiles, but the students managed to understand the absence of programming error and the non-deterministic substitutions for simple modelling.
- The platform has demonstrated a certain robustness during all these manipulations and has been enriched with the feedback collected so far. Several electronics/software errors were detected during the preparation of the course when designing exercises, others during these exercises.
- The IDE GUI was improved with the automation of the code generation process and the display of a carousel showing graphically the progress of the generation. The configuration of the board was also simplified, by displaying the position of the switches on the board and by filling the configuration file with default input and output names.
- CLEARSY Safety Platform is used to teach in Master 2^6 in universities and engineering schools. Electronic documentation[7] is used to structure the courses and is updated every 2 months. With 3 inputs and 2 inputs, the starter kit SK0 is for discovering the technology; another version of the board is planned for 2020 able to handle more I/O (up to 64).

8   Ready for Industry

The SK_0 board provides a good introduction to the programming of safety critical systems. However the framework proposed is mainly aimed at education and not perfectly fit for industry:

- the number of I/O is reduced (5)
- the programming schema is simple: read inputs, compute, set outputs. Identical algorithms are executed by I_1 and I_2.

The CLEARSY Safety Computer 0 (abbreviated as CS_0) (Fig. 12) was designed to offer more flexibility by providing:

- more I/O. The associated mother board brings 32 inputs and 32 outputs, all digital.
- and a programming model less constrained:
  - Safety functions are still programmed and proved in B, but are callable individually from the C program, in the main loop or associated to an interrupt vector.
  - Mandatory (watchdog-based) safety verification are still performed by the safety library but the developer is now responsible for calling in time the verification functions that keep the watchdogs alive.
  - Computation could be asymmetric between I_1 and I_2.

Fig. 12. The CS0 daughter board safety computer on the left, plugged on the mother board.

The CS_0 only embeds the 2 microcontrollers on a smart card format daughter board while the I/O and the power supply are located on a hosting mother board.

9 Conclusion and Perspectives

The CSSP provides a new way of practising formal methods by allowing students/engineers to connect formal models with the surrounding world. The CSSP is also used to create safety-critical systems, able to be certified at the highest safety levels[8,9,10].

As a consequence, the B0 modelling language has been (even more) restricted to allow an easier certification because of the simplicity of the tool chain. These restrictions oblige to have more verbose models (with more lines and more nesting levels). Even if these constraints could be released/removed in the future, the obtained proof automation level is a real improvement that would certainly ease its adoption in engineering processes. The invention of the CLEARSY Safety

---

[6] Second year of a Master's degree.
[7] Available at https://github.com/CLEARSY/CSSP-Programming-Handbook.
[8] Generic product certificate, CERTIFER 8891/200-1, 27th Feb 2017 SIL4.
[9] System certificate BUREAU VERITAS 6393741 3rd March 2017 SIL3.
[10] Generic product certificate BUREAU VERITAS 7092509, 23rd July 2019 SIL4.

Platform also paves the way for a broader use of the B formal method, in the railways and in other safety-related domains like energy or autonomous vehicles.

Acknowledgements. The work and results described in this article were partly funded by BPI-France (Banque Publique d'Investissement) and Métropole Aix-Marseille as part of the project LCHIP (Low Cost High Integrity Platform) selected for the call AAP-21.

References

1. Abrial, J.: Modeling in Event-B - System and Software Engineering. Cambridge University Press, Cambridge (2010)
2. Boulanger, J.: Formal Methods: Industrial Use from Model to the Code. Wiley, Hoboken (2013)
3. Gashi, I., Povyakalo, A., Strigini, L.: Diversity, safety and security in embedded systems: modelling adversary effort and supply chain risks. In: Proceedings of 2016 12th European Dependable Computing Conference (EDCC), Gothenburg, pp. 13–24 (2016)
4. Lecomte, T.: Safe and reliable metro platform screen doors control/command systems. In: Cuellar, J., Maibaum, T., Sere, K. (eds.) FM 2008. LNCS, vol. 5014, pp. 430–434. Springer, Heidelberg (2008). https://doi.org/10.1007/978-3-540-68237-0_32
5. Lecomte, T.: Applying a formal method in industry: a 15-year trajectory. In: Alpuente, M., Cook, B., Joubert, C. (eds.) FMICS 2009. LNCS, vol. 5825, pp. 26–34. Springer, Heidelberg (2009). https://doi.org/10.1007/978-3-642-04570-7_3
6. Lecomte, T.: Double cœur et preuve formelle pour automatismes sil4. 8E-Modèles formels/preuves formelles-sûreté du logiciel (2016)
7. Lecomte, T., Deharbe, D., Prun, E., Mottin, E.: Applying a formal method in industry: a 25-year trajectory. In: Cavalheiro, S., Fiadeiro, J. (eds.) SBMF 2017. LNCS, vol. 10623, pp. 70–87. Springer, Cham (2017). https://doi.org/10.1007/978-3-319-70848-5_6
8. Sabatier, D.: Using formal proof and B method at system level for industrial projects. In: Lecomte, T., Pinger, R., Romanovsky, A. (eds.) RSSRail 2016. LNCS, vol. 9707, pp. 20–31. Springer, Cham (2016). https://doi.org/10.1007/978-3-319-33951-1_2

Modelling Hybrid Programs with Event-B

Meryem Afendi[1](✉), Régine Laleau[1], and Amel Mammar[2]

[1] Université Paris-Est Créteil, LACL, Créteil, France
{meryem.afendi,laleau}@u-pec.fr
[2] SAMOVAR, Institut Polytechnique de Paris, Télécom SudParis, Évry, France
amel.mammar@telecom-sudparis.eu

Abstract. Hybrid systems are one of the most common mathematical models for Cyber-Physical Systems (CPSs). They combine discrete dynamics represented by state machines or finite automata with continuous behaviors represented by differential equations. The measurement of continuous behaviors is performed by sensors. When these sensors have a continuous access to these measurements, we call such model an *Event-Triggered* model. The properties of this model are easier to prove, while its implementation is difficult in practice. Therefore, it is preferable to introduce a more realistic model, called *Time-Triggered* model, where the sensors take periodic measurements. Contrary to *Event-Triggered* models, *Time-Triggered* models are much easier to implement, but much more difficult to verify. Based on the differential refinement logic (dR$\mathcal{L}$), a dynamic logic for refinement relations on hybrid systems, it is possible to prove that a *Time-Triggered* model refines an *Event-Triggered* model. The major limitation of such logic is that it is not supported by any prover. In this paper, we propose a correct-by-construction approach that implements the reasoning on hybrid programs particularly the reasoning of dR$\mathcal{L}$ in Event-B to take advantage of its associated tools.

Keywords: Cyber-Physical Systems · Hybrid systems · Event-B · Refinement · Differential refinement logic

1 Introduction

Recent progress in the industrial sector have allowed the development of a new production model based on digital network architectures to give birth to a fourth industrial revolution ("industry 4.0" or "industry of the future"). Cyber Physical Systems (CPSs) [2] are one of the main technologies in this industry and form the basis of future technologies. The domain of these systems has rapidly become a source of innovation with applications in many sectors: health, transport, smart grid, etc. This type of systems allows to connect the discrete virtual world and the continuous physical world via a network of sensors and actuators. One of

This work was supported in part by the DISCONT project [1] funded by the French National Research Agency (ANR).

© Springer Nature Switzerland AG 2020
A. Raschke et al. (Eds.): ABZ 2020, LNCS 12071, pp. 139–154, 2020.
https://doi.org/10.1007/978-3-030-48077-6_10

the most common architectures in CPSs is a discrete software controller that represents the computation part and controls the physical part through a loop involving sensors and actuators.

A common mathematical model for CPSs is that of hybrid systems that combine discrete behavior represented by state machines or finite automata with continuous behavior described by differential equations. The development of techniques and tools to effectively design hybrid systems has drawn the attention of many researchers. Traditional approaches are based on simulation tools like Matlab/Simulink or Stateflow. Since these tools are time-consuming and produce results tainted with uncertainty, traditional approaches can be very expensive and difficult to apply. To overcome these limitations, several formal approaches have been proposed. These approaches can be grouped into two categories: *model-checking-based* approaches and *proof-based* approaches. Model-checking-based approaches use hybrid automata to model hybrid systems and algorithmic analysis methods to prove their safety. They are based on the calculation of the set of reachable states for hybrid automata. These approaches suffer from the classical problems related to the state space explosion and boundedness of the considered variables issues. Proof-based approaches use deductive verification to prove the properties of hybrid systems. One of the strong points of these approaches is that they support the description of any kind of hybrid systems. However, they require significant effort and a high expertise in modelling and proof phases. This is the main reason why these approaches do not yet scale to industrial applications.

In CPSs, the measurement of continuous behaviors is performed by sensors. Ideally sensors have a continuous access to these measurements, this can be captured by an abstract model of CPSs, called *Event-Triggered* system by Kopetz in [3]. However, implementing such models is difficult in practice. Therefore, it is preferable to introduce a more realistic model, called *Time-Triggered* system in [3], where the sensors take periodic measurements. Contrary to *Event-Triggered* models, properties on *Time-Triggered* models are difficult to verify. Platzer et al. [4,5] use this approach to model hybrid systems. They have proved that a *Time-Triggered* model can be a refinement of an *Event-Triggered* model, by using an extension of the differential dynamic logic (d$\mathcal{L}$), called the differential refinement logic (dR$\mathcal{L}$). However dR$\mathcal{L}$ is not supported by any prover and dR$\mathcal{L}$ formulas can only be manually proved, which heavily restricts its use, especially in an industrial context. In this paper we propose an approach to model *Event-Triggered* systems and *Time-Triggered* systems in Event-B to take advantage of its well-defined refinement process and of its support tools. We also reused the work proposed in [6] that allows to model differential equations in Event-B.

This paper is organised as follows. Section 2 briefly describes d$\mathcal{L}$, dR$\mathcal{L}$ and Event-B. Section 3 presents a state of the art of some proof based-approaches for CPS modelling. Section 4 presents *Event* and *Time-Triggered* systems and their modelling in dR$\mathcal{L}$. Section 5 then introduces our proposed approach and discusses the difference between modelling *Event* and *Time-Triggered* systems in dR$\mathcal{L}$ and Event-B. Finally, Sect. 6 concludes and presents some future work.

2   Background

2.1   Differential Dynamic Logic d$\mathcal{L}$

This section describes a first-order dynamic logic in the domain of real (IR) introduced by A. Platzer to specify hybrid systems and verify their correctness using its associated proof calculus [4]. d$\mathcal{L}$ formulas are built using logical symbols of first-order logic and the modalities [ ] (Box-modality) and $\langle \rangle$ (Diamond-modality). Formula $[\alpha]\phi$ is true iff after all runs of the hybrid program α, formula ϕ holds. $\langle \alpha \rangle \phi$ is true iff there is at least one run of the hybrid program α, after which formula ϕ holds. The major advantage of d$\mathcal{L}$ is its ability to handle differential equations, even those with non-polynomial solutions. Moreover, d$\mathcal{L}$ and its associated proof calculus are supported by two automatic formal verification tools, KeYmaera [7] and its successor KeYmaera X [8].

In d$\mathcal{L}$, hybrid systems are given operationally as hybrid programs (HPs). These latter describe both discrete and continuous behaviors of hybrid systems using sequential composition (;), non-deterministic choice ($\cup$), non-deterministic repetition ($*$), discrete assignments ($:=$), continuous evolution ($'$), etc. Most HPs are defined using the notation, $(ctrl; plant)^*$, where $ctrl$ denotes the execution of the controller (discrete evolution), followed by the physical part $plant$ (continuous evolution). This sequence is non-deterministically repeated as denoted with the star ($*$).

Finally, in order to establish a safety property, $safeReq$, for a system, a typical formula expressing safety relative to initial conditions needs to be proved, $init \rightarrow [(ctrl; plant)^*](safeReq)$ that means: if the initial conditions ($init$) hold, then, after all runs of the hybrid program $safeReq$ holds.

2.2   Differential Refinement Logic dR$\mathcal{L}$

dR$\mathcal{L}$ is a logic with first-class support for refinement relations on hybrid systems [5]. It extends d$\mathcal{L}$ by introducing a refinement operator ($\leq$) for HPs. In addition to d$\mathcal{L}$ formulas, dR$\mathcal{L}$ introduces formulas of the form $\alpha \leq \beta$, α refines β, with α and β denoting HPs. According to [5], formula $\alpha \leq \beta$ is true in a state s iff all states reachable from s by following the transitions of α could also be reached from state s by following transitions of β.

dR$\mathcal{L}$ preserves the safety properties of refined hybrid programs by showing that if $\alpha \leq \beta$ and $[\beta]\phi$, then the formula ϕ is true in all states reachable from s by following the transitions of α ($[\alpha]\phi$). There is a similar rule for diamond modalities ($\langle \rangle$), which states that if α refines β, and there is at least one transition on α to a state where ϕ is true, then $\langle \beta \rangle \phi$ is true. Moreover, dR$\mathcal{L}$ establishes that a *Time-Triggered* system refines an *Event-Triggered* system using its associated proof calculus (Sect. 4).

2.3   Event-B

Event-B [9] is a formal method based on set theory, first-order logic and predicate logic. An Event-B model is composed of a set of machines and contexts. An

Event-B context consists of sets and constants with their axioms. An Event-B machine represents the dynamic behavior of a given system, and it may see one or more Event-B contexts. To any Event-B model, a set of proof obligations (POs) is associated. These POs must be proved to verify the correctness of a given Event-B model. They can be automatically generated using for example the Rodin platform [10], which is an Eclipse-based IDE for Event-B. This platform allows to add new features as Eclipse plug-ins. For example, the Theory plug-in [11] is a Rodin extension that allows to define new data types like *REAL*, new operators, etc. Event-B has a key feature that consists in using abstract modelling to represent the abstract behavior of a given system and the refinement to demonstrate compliance between an abstract model and a concrete one.

3   State of the Art

In this section, we focus on proof-based approaches defined to specify and verify hybrid systems using Event-B, the dR$\mathcal{L}$ approach will be discussed in Sect. 4. We briefly describe three main approaches. The approach presented in [12] proposes a new formal method, called Hybrid Event-B, to add continuous aspects to traditional discrete Event-B. It defines two kinds of events: *mode events* and *pliant events*. *mode events* represent the traditional discrete Event-B events. *pliant events* specify the continuous evolution of continuous measurements. As dR$\mathcal{L}$, Hybrid Event-B is not supported by any prover.

The authors of [13] propose an approach supported by the Rodin toolset to model hybrid systems using continuous functions over real intervals. Preserving the properties of these functions is the key for ensuring the correction of refined machines. This approach uses the Event-B refinement to reduce the non-determinism in continuous behaviors and introduce periodic control.

Finally, the approach proposed by Dupont et al. in [6] uses the Theory plug-in of Event-B to define theories that handle continuous aspects of hybrid systems. The behavior of CPSs is specified by the following three Event-B models:

– **System** **model** is used to describe the continuous evolution of the physical part of a hybrid system. Its machine contains two events:
  - the *Progress* event models the continuous evolution of time by using a positive real variable $t \in (TIME = RRealPlus)$[1]. The lt symbol corresponds to the operator $(<)$ in the *RReal* theory [6].

Progress		
THEN	act1: t : $\mid t' \in TIME \land (t \mapsto t' \in lt)$	**END**

  - the *Behave* event models the physical part's evolution represented by the physical state variable $plantV$. While modelling a car, $plantV$ will be replaced by the car's position p and the car's velocity v. $plantV$ evolves according to the differential equation $e \in DE(S = RReal * RReal)$[2]

---

[1] *RRealPlus* represents $\mathbb{R}^+$ in the *RReal* theory developed by Dupont et al.

[2] *RReal* represents $\mathbb{R}$ in the *RReal* theory developped by Dupont et al.

defined as a parameter of the *Progress* event, where $DE(S)$ is a set of differential equations defined on S. This differential equation must have a solution in the interval $[t, \infty[$ that is represented by Guard *grd2*.

Behave
ANY   e
WHERE
    grd1: $e \in DE(S)$
    grd2: $Solvable(Closed2Infinity(t), e)$
THEN
    act1: $plantV$: | $plantV' \in TIME \rightarrow S \; \wedge$
                $AppendSolutionBAP(e, \; TIME, \; Closed2Open(Rzero, \; t),$
                          $Closed2Infinity(t), \; plantV, \; plantV')$
END

- *State_System* **model** refines the previous model by adding the evolution of the discrete part (the controller). It introduces a new variable named x_s to model the possible states of the controller. It also introduces a new event, named *Transition*, to update the controller's state by assigning a non deterministic value to x_s. The possible values that can be assigned to this variable are defined in the associated context as elements of a set $STATES$ defined in the same context.

- *Controlled_System* **model** refines the *State_System* model by adding two new events that allow the interaction between the physical part and the discrete part:
  - *Sense* event allows to modify the controller's state according to the physical part's state. It introduces a parameter p which depends on x_s, t and $plantV(t)$ ($p \in P(STATES \times TIME \times S)$). This parameter allows to define the system safety envelope according to its discrete state.
  - *Actuate* event refines the *Behave* event by adding a constraint on the controller's state. This constraint is represented by the following formulas: $s \subseteq STATES$ and $x\_s \in STATES$.

4   Event and Time-Triggered Systems

In order to design a model that better corresponds to real CPSs it is preferable to start with an abstract one, called *Event-Triggered* model, where the controller interrupts the physical part when a particular event occurs, and then introduce a more realistic model, called *Time-Triggered* model, where the controller interrupts periodically the physical part [3]. *Event-Triggered* models describe an ideal behavior where the time is continuous and the sensors have continuous access to continuous measurements which is not always possible in practice. *Time-Triggered* models describe a more realistic behavior where the sensors take periodic measurements. Therefore, the controller of a *Time-Triggered* system must make a choice that will be safe until the next sensor's update, which makes this type of systems difficult to prove compared to *Event-Triggered* systems.

dR$\mathcal{L}$ allows to specify and prove that a *Time-Triggered* system refines an *Event-Triggered* system. It introduces two generic templates [5], *Model 1* and *Model 2*, to model and prove these two types of systems. The control part of these two generic templates has only two modes: the *normal* mode which is triggered if the system safety envelope, denoted *safe*, is satisfied, and the *evade* mode which is triggered otherwise. As already mentioned, the major limitation of dR$\mathcal{L}$ is that it is not supported by any prover. This limitation represents a strong restriction on its application to more complex hybrid systems. This paper proposes to deal with this restriction through the use of Event-B and its support tools.

4.1   Event-Triggered Model

Model 1: Event-triggered Generic Model

$$event^* \equiv (ctrl_{Ev}; plant_{Ev})^* \tag{1.1}$$
$$ctrl_{Ev} \equiv (ctrlV := evade\_value) \cup (ctrlV := *; ?safe(plantV)) \tag{1.2}$$
$$plant_{Ev} \equiv t := 0; \; plantV_0 := plantV;$$
$$(plantV' = f\_evol(ctrlV), t' = 1 \; \& \; evt\_trig(plantV)$$
$$\wedge \; dom\_evol(plantV)) \tag{1.3}$$
$$\cup (plantV' = f\_evol(ctrlV), t' = 1 \; \& \; \sim evt\_trig(plantV)$$
$$\wedge \; dom\_evol(plantV)) \tag{1.4}$$

---

where:

> $ctrlV$: the control variable (acceleration in the case of a car).
>
> $plantV$: the state variable of the system.
>
> $safe(plantV)$: defines the system safety envelope. It is calculated from the safety requirement that the system must satisfy.
>
> $plantV' = f\_evol(ctrlV)$: defines the system ODE that describe the continuous evolution of the system.
>
> $evt\_trig(plantV)$: the predicate that defines the boundary of the safety envelope. When this latter becomes false, the controller triggers the *evade* mode. It must define a closed domain.
>
> $\sim evt\_trig(plantV)$: topological closure of the complement of $evt\_trig(plantV)$.
>
> $dom\_evol(plantV)$: defines the evolution domain of the system. It is a set of constraints on the state variable.
>
> $plantV_0$: represents the initial value of $plantV$.

N.B: the variables t and $plantV_0$ have no effect on the state of this model. They will be used in the second model.

Model 1 represents the generic model associated with a controller triggered by events. When the formula *safe* is satisfied, the system can evolve continuously according to formula **(1.3)** until it reaches the boundary of the domain $evt\_trig(plantV)$. Once the system reaches the boundary of this domain, the controller must then switch to the *evade* mode by affecting a deterministic value *evade_value* to the control variable *(ctrlV)*. After the switch of the controller to the *evade* mode, the system no longer satisfies the formula *safe*. Therefore, it can no longer evolve in the domain $evt\_trig(plantV)$ that's why dR$\mathcal{L}$ defines formula **(1.4)**. This latter allows the system to evolve continuously when it is in the *evade* mode. To prove the safety of this model, dR$\mathcal{L}$ provides the following proof rule where Γ represents other assumptions not affected by the program *event*:

$$evt\_trig(plantV) \wedge \Gamma \vdash [event](evt\_trig(plantV) \wedge \Gamma)$$

This proof states that *Model 1* is safe if its associated hybrid program *event* always satisfies the loop invariant $evt\_trig(plantV)$ which includes the formula $safe(plantV)$.

4.2 Time-Triggered Model

Model 2: Time-triggered generic model

$$time^* \equiv (ctrl_t; plant_t)^* \qquad \text{(2.1)}$$
$$ctrl_t \equiv (ctrlV := evade\_value)$$
$$\cup \ (ctrlV := *; ?safe_\epsilon(plantV, ctrlV)) \qquad \text{(2.2)}$$
$$plant_t \equiv t := 0; \ plantV_0 := plantV; (plantV' = f\_evol(ctrlV),$$
$$t' = 1 \ \& \ t \le \epsilon \wedge dom\_evol(plantV)) \qquad \text{(2.3)}$$

---

where
   ϵ: maximum time between two sensors updates.
   t: allows to know if the duration ϵ is reached or not.

Model 2 represents the generic model associated with a *Time-Triggered* system. The controller of such system reacts at least every ϵ seconds. To express this constraint, dR$\mathcal{L}$ replaces formulas **(1.3)** and **(1.4)** by a single one **(2.3)**. Formula *safe* is also replaced by formula $safe_\epsilon$, which depends on both the current choice of *ctrlV* and the time duration ϵ, in addition to the current state *plantV*, in order to guarantee that the controller will make a choice that will be safe for up to ϵ time. To prove that *Model 2* satisfies a safety property ϕ, dR$\mathcal{L}$ has introduced the following proof rule ($[\le]$) where Δ represents other obligations in the context not affected by the proof rule.

$$\frac{\Gamma \vdash [event^*]\phi, \Delta \quad \Gamma \vdash (time^* \leq event^*), \Delta}{\Gamma \vdash [time^*]\phi, \Delta}[\leq]$$

This proof consists of two sub-goals: the first one is to prove that *Model 1* satisfies the system safety property ϕ, and the second aims at verifying that *Model 2* refines *Model 1*.

4.3   Time-Triggered Model Refines Event-Triggered Model

To prove that a *Time-Triggered* system refines an *Event-Triggered* system, dR$\mathcal{L}$ provides three proof obligations:

- **PO 1:** $evt\_trig(plantV) \wedge \Gamma \wedge safe_\epsilon(plantV, ct\bar{r}lV) \vdash safe(plantV)$

  where:
  $ct\bar{r}lV$: represents a non-deterministic choice of the control variable.

  This proof expresses that the safety envelope of *Model 2* implies that of *Model 1*, which means that the discrete controller refines the continuous one.

- **PO 2:** $evt\_trig(pla\bar{n}tV_0) \wedge \Gamma \wedge safe_\epsilon(pla\bar{n}tV_0, ct\bar{r}lV) \wedge 0 \leq t \leq \epsilon$
  $\wedge \; dom\_evol(pla\bar{n}tV) \wedge plantV = S_{pla\bar{n}tV_0, ct\bar{r}lV}(t) \vdash evt\_trig(pla\bar{n}tV)$

  where:
  $pla\bar{n}tV_0$: set of physical state variables values at instant $t = 0$.
  $pla\bar{n}tV$: set of physical state variables values at instant t.
  $S_{pla\bar{n}tV_0, ct\bar{r}lV}(t)$: solutions of the ODE associated with $plantV_0$, given a control variable choice $ct\bar{r}lV$.

  This proof expresses that the non-deterministic choice of $ctrlV := *$ expressed by $ct\bar{r}lV$ guarantees that the system will not cross the boundary of $evt\_trig(plantV)$ within time ϵ.

- **PO 3:** $evt\_trig(pla\bar{n}tV_0) \wedge \Gamma \wedge 0 \leq t \leq \epsilon \wedge dom\_evol(pla\bar{n}tV)$
  $\wedge \; pla\bar{n}tV = S_{pla\bar{n}tV_0, evade\_value}(t) \vdash evt\_trig(pla\bar{n}tV)$

  This proof is similar to the previous one except that here the control choice is deterministic $ctrlV := evade\_value$.

5   Modelling Hybrid Programs with Event-B

The objective of the DISCONT project [1] is to elaborate a correct-by-construction method, based on Event-B, to specify hybrid systems models. Two approaches are considered. The first one, developed by Dupont et al. [6], is based on a translation of hybrid automata in Event-B extended by theories that handle differential equations and continuous functions (derivation, Lipschitz condition, etc.). In our approach we propose to model the high-level structure of hybrid programs, (ctrl;plant)*, in Event-B, and more precisely the generic templates defined for modelling *Event* and *Time-Triggered* systems in dR$\mathcal{L}$.

One of our objectives is to use the Event-B refinement and its associated tools to demonstrate the compliance between these two models, and also compare the refinement proof obligations generated in Event-B with those provided by dR$\mathcal{L}$. The approach consists of three models as depicted in Fig. 1 where *System_M* and *System_Ctx* are those of [6]. We also reuse the Event-B theories that handle continuous aspects of hybrid systems. The whole models can be downloaded from: https://cloud.lacl.fr/index.php/s/K75Lt28ApPbkY7z.

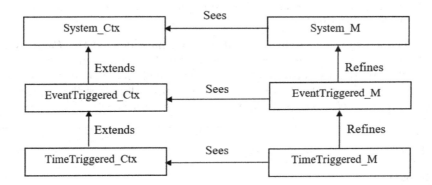

Fig. 1. Structure of the Event-B specification.

5.1   Event and Time-Triggered in Event-B

Event-Triggered Model is a generic model designed to specify and prove *Event-Triggered* systems in Event-B. It is based on the first generic template of dR$\mathcal{L}$, *Model 1*. As we mentioned above, dR$\mathcal{L}$ models an *Event-triggered* system by adding the constraint $evt\_trig(plantV)$ to the system evolution domain. Since Event-B models the transitions between discrete states as events, we do not need to use this constraint in Event-B. Moreover, through the use of Event-B, we can see the different transitions of a given system.

The *Event-Triggered* model is composed of an Event-B context named *Event-Triggered_Ctx* and an Event-B machine named *EventTriggered_M*. *EventTriggered_Ctx* defines a set of constants and axioms required to model an *Event-Triggered* system, such as the formula *safe* that represents the safety envelope for the modeled system. As in dR$\mathcal{L}$, the formula *safe* depends on the current physical state variable as well as the control variable since it may contain some limits on how this latter may be set. The domain of this formula must be included in that of $evt\_trig(plantV)$ formula. Moreover, *safe* must be initially satisfied. In that case, proving the safety of an *Event-Triggered* model consists in ensuring that the specific choice of the *evade* mode is safe. Machine *EventTriggered_M* refines that of the abstract model *System* by adding two new variables:

– *ctrlV* represents the control variable and belongs to *RReal*. The current value of this variable corresponds to the current controller's state.
– *exec* is a flag used to model the alternation between the control part and the physical part as represented in the high-level structure of hybrid programs, $(ctrl; plant)^*$. Therefore, *exec* can take two values *ctrl* and *plant*. In Event-B, the time must be explicitly handled. To be sure that this explicit time will only be updated after the execution of the controller and the physical part, we added a third value, *prg*, to *exec*. Moreover, we refined the *Progress* event of the Machine *System_M* (page 4) to add the constraint *exec* = *prg* as a guard and *exec* := *ctrl* as an action. Therefore, our model follows the following structure: $init; (ctrl; plant; prg)^*$, where *init* represents the *INITIALISATION* event.

To model the evolution of the physical part, we have defined the *Plant* event. This latter refines the *Behave* event by replacing the abstract differential equation *e* with that defined for a function denoted *f_evol_plantV* in order to model $plantV' = f\_evol(ctrlV)$. The function *f_evol_plantV* describes the evolution of the state variable *plantV* according to the system discrete state.

Plant
REFINES
  Behave
WHERE
  grd1: $ode(f\_evol\_plantV(ctrlV),\ plantV(t),\ t) \in DE(S)$
  grd2: $Solvable(Closed2Infinity(t), ode(f\_evol\_plantV(ctrlV), plantV(t), t))$
  grd3: $exec = plant$
WITH
    e: $e = ode(f\_evol\_plantV(ctrlV),\ plantV(t),\ t)$
THEN
  act1: $plantV :|\ plantV' \in (TIME \to S)\ \wedge$
     $AppendSolutionBAP(ode(f\_evol\_plantV(ctrlV),\ plantV(t),\ t),$
      $TIME, Closed2Open(Rzero, t), Closed2Infinity(t), plantV, plantV')$
  act2: $exec := prg$
END

Regarding the evolution of the control part, we have added two new events:

– *Ctrl_normal* event representing the *normal* mode. It is triggered when it is the controller's turn (*exec* = *ctrl*) and the formula *safe* is true. It assigns a non-deterministic value, defined in the *ANY* clause, to the control variable *ctrlV* and gives the turn to the physical part (*exec* := *plant*).
– *Ctrl_evade* event representing the *evade* mode. It assigns the parameter *evade_value* to the control variable *ctrlV* and gives the turn to the physical part (*exec* := *plant*). This event can be triggered even if the system has not yet reached the boundary of *evt_trig(plantV)*, i.e. the system still satisfies the formula *safe*. However, we keep the guarantee that it will be triggered exactly when the system reaches the boundary of *evt_trig(plantV)* since the controller is continuous.

Ctrl_normal
ANY $nrml\_value$
WHERE
    grd1: $nrml\_value \in RReal$
    grd2: $exec = ctrl$
    grd3: $safe(plantV(t) \mapsto$
        $nrml\_value) = TRUE$
THEN
    act1: $ctrlV := nrml\_value$
    act2: $exec := plant$
END

Ctrl_evade
ANY $evade\_value$
WHERE
    grd1: $exec = ctrl$
    grd2: $evade\_value \in RReal$
THEN
    act1: $ctrlV := evade\_value$
    act2: $exec := plant$
END

Time-Triggered Model refines the previous model to get a system corresponding to that described by *Model 2*. As mentioned in the previous section, the sensors of a *Time-Triggered* system take periodic measurements of physical state variables and its controller executes each time those sensors updates are taken. Moreover, the longest time between sensors updates is bounded by a symbolic duration named ϵ. Therefore, the controller can execute at least every ϵ time. For this purpose, we have calibrated a new variable named d (variable t in dR$\mathcal{L}$) to know whether the duration ϵ is reached or not. This variable is reset (set to $Rzero$) before each execution of the physical part and evolves according to a function f_evol_d defined in the associated context. We have also added the constraint $d(t') \leq \epsilon$ to the first action of the *Progress* event to be sure that the sensors updates occurs at least every ϵ. Since the controller of a *Time-Triggered* system must make a choice that will be safe for up to ϵ time, we defined a new safety envelope named $safeEpsilon$ ($safe_\epsilon(plantV, ctrlV)$ in dR$\mathcal{L}$). As in dR$\mathcal{L}$, we have replaced $safe$ with $safeEpsilon$ by defining a new event named *Ctrl_normal_time*. This latter refines the *Ctrl_normal* event and is triggered when a given value, $nrml\_value$, satisfies the formula $safeEpsilon$. In that case, we assign this value to $ctrlV$ and give the turn to the physical part.

5.2   Application

To apply our approach to a concrete system, we define two concrete models, *Concrete_System_Event-triggered* model and *Concrete_System_Time-triggered* model. The first model refines the *Event-Triggered* model through replacing $plantV$ by the system physical state variables, defining the system safety properties as invariants in addition to the associated evolution function f_evol_plantV, then the formula $safe$ is instanciated to define the system safety envelope. The second model, can either refine the first one or the *Time-Triggered* model. If we choose the first alternative, the refined model will then inherit the system safety properties but on the other hand we must add the notion of control period *epsilon*.

5.3   Proof of Refinement

In Event-B, two proof obligations are generated to prove that a concrete Event-B machine refines an abstract one:

- *Guard strengthening (GRD):* ensures that a concrete guard is stronger than the corresponding abstract one.
- *Action simulation (SIM):* ensures that each concrete action is not contradictory to the corresponding abstract one.

As mentioned earlier, we replaced the safety envelope formula *safe* by the formula *safeEpsilon* in the *Ctrl_normal_time* event. In this case, the following *Guard strengthening (GRD)* proof obligation has been generated:

$$(exec = ctrl \land safeEpsilon(plantV(t) \mapsto nrml\_value) = TRUE)$$
$$\Rightarrow safe(plantV(t)) = TRUE$$

To prove that the concrete machine, *TimeTriggered_M*, refines the abstract one, *EventTriggered_M*, we must prove that, during a control period ϵ, the safety formula *safeEpsilon*, defined in the concrete model, implies the safety formula *safe* defined in the abstract one. This proof is similar to the *PO 1* provided by dR$\mathcal{L}$. *PO 2* and *PO 3* are not generated as refinement POs by the proof obligation generator of Event-B, though they are needed to prove the refinement relation between our two generic models, *Time-Triggered* model and *Event-Triggered* model. Therefore, they must be added manually as Event-B proof obligations. Since we model the evolution of the physical state variables using a single event, *Plant* in the *Event-Triggered* model and *Plant_time* in the *Time-Triggered* model, we will then replace the equations of the dR$\mathcal{L}$ POs, $plantV = S_{plantV_0, ctrlV}(t)$ and $plantV = S_{plantV_0, evade\_value}(t)$ by the guard $exec = prg$. *init* represents the initial conditions of the modeled system and *plantV(t0)* represents the initial value of the physical state variable *plantV*. In Event-B, the proof obligations are as follows:

- PO 2:

$$safeEpsilon(plantV(t0) \mapsto nrml\_value) = TRUE \land evt\_trig(plantV(t0))$$
$$\land init \land (Rzero \mapsto t \in leq) \land (t \mapsto epsilon \in leq) \land exec = prg$$
$$\Rightarrow evt\_trig(plantV)$$

- PO 3:

$$evt\_trig(plantV(t0)) \land init \land (Rzero \mapsto t \in leq) \land (t \mapsto epsilon \in leq)$$
$$\land exec = prg \Rightarrow evt\_trig(plantV)$$

These two proof goals are based on the safety envelope of the system and the choices of the control variable. When the safety envelope of the system is satisfied, the controller can non-deterministically choose between the *normal* mode or the *evade* mode. In the case of a *Event-Triggered* system, we have the guarantee that the controller is able to switch to the *evade* mode exactly when the safety envelope is no longer satisfied. While in a *Time-Triggered* system, we must prove that *nrml_value* and *evade_value* guarantee that the system will not exceed the domain of the safety envelope within time ϵ.

5.4   Case Study

To validate our approach, we chose the *Stop Sign* case study [14] which deals with a stop sign controller whose objective is to ensure the stopping of a car before a stop signal SP. The control strategy is to adjust the velocity of the car by accelerating or braking, without ever backing down. The continuous behavior of this system is modeled by the position and the velocity of the car specified respectively by the state variables p and v, as well as its acceleration represented by the control variable $ctrlV$. This continuous behavior evolves according to linear differential equations, $p' = v, v' = ctrlV \equiv (\frac{dp}{dt} = v, \frac{dv}{dt} = ctrlV)$, which describe the evolution of the position and the velocity over time. The system can behave according to the following two discrete states: State *accelerate* and State *braking*. State *accelerate* is triggered when the car is very far from the stop signal SP. In this case, the car velocity can evolve according to a non-deterministic value assigned to the control variable $ctrlV$. This value must never exceed the physical limits of the car expressed by A (maximum limit of acceleration) and B (maximum limit of braking). State *braking* is triggered when the car is very close to the stop signal SP. In this case, we must decrease the car velocity by assigning $-B$ to $ctrlV$. To model this system using our approach, we followed the schema depicted in Fig. 2. The whole models of this development can be downloaded from https://cloud.lacl.fr/index.php/s/aiKiPxkrfmWpakR.

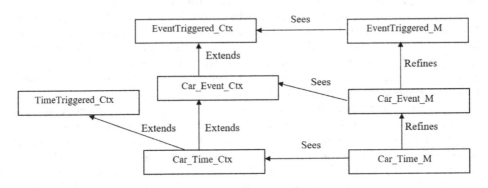

Fig. 2. *Stop sign* case study development schema.

Machine *Car_Event_M* refines Machine *EventTriggered_M* through replacing the generic state variable *plantV* by the physical state variables associated with the *Stop sign* case study, p and v. This replacement is done using the operator *bind* defined in the differential equations theory [6]. The physical part is modeled by the *Plant_event_car* event. This latter refines the *Plant* event by adding a witness that replaces the evolution of the generic state variable *plantV* by the evolution of the position p and the velocity v represented by the *f_evol_plantV* defined in the associated context. State *accelerate* is modeled using the *Ctrl_acceleration_car* event that refines the *Ctrl_normal* event.

State *braking* is modeled through replacing the value *evade_value* by $-B$ in the *CtrLevade* event.

Machine *Car_Time_M* refines Machine *Car_Event_M* in order to preserve the system safety property, $p \leq SP$ and $0 \leq v$. If we prove that *safeTime* implies *safe*, and the *Car-Event-Triggered* model satisfies the property $p \leq SP$, so we can say that the *Car-Event-Triggered* model also satisfies this property. *Car_Time_M* is based on the Machine *TimeTriggered_M*, therefore we added the variable d and its evolution.

To prove that *Car_Time_M* refines *Car_Event_M* we must prove the three associated POs presented in Sect. 5.3. As we mentioned above, the choice of the parameters *nrml_value* and *evade_value* is the key to prove the safety of the system which can be done by using external mathematical tools for a parametric analysis since the differential equation of the *Stop Sign* case study is linear.

5.5   Comparing Event-B Refinement with Differential Refinement Logic

Type of Refinement. Event-B refinement is based on the execution traces starting from the initial state. Therefore, to prove that a concrete Event-B machine refines an abstract one, we have to establish that the set of execution traces of the concrete one is included in that corresponding to the abstract one. The refinement of dR$\mathcal{L}$ is based on reachable states. In hybrid automata and hybrid programs, a state is defined by a couple $(x\_s, plantV)$ composed of the current discrete state x_s and the current value of the continuous variable $plantV$. Therefore, to prove that a hybrid program α refines another hybrid program β ($\alpha \leq \beta$), we have to establish that the set of reachable states from a state s following the transitions of α is included in the set of reachable states from the same state s following some transitions of β.

Both Event-B refinement and dR$\mathcal{L}$ allow preserving the safety properties of the refined model. This is ensured in dR$\mathcal{L}$ through combining refinement relations and modalities. Despite the several features of dR$\mathcal{L}$'s refinement, computing reachable states for non linear system requires solving non-linear real arithmetic problems which is difficult in general [15]. Moreover, dR$\mathcal{L}$ refinement does not preserve the safety properties on the traces, but it is less constrained than the Event-B refinement.

Proofs Complexity. As we mentioned earlier, dR$\mathcal{L}$ has introduced a refinement strategy based on comparing reachable states for hybrid programs. Using this refinement strategy, one can start with an ideal system where the controller has continuous control over the system behavior (*Event-Triggered* system), then introduces a more realistic system where the controller interrupts the physical part at least every ϵ time (*Time-Triggered* system). The main advantage of dR$\mathcal{L}$ is that it uses differential equations to describe the continuous evolution of a given hybrid system by employing differential invariants, differential cuts, and differential refinement techniques. Moreover, the refinement relation between *Time* and *Event-Triggered* systems have been successfully proved using the dR$\mathcal{L}$'s refinement proof rules. Despite these advantages, dR$\mathcal{L}$ is not supported by any prover, which makes the proofs difficult to achieve in the case of complex systems especially for systems with more than two modes. Through using Event-B, we can overcome this limitation since its support tools aid in discharging proof obligations either automatically or with the interactive prover. Therefore using our approach, we can model an hybrid system with more than two modes.

The major limitation in using Event-B to model and verify hybrid systems is the absence of support for the continuous aspects of CPSs, such as continuous time and differential equations. As we mentioned, the approach proposed in [6] has tried to overcome this limitation by defining an Event-B theory that includes different kinds of differential equations. Using the abstract model of this approach, it becomes possible to represent the reasoning on hybrid programs in Event-B.

6   Conclusion and Future Work

In this paper, we have presented a proof-based approach that uses Event-B and its refinement technique to specify and verify *Event-Triggered* systems and *Time-Triggered* systems. We have defined two generic templates for these systems, directly inspired from the dR$\mathcal{L}$ specification, that represent hybrid systems as hybrid programs. dR$\mathcal{L}$ proof obligations have been defined to establish the refinement of the *Event-Triggered* template by the *Time-Triggered* template. Then we have compared the Event-B refinement with the dR$\mathcal{L}$ refinement and the generated POs. This led us to define new refinement POs in Event-B. One of the main advantages of Event-B is its support tools (provers, model-checkers, ...) to discharge POs, contrary to dR$\mathcal{L}$.

To demonstrate the usability of our approach, we have experimented it on a *Stop Sign* case study. In this case study, the differential equations that represent the evolution of the physical part are linear and can be easily solved. To handle more difficult differential equations we need to use an external tool like Mathematica [16], a symbolic mathematical computation system. Moreover, our approach is still in the abstract level where all transitions are instantaneous. It does not take into account the duration between the sending of continuous measurements by the sensors and their processing by the controller as well as the duration between the sending of actions by the controller and their execution. As future work, we plan to define a refinement of the *Time-Triggered* model to model these durations. We also plan to integrate Mathematica as a back-end tool in the Rodin platform to resolve differential equations.

References

1. ANR-17-CE25-0005: DISCONT ANR project (2017). https://discont.loria.fr

2. Lee, E.A.: Cyber physical systems: design challenges. In: 2008 11th IEEE International Symposium on Object and Component-Oriented Real-Time Distributed Computing (ISORC), pp. 363–369. IEEE (2008)
3. Kopetz, H.: Event-triggered versus time-triggered real-time systems. In: Karshmer, A., Nehmer, J. (eds.) Operating Systems of the 90s and Beyond. LNCS, vol. 563, pp. 86–101. Springer, Heidelberg (1991). https://doi.org/10.1007/BFb0024530
4. Platzer, A.: Differential dynamic logic for hybrid systems. J. Autom. Reasoning **41**(2), 143–189 (2008). https://doi.org/10.1007/s10817-008-9103-8
5. Loos, S.M., Platzer, A.: Differential refinement logic. In: 2016 31st Annual ACM/IEEE Symposium on Logic in Computer Science (LICS), pp. 1–10. IEEE (2016)
6. Dupont, G., Aït-Ameur, Y., Pantel, M., Singh, N.K.: Proof-based approach to hybrid systems development: dynamic logic and Event-B. In: Butler, M., Raschke, A., Hoang, T.S., Reichl, K. (eds.) ABZ 2018. LNCS, vol. 10817, pp. 155–170. Springer, Cham (2018). https://doi.org/10.1007/978-3-319-91271-4_11
7. Platzer, A., Quesel, J.-D.: KeYmaera: a hybrid theorem prover for hybrid systems (system description). In: Armando, A., Baumgartner, P., Dowek, G. (eds.) IJCAR 2008. LNCS (LNAI), vol. 5195, pp. 171–178. Springer, Heidelberg (2008). https://doi.org/10.1007/978-3-540-71070-7_15
8. Fulton, N., Mitsch, S., Quesel, J.-D., Völp, M., Platzer, A.: KeYmaera X: an axiomatic tactical theorem prover for hybrid systems. In: Felty, A.P., Middeldorp, A. (eds.) CADE 2015. LNCS (LNAI), vol. 9195, pp. 527–538. Springer, Cham (2015). https://doi.org/10.1007/978-3-319-21401-6_36
9. Abrial, J.-R.: Modeling in Event-B: System and Software Engineering. Cambridge University Press, Cambridge (2010)
10. Abrial, J.-R., Butler, M., Hallerstede, S., Hoang, T.S., Mehta, F., Voisin, L.: Rodin: an open toolset for modelling and reasoning in Event-B. Int. J. Softw. Tools Technol. Transf. **12**(6), 447–466 (2010). https://doi.org/10.1007/s10009-010-0145-y
11. Butler, M., Maamria, I.: Mathematical extension in Event-B through the Rodin theory component (2010)
12. Banach, R., Butler, M., Qin, S., Verma, N., Zhu, H.: Core hybrid Event-B I: single hybrid Event-B machines. Sci. Comput. Programm. **105**, 92–123 (2015)
13. Butler, M., Abrial, J.-R., Banach, R.: Modelling and refining hybrid systems in Event-B and Rodin (2016)
14. Quesel, J.-D., Mitsch, S., Loos, S., Aréchiga, N., Platzer, A.: How to model and prove hybrid systems with KeYmaera: a tutorial on safety. Int. J. Softw. Tools Technol. Transf. **18**(1), 67–91 (2015). https://doi.org/10.1007/s10009-015-0367-0
15. Chen, X.: Reachability analysis of non-linear hybrid systems using Taylor models. Ph.D. thesis, Fachgruppe Informatik, RWTH Aachen University (2015)
16. Wolfram, S.: The Mathematica Book, 5th edn. Wolfram Media, Champaign (2003)

Event-B-Supported Choreography-Defined Communicating Systems
Correctness and Completeness

Sarah Benyagoub[1], Yamine Aït-Ameur[1], and Klaus-Dieter Schewe[2(✉)]

[1] Université de Toulouse, IRIT/INPT-ENSEEIHT, Toulouse, France
{sarah.benyagoub,yamine}@enseeiht.fr
[2] Zhejiang University, UIUC Institute, Haining, China
kd.schewe@intl.zju.edu.cn, kdschewe@acm.org

Abstract. Choreographies prescribe the rendez-vous synchronisation of messages in a communicating system. Such a system is called *realisable*, if the traces of the prescribed communication coincide with those of the asynchronous system of peers, where the communication channels either use FIFO queues or multiset mailboxes. It has recently been shown that realisability can be characterised by two necessary conditions that together are also sufficient, whereas in general the *synchronisability* of communicating peers is undecidable. The sufficiency of the conditions permits the construction of correct communicating systems; their necessity shows that all choreography-defined communicating system can be obtained in this way. This article provides an integrated framework based on Event-B for such a construction with a major emphasis on Rodin-based proofs of correctness and completeness.

Keywords: Event-B · Choreography · Realisability · Correctness proof · Completeness proof

1 Introduction

In a communicating system peers communicate asynchronously through messages. If the computations performed by the peers are disregarded and only the sequences of messages sent and received are considered, the system becomes a system of communicating FSMs with a semantics defined by the traces of sent messages. In addition, only those traces may be taken into account in which all sent messages have also been received.

Such a trace semantics can be defined in various ways using channels organised as FIFO queues for each pair of peers [6] or just for each receiver [1]. Alternatively, channels may be organised as multisets [8]. Naturally, one may also consider the possibility of messages being lost [7]. The *synchronisability* problem for such communicating systems is to decide whether the traces remain the same,

© Springer Nature Switzerland AG 2020
A. Raschke et al. (Eds.): ABZ 2020, LNCS 12071, pp. 155–168, 2020.
https://doi.org/10.1007/978-3-030-48077-6_11

if a rendez-vous synchronisation of sending and receiving of messages is considered. This was proven to be undecidable in general [9]. The picture changes in the presence of *choreographies* which prescribe the rendez-vous synchronisation [2]. In this case the peers are projections of a choreography, and synchronisability becomes *realisability* of the given choreography. Recently it was shown that in this case the rendez-vous composition of the projected peers coincides with the choreography, and *language synchronisability* based only on the message traces concides with *synchronisability* based in addition on the stable configurations reached [10]. This further enabled the characterisation of realisability of choreographies by two necessary conditions on a communication choreography, which together are sufficient.

A constructive Event-B-based approach to develop realisable choreographies and consequently communicating systems was brought up in [4,5]. The general idea is to exploit construction operators, by means of which realisable choreographies can be built out of a primitive base [11]. This already contains a hint on the sufficient conditions used in the associated proofs that were conducted using Rodin [3]. As the sufficiency proof in [10] removes some unnecessary assumptions, this approach becomes general. More importantly, the necessity of the conditions shows that all choreography-defined communicating systems can be obtained in this way. *In this paper we continue this route and show that also the necessity proof for realisable choreographies can be supported by Event-B and Rodin. This further gives us means for repairing choreographies.*

The remainder of this article is organised as follows. Section 2 is dedicated to theoretical foundations, where we review the fundamental definitions around peer-to-peer (P2P) systems and choreographies as well as the theory of realisable choreographies developed in [10]. Different to previous work we concentrate only on the most restrictive composition using a single message queue per peer. In Sect. 3 we briefly review our previous work on the Event-B-based construction of choreography-defined P2P systems with a slight extension of the Rodin-based proofs based on our newer insights. Section 4 is the core of this paper emphasising the necessary conditions for realisable choreographies and the Rodin-based proof. We conclude with a brief summary and outlook in Sect. 5.

2   Theoretical Background of Realisable Choreographies

Let M and P be finite, disjoint sets, elements of which are called *messages* and *peers*, respectively. Each message $m \in M$ has a unique *sender* $s(m) \in P$ and a unique *receiver* $r(m) \in P$ with $s(m) \neq r(m)$. We use the notation $i \xrightarrow{m} j$ for a message m with $s(m) = i$ and $r(m) = j$. We also use the notation $!m^{i \to j}$ and $?m^{i \to j}$ for the *event* of sending or receiving the message m, respectively. Write M_p^s and M_p^r for the sets of messages, for which the sender or the receiver is p, respectively.

Let $s(M)$ and $r(M)$ denote the sets of send and receive events defined by a set M of messages. A *P2P system* over M and P is a family $\{\mathcal{P}_p\}_{p \in P}$ of finite state machines (FSMs) $\mathcal{P}_p$ over an alphabet $\Sigma_p = s(M_p^s) \cup r(M_p^r)$. By abuse of terminology $\mathcal{P}_p$ is also called a *peer*.

We write $\mathcal{P}_p = (Q_p, \Sigma_p, q_{0,p}, F_p, \delta_p)$, where Q_p is the finite set of states of the FSM, $q_{0,p} \in Q_p$ is the start state, $F_p \subseteq Q_p$ is the set of final states, and δ_p is the transition function, i.e. $\delta_p : Q_p \times \Sigma_p \to Q_p$. Without loss of generality we may concentrate on deterministic FSMs (see [10, Prop.1]).

2.1 Composition of Peers

A composition of a P2P system over M and P will be another automaton, the alphabet of which will be either M or $s(M) \cup r(M)$.

The *rendez-vous composition* of a P2P system $\{\mathcal{P}_p\}_{1 \leq p \leq n}$ with $\mathcal{P}_p = (Q_p, \Sigma_p, q_{0p}, Q_p, \delta_p)$ is the FSM $\mathcal{C}_{rv} = (Q, M, q_0, Q, \delta)$ with $Q = Q_1 \times \cdots \times Q_n$, $q_0 = (q_{01}, \ldots, q_{0n})$, and $\delta((q_1, \ldots, q_n), i \xrightarrow{m} j) = (q_1', \ldots, q_n')$ holds if $\delta_i(q_i, !m^{i \to j}) = q_i'$ and $\delta_j(q_j, ?m^{i \to j}) = q_j'$ hold, and $q_x = q_x'$ for all $x \notin \{i, j\}$.

The *mailbox composition* of a P2P system $\{\mathcal{P}_p\}_{1 \leq p \leq n}$ with $\mathcal{P}_p = (Q_p, \Sigma_p, q_{0p}, Q_p, \delta_p)$ is the automaton $\mathcal{C}_m = (Q, \Sigma, q_0, Q, \delta)$ satisfying the following conditions:

- The set of states is $Q = Q_1 \times \cdots \times Q_n \times (c_j)_{1 \leq j \leq n}$, where each c_j is a finite queue with elements in M.
- The alphabet is $\Sigma = s(M) \cup r(M)$.
- The initial state is $q_0 = (q_{01}, \ldots, q_{0n}, ([\,])_{1 \leq j \leq n})$, i.e. initially all channels are empty.
- The transition function δ is defined by $\delta((q_1, \ldots, q_n, (c_j)_{1 \leq j \leq n}), e) = (q_1', \ldots, q_n', (c_j')_{1 \leq j \leq n})$ if there exists i such that $\delta_i(q_i, e) = q_i'$ holds, $q_x = q_x'$ for all $x \neq i$, and
  - either $e = !m^{i \to j}$ for some j, $c_j' = c_j {^\frown} [i \xrightarrow{m} j]$, and $c_k = c_k'$ for all $k \neq j$
  - or $e = ?m^{j \to i}$ for some j and $c_i = [j \xrightarrow{m} i] {^\frown} c_i'$ and $c_k = c_k'$ for all $k \neq i$.

As above we call a state $(q_1, \ldots, q_n, (c_j)_{1 \leq j \leq n})$ *stable* if and only if all channels c_j are empty.

Peers as well as any composition of a P2P system are defined by automata, so their semantics is well defined by the notion of language accepted by them. It is common to consider just sequences of sending events, i.e. for a word $w \in M^*$ let $\sigma(w)$ denote its restriction to its sending events. Formally, we have $\sigma(\epsilon) = \epsilon$, $\sigma(i \xrightarrow{m} j) = !m^{i \to j}$, and $\sigma(w_1 \cdot w_2) = \sigma(w_1) \cdot \sigma(w_2)$, where $\cdot$ denotes concatenation. Analogously, for words in $(s(M) \cup r(M))^*$ we have $\sigma(\epsilon) = \sigma(?m^{i \to j}) = \epsilon$, $\sigma(!m^{i \to j}) = !m^{i \to j}$, and $\sigma(w_1 \cdot w_2) = \sigma(w_1) \cdot \sigma(w_2)$.

If $\mathcal{L}$ is the language accepted by an FSM $\mathcal{A}$ with alphabet M or $\Sigma = s(M) \cup r(M)$, then $\mathcal{L}(\mathcal{A}) = \sigma(\mathcal{L})$ is the *trace language* of $\mathcal{A}$. This applies for the cases where $\mathcal{A}$ is a peer $\mathcal{P}_p$ or a composition $\mathcal{C}_{rv}$ or $\mathcal{C}_m$. We use the notation $\mathcal{L}_0(\mathcal{P}) = \mathcal{L}(\mathcal{C}_{rv})$, $\mathcal{L}_\omega(\mathcal{P}) = \mathcal{L}(\mathcal{C}_m)$.

If we restrict final states to be stable, we obtain a different language $\hat{\mathcal{L}}(\mathcal{C}_m) \subseteq \mathcal{L}(\mathcal{C}_m)$, which we call the *stable trace language* of $\mathcal{C}_m$.

A P2P system $\mathcal{P} = \{\mathcal{P}_p\}_{1 \leq p \leq n}$ is called *language-synchronisable*, if $\mathcal{L}_0(\mathcal{P}) = \mathcal{L}_\omega(\mathcal{P})$ holds. $\mathcal{P} = \{\mathcal{P}_p\}_{1 \leq p \leq n}$ is called *synchronisable*, if $\mathcal{L}_0(\mathcal{P}) = \mathcal{L}_\omega(\mathcal{P}) = \hat{\mathcal{L}}_\omega(\mathcal{P})$ holds.

2.2   Choreography-Defined P2P Systems

Let us now look into choreographies. We define a *choreography* by an FSM $\mathcal{C} = (Q, M, q_0, F, \delta)$, where M is again a set of messages. As before we ignore final states and assume $F = Q$. Then every rendez-vous composition of a P2P system $\mathcal{P} = \{\mathcal{P}_p\}_{1 \leq p \leq n}$ defines a choreography.

Let $\mathcal{C} = (Q, M, q_0, Q, \delta)$ be a choreography with messages M and peers P. For $p \in P$ the *projection* $\pi_p(\mathcal{C})$ is the FSM $(Q, \Sigma, q_0, Q, \delta_p)$ with $\Sigma = s(M) \cup r(M)$ and $\delta_p(q, e) = q'$ if $e = !m^{p \to j}$ for some j with $\delta(q, i \xrightarrow{m} j) = q'$, $e = ?m^{i \to p}$ for some i with $\delta(q, i \xrightarrow{m} p) = q'$ or $e = \epsilon$ for $\delta(q, i \xrightarrow{m} j) = q'$ with $p \notin \{i, j\}$.

The *peer* $\mathcal{P}_p$ *defined by* $\mathcal{C}$ is the FSM without ϵ-transitions corresponding to $\pi_p(\mathcal{C})$. A P2P system $\mathcal{P} = \{\mathcal{P}_p\}_{1 \leq p \leq n}$ is *choreography-defined* if there exists a choreography with peers $\mathcal{P}_p$ for all p.

There is a close relationship between rendez-vous compositions and choreography-defined P2P systems. In [10] we proved that each choreography $\mathcal{C}$ coincides (up to isomorphism) with the rendez-vous composition of its peers. Thus, not all P2P systems are choreography-defined. In fact, if a P2P system is choreography-defined, then it must consist of the peers defined by its rendez-vous composition.

For choreography-defined P2P systems the synchronisability problem is much simpler than in the general case. In [10] we proved that a choreography-defined P2P system $\mathcal{P} = \{\mathcal{P}_p\}_{1 \leq p \leq n}$ is synchronisable if and only if it is language-synchronisable.

Therefore, we may focus only on language-synchronisability: if a trace is accepted, then it will be accepted in a stable configuration. We may also identify the rendez-vous composition with the given choreography. Therefore, a choreography $\mathcal{C}$ is called *realisable*, if $\mathcal{L}_0(\mathcal{P}) = \mathcal{L}_\omega(\mathcal{P})$ holds for the P2P system $\mathcal{P}$ defined by the projections of $\mathcal{C}$.

2.3   Characterisation of Realisability

The main result from [10] states that there are two necessary conditions for realisability, which together are sufficient. The sequence condition expresses that if two messages appear in a sequence, the sender of the second message must coincide with either the sender or the receiver of the preceding message. The choice condition expresses that if there is a choice of continuation with two different messages, then these messages must have the same sender.

Sequence Condition. Whenever there are states $q_1, q_2, q_3 \in Q$ with $\delta(q_1, i \xrightarrow{m_1} j) = q_2$ and $\delta(q_2, k \xrightarrow{m_2} \ell) = q_3$ for non-independent messages $i \xrightarrow{m_1} j$ and $k \xrightarrow{m_2} \ell$, we must have $k \in \{i, j\}$.

Choice Condition. Whenever there are states $q_1, q_2, q_3 \in Q$ with $\delta(q_1, i \overset{m_1}{\rightarrow} j) = q_2$, $\delta(q_1, k \overset{m_2}{\rightarrow} \ell) = q_3$ and $q_2 \neq q_3$ for non-independent messages $i \overset{m_1}{\rightarrow} j$ and $k \overset{m_2}{\rightarrow} \ell$, we must have $k = i$.

Both conditions establish constraints on δ for two messages $i \overset{m_1}{\rightarrow} j$ and $k \overset{m_2}{\rightarrow} \ell$, but in both cases we need to exclude that these two messages are *independent* in the sense that they may appear in any order, i.e. we request that if there are states q_1, q_2, q_3 with $\delta(q_1, i \overset{m_1}{\rightarrow} j) = q_2$ and $\delta(q_2, k \overset{m_2}{\rightarrow} \ell) = q_3$, then we cannot have both $\delta(q_1, k \overset{m_2}{\rightarrow} \ell) = q_2$ and $\delta(q_2, i \overset{m_1}{\rightarrow} j) = q_3$. The following theorem is the main result in [10].

Theorem 1. *A choreography $\mathcal{C}$ is a realisable with respect to P2P, queue or mailbox composition if and only if it satisfies the sequence and choice conditions.*

3   Correctness by Construction

We now address the construction of realisable choreographies. For this we will first introduce several composition operators in Subsect. 3.1. We can easily define conditions on the constructors to ensure that all choreographies obtained by composition will satisfy the choice and sequence conditions and thus are realisable by Theorem 1. However, following [3,4] we will actually redo the (sufficiency) proof using specifications of the constructors in Event-B and the Rodin prover.

3.1   Composition Operators

In the following we use the notation CP to refer to a choreography, and we add indices to distinguish different choreographies, whenever the need arises. Without loss of generality we also introduce distinguished final states q_{CP}^f, which ease the proofs. We define three composition operators: *sequence* composition $\otimes_{(\gg, q_{CP}^f)}$, *branching* composition $\otimes_{(+, q_{CP}^f)}$, and *loop* composition $\otimes_{(\circlearrowleft, q_{CP}^f)}$.

Each expression of the form $\otimes_{(op, q_{CP}^f)}(CP, CP_b)$ assumes that the initial state of CP_b is fused with the final state s_{CP}^f. Informally, we can say that CP_b is appended to CP at state s_{CP}^f.

Definition 1 (Sequential Composition). Given a choreograhy CP with final state $q_{CP} \in Q_{CP}^f$ and a choreography CP_b with a single transition $\delta_{CP_b}(q_{CP_b}, l_{CP_b}) = q'_{CP_b}$, the *sequential* composition $CP_{\gg} = \otimes_{(\gg, s_{CP})}(CP, CP_b)$ is defined by $Q_{CP_{\gg}} = Q_{CP} \cup \{q'_{CP_b}\}$, $M_{CP_{\gg}} = M_{CP} \cup \{m_{CP_b}\}$, $Q_{CP_{\gg}}^f = (Q_{CP}^f \setminus \{q_{CP}\}) \cup \{q'_{CP_b}\}$ and $\delta_{CP_{\gg}} = \delta_{CP} \cup \{((q_{CP}, l_{CP_b}), q'_{CP_b})\}$.

Definition 2 (Branching Composition). Given a choreography CP with final state $q_{CP} \in Q_{CP}^f$ and a family of choreographies $\{CP_{bi}\}_{1 \leq i \leq n}$, each comprising a single transition $\delta_{CP_{bi}}(q_{CP_{bi}}, l_{CP_{bi}}) = q'_{CP_{bi}}$, the *branching* composition $CP_+ = \otimes_{(+, q_{CP})}(CP, \{CP_{bi}\})$ is defined by

$-\ Q_{CP_+} = Q_{CP} \cup \{q'_{CP_1}, \ldots, q'_{CP_{bn}} \mid \delta_{CP_{bi}}(q_{CP_{bi}}, l_{CP_{bi}}) = q'_{CP_{bi}}\},$

$-\ M_{CP_+} = M_{CP} \cup \{l_{CP_{b_i}}, \ldots, l_{CP_{bn}}\}$

$-\ \delta_{CP_+} = \delta_{CP} \cup \{((q_{CP}, l_{CP_{bi}}), q'_{CP_{bi}}) \mid 1 \le i \le n\}, and$

$-\ Q^f_{CP_+} = (Q^f_{CP} \setminus \{q_{CP}\}) \cup \{q'_{CP_{b1}}, \ldots, q'_{CP_{bn}}\}.$

Definition 3 (Loop Composition). Given a choreography CP with final state $q_{CP} \in Q^f_{CP}$ and a choreography CP_b with a single transition $\delta_{CP_b}(q_{CP_b}, l_{CP_b}) = q'_{CP_b}$ such that $q'_{CP_b} \in Q_{CP}$ holds, the *loop* composition $CP_{\circlearrowleft} = \otimes_{(\circlearrowleft,\, s_{CP})}(CP, CP_b)$ is defined by $Q_{CP_{\circlearrowleft}} = Q_{CP}, M_{CP_{\circlearrowleft}} = M_{CP} \cup \{l_{CP_b}\}$, $\delta_{CP_{\circlearrowleft}} = \delta_{CP} \cup \{((q_{CP}, l_{CP_b}), q'_{CP_b})\}$, and $Q^f_{CP_{\circlearrowleft}} = Q^f_{CP}$.

Clearly, according to Theorem 1 we must require that in a sequence the sender of the added message equals the sender or receiver of any message associated with a transition to $q_{CP} \in Q^f_{CP}$. The same must hold for the new messages introduced by a branching composition. In addition, the senders associated with the new messages must be pairwise different. In case of a loop composition we must in addition require that the sender of any message associated with a transition from $q'_{CP_b} \in Q_{CP}$ equals the sender or receiver of the newly introduced message.

3.2 Correctness Proof

We use Event-B to prove the correctness of the compositions thereby giving an alternative Rodin-based proof of Theorem 1. An Event-B model (see Table 1) is defined to encode this incremental process.

Table 1. An excerpt of the LTS_model.

```
INITIALISATION ≜
EVENTS
    Add_Seq ≜
        Any Some_cp_b
        Where
            grd1: Some_cp_b ∈ cps_b
            grd2: MESSAGE(Some_cp_b) ≠ End
            grd3: Some_cp_b ∈ ISeqF
            grd4: SOURCE_STATE(Some_cp_b) ∈ CP_Final_states
            ...
        Then
            act1: BUILT_CP := BUILT_CP ∪ {Some_cp_b}
            act3: CP_Final_states := (CP_Final_states ∪
                    {DESTINATION_STATE(Some_cp_b)})\
                    {SOURCE_STATE(Some_cp_b)}
            ...
    End
    Add_Choice ≜ ...
    Add_Loop ≜ ...
    Add_End ≜ ...
End
```

Once initialisation ($INITIALISATION$) is performed, three events ($Add\_Sequence$, Add_Choice and Add_Loop) for sequence, choice and loop are interleaved to build a choreography CP. All these events are guarded by the identified

Table 2. An excerpt of the LTS_CONTEXT.

```
LTS_CONTEXT
SETS PEERS, MESSAGES , CP_STATES.
CONSTANTS CPs_B, DC, ISeqF, NDC, ...
AXIOMS
    axm1: CPs_B ⊆ CP_STATES × PEERS × MESSAGES × PEERS × CP_STATES × ℕ
        − Determinstic CP definition DC
    axm2_Cond1: NDC ⊆ CPs_B
    axm3_Cond1: ∀Trans2, Trans1·(Trans1 ∈ CPs_B ∧ Trans2 ∈ CPs_B ∧
        SOURCE_STATE(Trans1) = SOURCE_STATE(Trans2) ∧
        LABEL(Trans1) = LABEL(Trans2) ∧
        DESTINATION_STATE(Trans1) ≠ DESTINATION_STATE(Trans2))
            ⇒ {Trans1, Trans2} ⊆ NDC
    axm4_Cond1: DC = CPs_B \ NDC
        − Independent sequence freeness definition ISEQF
    axm5_Cond2: ISeqF ⊆ CPs_B
    axm6_Cond2: ∀ cp_b · ( cp_b ∈ CPs_B ∧
        (PEER_SOURCE(cp_b) = LAST_SENDER_PEERS(SOURCE_STATE(cp_b)) ∨
        PEER_SOURCE(cp_b) = LAST_RECEIVER_PEERS(SOURCE_STATE(cp_b))))
            ⇒ {cp_b} ⊆ ISeqF
        − Parallel Choice freeness PCF
    axm7_Cond3: PCF ⊆ CPs_B
    axm8_Cond3: ∀ cp_b· (cp_b ∈ CPs_B ∧
        {PEER_SOURCE(cp_b)} = BRANCHES_PEERS_SOURCE(cp_b) )
            ⇒ {cp_b} ⊆ PCF
        ...
End
```

conditions deterministic, sequence and choice conditions defined in the context *LTS_CONTEXT* of Table 2.

In this context (see Table 2), we introduce using sets and constants, the whole basic definitions of messages, choreography states, basic choreographies (i.e. choreographies with a single transition as used in the definitions of the composition operators), etc. A set of axioms is used to define the relevant properties of these definitions. For example, in Axiom $axm1$, a choreography CP is defined as a set of transitions with a source and target state, a message and a source and target peers. Axiom $axm3\_Cond1$ defines that a non-deterministic choreography is using the NDC set. This NDC set characterises all the non-deterministic choices in a choreography CP. Note that Axiom $axm4\_Cond1$ defines the assumed deterministic choice condition. The capture of sequence conditions is given by Axioms $axm5\_Cond2$ and $axm6\_Cond2$. It compares the source peer $PEER\_SOURCE(cp\_b)$ with the sender peer $LAST\_SENDER\_PEERS$ or with the receiver peer $LAST\_RECEIVER\_PEERS$ of the last transition of the choreography.

Similarly, to define the choice condition, in Axioms $axm7\_Cond3$ and $axm8\_Cond3$ the sender peers $PEER\_SOURCE(Trans)$ of the transitions involved in a branch are compared.

The correctness proof rebuilds the three decisive parts of the proof of Theorem 1:

1. It shows that the trace language of the choreography coincides with the one of the rendez-vous composition of its projected peers. This property was called *equivalence* in earlier work [2].

2. It shows the language synchronisability between the rendez-vous composition and the mailbox composition, which was referred to as *synchonisability* in [2].
3. It shows that all accepted sequence of messages of the mailbox composition system are accepted in a state, where the mailboxes are empty. This property was called *well-formedness* in earlier work [2].

4   Completeness Proof: A Correct-by-Construction Approach with Event-B

To prove that all the choreographies CP built using the previously defined events, encoding the composition operators, we rely on refinement offered by Event-B. As indicated above we can decomposed the realisability property into three properties, namely *equivalence, synchronisability* and *well-formedness*. The Event-B context in Table 3 defines these properties.

Table 3. An excerpt of the LTS_SYNC_CONTEXT.

```
LTS_SYNC_CONTEXT, EXTENDS LTS_CONTEXT
SETS ACTIONS. CONSTANTS CPs_B , EQUIV, ...
AXIOMS
    axm1: CPs_SYNC_B ⊆ CP_STATES × ACTIONS × MESSAGES × PEERS×
                        PEERS × ACTIONS × MESSAGES × CP_STATES × ℕ
    axm2: CPs_ASYNC_B ∈ (A_STATES × ETIQ × ℕ) ↠ A_STATES
          − Equivalence of CP and Synchronous projection
    axm_1.a: EQUIV ∈ CPs_B ↣ CPs_SYNC_B
    axm_1.a1: EQUIV = { Trans ↦ S_Trans | Trans ∈ CPs_B ∧ S_Trans ∈ CPs_SYNC_B ∧
                        SOURCE_STATE(Trans) = S_SOURCE_STATE(S_Trans) ∧
                        DESTINATION_STATE(Trans) = S_DESTINATION_STATE(S_Trans) ∧
                        PEER_SOURCE(Trans) = S_PEER_SOURCE(S_Trans) ∧
                        PEER_DESTINATION(Trans) = S_PEER_DESTINATION(S_Trans) ∧
                        MESSAGE(Trans) = S_MESSAGE(S_Trans) ∧
                        INDEX(Trans) = S_INDEX(S_Trans) }
          − Synchronisability property
    axm_1.b: SYNCHRONISABILITY ∈ CPs_SYNC_B ↣ R_TRACE_B
    axm_1.b1: SYNCHRONISABILITY = {S_Trans↦ R_Trans | S_Trans ∈ CPs_SYNC_B ∧
                        R_Trans ∈ R_TRACE_B ∧ S_INDEX(S_Trans) = R_INDEX(R_Trans) ∧
                        S_SOURCE_STATE(S_Trans) = R_SOURCE_STATE(R_Trans) ∧
                        S_PEER_SOURCE(S_Trans) = R_PEER_SOURCE(R_Trans) ∧
                        S_MESSAGE(S_Trans) = R_MESSAGE(R_Trans) ∧
                        S_PEER_DESTINATION(S_Trans) = R_PEER_DESTINATION(R_Trans) ∧
                        S_DESTINATION_STATE(S_Trans) = R_DESTINATION_STATE(R_Trans)}
          − Well formedness property
    axm_1.c: WF ∈ A_TRACES ↠ QUEUE
    axm_1.c1: ∀ A_TR,queue · ( A_TR ∈ A_TRACES ∧ queue ∈ QUEUE ∧ queue = ∅ )
    ⇒ A_TR↦ queue ∈ WF
          ...
End
```

Each property is formalised by a set of choreographies satisfying the corresponding property. These definitions use the rendez-vous composition CP_{rv} defined as set CPs_SYNC_B and the mailbox composition CP_m defined as set CPs_ASYNC_B in context $LTS\_SYNC\_CONTEXT$ of Table 3.

4.1   An Event-B Context for the Realisability Property

The definition of the state-transitions system corresponding to the synchronous projection is given by the set CPs_SYNC_B defined by Axiom $axm1$ in Table 3. Actions (send ! and receive ?) are introduced. Then two other important axioms, namely $axm\_1.a$ and $axm\_1.a1$, are given to define the equivalence between a choreography CP and its synchronous projection. The $EQUIV$ relation is introduced. It characterises the set of CPs that are equivalent to their synchronous projection. Axiom $axm\_1.a1$ formalises *equivalence*. The properties related to *synchronisability* are captured by Axioms $axm\_1.b$ and $axm\_1.b1$. Well-formedness is captured by Axioms $axm\_1.c$ and $axm\_1.c1$.

4.2   Refinement

We exploit the characterisation of realisability by three properties in a refinement strategy, which establishes the necessity step-by-step. These properties are introduced as invariants and inductively proven for each composition operator (sequence, choice and loop). That is, two refinements of the initial machine of Table 1 are defined:

- The first refinement introduces the equivalence property by defining the (synchronous) rendez-vous projection of the initial choreography CP.
- Synchronisability and well-formedness properties are proven in the second refinement.

Below we present a sketch of this development focusing on the definition of the sequence operator. The complete development can be accessed from http://yamine.perso.enseeiht.fr/ABZ2020EventBModels.pdf.

Table 4. An excerpt of the LTS_Synchronous_model.

```
INITIALISATION
    . . .
EVENTS
    Add_Seq Refines Add_Seq ≜
        Any
            S_Some_cp_b, Some_cp_sync_b
        Where
            grd1: Some_cp_sync_b ∈ cps_sync_b
            grd3: S_SOURCE_STATE(Some_cp_sync_b) ∈ CP_Final_states
            grd4: S_Some_cp_b ∈ ISeq
            grd8: MESSAGE(S_Some_cp_b) ≠ End
            grd9: MESSAGE(S_Some_cp_b) = S_MESSAGE(Some_cp_sync_b)
            . . .
        With Some_cp_b: Some_cp_b = S_Some_cp_b
        Then
            act1: BUILT_CP := BUILT_CP ∪ {S_Some_cp_b}
            act2: BUILT_SYNC := BUILT_SYNC ∪ {Some_cp_sync_b}
            . . .
    End
```

First Refinement: Equivalence. The first refinement introduces the synchronous projection of the $BUILT\_CP$ defined by variable $BUILT\_SYNC$ in Table 4.

The event Add_Seq or sequence operator (Table 4) refines the same event of the root model of Table 1. It introduces the $BUILT\_SYNC$ set corresponding to the synchronous projection as given in Sect. 2.1. Here, again, the Add_Seq applies only if the conditions in the guards hold. The $With$ clause provides a witness to glue $Some\_cp\_b$ CP with its synchronous version.

Second Refinement: Synchronisability and Well-Formedness. The second refinement introduces the asynchronous projection with sending and receiving peers actions.

Well-formedenss and synchronisability remain to be proven in order to complete realisability preservation. At this level each event corresponding to a composition operator is refined by three events: one to handle sending of messages (Add_Seq_send), one for receiving messages ($Add\_Seq\_receive$), and a third one ($Add\_Seq\_send\_receive$) refining the abstract Add_seq event. Queues are introduced as well.

Table 5 defines these events. Sending and receiving events are interleaved in an asynchronous manner. Once a pair of send and receive events has been triggered, the event $Add\_Seq\_send\_receive$ records that the emission-reception is completed. This event increases the number of received messages (Action $act5$). Traces are updated accordingly by the events, they are used for proving the invariants.

4.3 Completeness Proof

The proof of completeness consists in proving a choreography is realisable if and only if it is built using the defined composition operators. The Rodin-based proofs exploits that realisability can be equivalently expressed by *equivalence*, *synchronisability* and *well-formedness*. Note that the proof strategy with the sufficiency and necessity parts is quite similar, as the same development and refinement steps are used in both cases. The main difference resides in the definition of two invariants, which correspond to each direction of the implication corresponding to the necessity and sufficiency conditions.

Sufficiency. Sufficiency consists in proving that, if a choreography is built using the defined composition operator, then it is realisable. This property has been proven by proving the invariants described in Table 6.

These invariants state that for each CP built using the composition operators, the obtained CP fulfils *Equivalence*, *Synchronisability* and *WF* by set belonging property. Table 6 introduces the *equivalence* property through invariant inv_1.a. The invariant requires equivalence between a CP and its synchronous projection. inv_1.b and inv_1.c introduce respectively the *synchronisability* and *well-formedness* properties.

Table 5. An excerpt of the LTS_Asynchronous_model.

```
Event Add_Seq_Send ≙
    Any
        send, lts_s, lts_d, msg, index
    Where
        grd1: ∃send_st_src, send_st_dest · ((lts_s ↦ send_st_src) ∈ A_GS ∧ ((send_st_src ↦
            (Send ↦ msg ↦ lts_d) ↦ index) ↦ send_st_dest) ∈ CPs_ASYNC_B∧ ...
        ...
    Then
        act1: A_TRACE := A_TRACE ∪ {Reduces_Trace_states ↦ St_Num ↦
            Send ↦ lts_s ↦ msg ↦ lts_d ↦ Reduces_Trace_states ↦
            (St_Num + 1) ↦ A_Trace_index}
        act2: queue, back := queue ∪ {lts_d ↦ msg ↦ back}, back + 1
        act3: A_GS := A_Next_States({send} ↦ A_GS ↦ queue)
        ...
End
Event Add_Seq_Receive ≙
    Any
        send, receive, lts_s, lts_d, msg, index
    Where
        grd1: queue ≠ ∅ ∧ lts_d ↦ msg ↦ front ∈ queue
        grd2: ∃receive_st_src, receive_st_dest · (((lts_d ↦ receive_st_src) ∈ A_GS)∧
            ((receive_st_src ↦ (Receive ↦ msg ↦ lts_s) ↦ index) ↦ receive_st_dest)
            ∈ CPs_ASYNC_B ∧ ...
        ...
    Then
        act1: A_TRACE := A_TRACE ∪ {Reduces_Trace_states ↦ St_Num ↦
            Receive ↦ lts_s ↦ msg ↦ lts_d ↦ Reduces_Trace_states ↦ (St_Num + 1)
            ↦ A_Trace_index}
        act2: queue := queue \ {lts_d ↦ msg ↦ front}
        ...
End
Event Add_Seq_Send − Receive Refines Add_Seq ≙
    Any
        A_Some_cp_b, A_Some_cp_sync_b, Send_cp_async_b, Receive_cp_async_b, R_trace_b
    Where
        grd1: A_MESSAGE(Send_cp_async_b) = A_MESSAGE(Receive_cp_async_b)
        grd2: ACTION(Receive_cp_async_b) = Receive ∧ ACTION(Send_cp_async_b) = Send
        grd3: A_Some_cp_b ∈ ISeq
        grd4: MESSAGE(A_Some_cp_b) = A_MESSAGE(Send_cp_async_b)
        ...
        With S_Some_cp_b : S_Some_cp_b = A_Some_cp_b,
            Some_cp_sync_b : Some_cp_sync_b = A_Some_cp_sync_b
    Then
        act1: BUILT_CP := BUILT_CP ∪ {A_Some_cp_b}
        act2: BUILT_SYNC := BUILT_SYNC ∪ {A_Some_cp_sync_b}
        act3: BUILT_ASYNC := BUILT_ASYNC ∪ {Send_cp_async_b} ∪ {Receive_cp_async_b}
        act4: REDUCED_TRACE := REDUCED_TRACE ∪ {R_trace_b}
        ...
    End
    ...
End
```

Necessity. Necessity consists in proving that if a CP is realisable, then it is built using the defined composition operator.

This property has been established by proving the invariants described in Table 7. Invariant `inv2.a` states that any CP belonging to the *equivalence* set is a peer to peer CP, `inv2.b` states that any synchronisable CP belongs to the set of built CP and finally `inv2.c` states that all the well formed CP exchanging the ending message is built at the asynchronous level.

Proof Statistics. Table 8 gives the results of our experiments. We can observe that all the proof obligations (POs) have been proved. A large amount of these

Table 6. An excerpt of sufficient model invariants.

Invariants
> **inv1:** $BUILT\_SYNC \subseteq CPs\_SYNC\_B$
> **inv2** $BUILT\_ASYNC \subseteq CP\_ASYNC\_B$
> **inv3** $REDUCED\_TRACE \subseteq R\_TRACE\_B$
> **inv4** $A\_TRACE \subseteq A\_TRACES$
> **inv_1.a:** $\forall Trans\cdot\exists S\_Trans\cdot(Trans \in BUILT\_CP \land S\_Trans \in BUILT\_SYNC \land$
> $\qquad BUILT\_CP \neq \varnothing) \Rightarrow$ **Trans** $\mapsto$ **S_Trans** $\in$ **EQUIV**
> **inv_1.b** $\forall S\_Trans\cdot\exists R\_Trans\cdot(S\_Trans \in BUILT\_SYNC \land R\_Trans \in$
> $\qquad REDUCED\_TRACE) \Rightarrow$
> $\qquad\qquad$ **S_Trans** $\mapsto$ **R_Trans** $\in$ **SYNCHRONISABILITY**
> **inv_1.c** $\forall A\_Trans\cdot(A\_Trans \in A\_TRACES \land MESSAGE(Last\_cp\_trans) = End \land$
> $\qquad A\_TRACE \neq \varnothing) \Rightarrow$ **A_Trans** $\mapsto$ **queue** $\in$ **WF**
> $\quad$ ...

Table 7. An excerpt of necessary and sufficient model invariants.

Invariants
> **inv2.a** $\forall Trans.\exists S\_Trans.($**Trans** $\mapsto$ **S_Trans** $\in$ **EQUIVALENCE** $\land BUILT\_CP \neq \varnothing)$
> $\qquad \Rightarrow Trans \in BUILT\_CP \land S\_Trans \in BUILT\_SYNCHRONE$
> **inv2.b** $\forall S\_Trans.\exists R\_Trans.($**S_Trans** $\mapsto$ **R_Trans** $\in$ **SYNCHRONISABILITY** $\land$
> $\qquad BUILT\_SYNCHRONE \neq \varnothing \land REDUCED\_TRACE \neq \varnothing)$
> $\qquad \Rightarrow S\_Trans \in BUILT\_SYNCHRONE \land R\_Trans \in REDUCED\_TRACE$
> **inv2.c** $\forall A\_Trans.($**A_Trans** $\mapsto$ **queue** $\in$ **WF**$) \Rightarrow (A\_Trans \in A\_TRACES \land$
> $\qquad queue = \varnothing \land MESSAGE(Last\_cp\_trans) = End\_message)$
> $\quad$ ...

Table 8. Rodin proofs statistics

Event-B model	Interactive proofs	Automatic proofs	Proof Obligations
Abstract context	06 (100%)	0 (0%)	06 (100%)
Synchronous context	02 (100%)	0 (0%)	02 (100%)
Asynchronous context	01 (33,33%)	02 (66,67%)	03 (100%)
Abstract model	28 (58,33%)	20 (41,67%)	48 (100%)
Synchronous model	43 (41,34%)	61 (58,65%)	104 (100%)
Asynchronous model	81 (41,32%)	115 (58,67%)	196 (100%)
Total	161 (100%)	198 (100%)	359 (100%)

POs has been proved automatically using the different provers associated to the Rodin platform. Interactive proofs of POs required to combine some interactive deduction rules and the automatic provers of Rodin. Few steps were required in most of the cases, and a maximum of 15 steps was reached.

5 Conclusion

In this article we extended the Event-B-based approach to the construction of realisable choreographies [4,5] based on recent new insights into choreography-defined P2P systems. In [10] we proved that under the presence of a choreography that prescribes the rendez-vous synchronisation of the peers there are two necessary conditions on realisable choreographies which together guarantee realisability. A consequence is decidability of realisability in the presence of a choreography. We removed unnecessary assumptions in the Event-B-based proofs and extended them to cover also necessity of the conditions. In doing so we demonstrated the power of the Rodin tool. All the models are accessible through http://yamine.perso.enseeiht.fr/ABZ2020EventBModels.pdf.

Naturally, using Event-B in this context provides an open invitation for a refinement-based approach taking choreographies to communicating systems that do not just emphasise the flow of messages. As we are now able to detect violations of a necessary condition, it allows us to find minimal repairs to the choreography to restore realisability. Such repairs have to be validated by a designer. In addition, we need a systematic investigation of refinements based on Event-B. In this context an analysis of the realisation of the messaging channels is due, for which we expect the most natural semantics using mailboxes to be the simplest to be realised. This refinement method provides an open invitation for the continuation of this research towards a verifiable method for the specification and refinement of correct P2P systems.

References

1. Basu, S., Bultan, T.: On deciding synchronizability for asynchronously communicating systems. Theor. Comput. Sci. **656**, 60–75 (2016)
2. Basu, S., Bultan, T., Ouederni, M.: Deciding choreography realizability. In: Field, J., Hicks, M. (eds.) Proceedings of the 39th ACM SIGPLAN-SIGACT Symposium on Principles of Programming Languages (POPL 2012), pp. 191–202. ACM, New York (2012)
3. Benyagoub, S., Aït-Ameur, Y., Ouederni, M., Mashkoor, A., Medeghri, A.: Formal design of scalable conversation protocols using Event-B: validation, experiments and benchmarks. J. Softw.: Evol. Process **23**(2), 129–145 (2019)
4. Benyagoub, S., Ouederni, M., Aït-Ameur, Y., Mashkoor, A.: Incremental construction of realizable choreographies. In: Dutle, A., Muñoz, C., Narkawicz, A. (eds.) NFM 2018. LNCS, vol. 10811, pp. 1–19. Springer, Cham (2018). https://doi.org/10.1007/978-3-319-77935-5_1
5. Benyagoub, S., Ouederni, M., Singh, N.K., Ait-Ameur, Y.: Correct-by-construction evolution of realisable conversation protocols. In: Bellatreche, L., Pastor, Ó., Almendros Jiménez, J.M., Aït-Ameur, Y. (eds.) MEDI 2016. LNCS, vol. 9893, pp. 260–273. Springer, Cham (2016). https://doi.org/10.1007/978-3-319-45547-1_21
6. Brand, D., Zafiropulo, P.: On communicating finite-state machines. J. ACM **30**(2), 323–342 (1983)
7. Chambart, P., Schnoebelen, P.: Mixing lossy and perfect FIFO channels. In: van Breugel, F., Chechik, M. (eds.) CONCUR 2008. LNCS, vol. 5201, pp. 340–355. Springer, Heidelberg (2008). https://doi.org/10.1007/978-3-540-85361-9_28
8. Clemente, L., Herbreteau, F., Sutre, G.: Decidable topologies for communicating automata with FIFO and bag channels. In: Baldan, P., Gorla, D. (eds.) CONCUR 2014. LNCS, vol. 8704, pp. 281–296. Springer, Heidelberg (2014). https://doi.org/10.1007/978-3-662-44584-6_20

9. Finkel, A., Lozes, É.: Synchronizability of communicating finite state machines is not decidable. In: Chatzigiannakis, I., et al. (eds.) 44th International Colloquium on Automata, Languages, and Programming (ICALP 2017), volume 80 of LIPIcs, pp. 122:1–122:14. Schloss Dagstuhl - Leibniz-Zentrum für Informatik (2017)
10. Schewe, K.-D., Aït-Ameur, Y., Benyagoub, S.: Realisability of choreographies. In: Herzig, A., Kontinen, J. (eds.) FoIKS 2020. LNCS, vol. 12012, pp. 263–280. Springer, Cham (2020). https://doi.org/10.1007/978-3-030-39951-1_16
11. Farah, Z., Ait-Ameur, Y., Ouederni, M., Tari, K.: A correct-by-construction model for asynchronously communicating systems. Int. J. Softw. Tools Technol. Transf. **19**(4), 465–485 (2016). https://doi.org/10.1007/s10009-016-0421-6

Formally Verified Architecture Patterns of Hybrid Systems Using Proof and Refinement with Event-B

Guillaume Dupont$^{(\boxtimes)}$, Yamine Aït-Ameur, Marc Pantel, and Neeraj K. Singh

INPT-ENSEEIHT/IRIT, University of Toulouse, Toulouse, France
{guillaume.dupont,yamine,marc.pantel,nsingh}@enseeiht.fr

Abstract. Cyber-Physical Systems (CPS) play a central role in modern days technology. From simple thermostat controllers to more advanced autonomous cars, their versatility makes them perfect candidates for many applications, in particular for safety critical ones. Thus, their certification is a key issue and formal methods are good candidates to assess safety and produce associated certificates. Hybrid systems show continuous-time dynamics depending on mode that is required in several stages of the architecture of Cyber-Physical Systems. Our work addresses the problem of formally verifying hybrid systems using refinement and proof with Event-B. Our previous work [14] presented formally verified generic architecture patterns for designing centralised hybrid systems, based on our generic approach [15]. We extend this work and give a formally verified architecture pattern aimed at modelling distributed hybrid systems, featuring multiple plants and multiple controllers. We validate the approach and illustrate the use of the defined pattern on an extension of a very common case study, borrowed from literature.

Keywords: Hybrid systems · Cyber-physical systems · Architecture design patterns · Event-B · Refinement · Proof

1 Introduction

Cyber-Physical Systems (CPS) can be described as complex systems that integrate both discrete and continuous features [19]. Such system generally consists of a discrete algorithm or *controller* that interacts with a continuous process or *plant* in order to control its behaviour. The controller can retrieve information from the plant through sensors and may alter its behaviour with actuators.

Because of this hybridation, CPS are often regarded as quite hard to trust. However, their versatility, adaptability and price made them unavoidable in our everyday life, from Internet of Things (IoT) to smart systems (e.g. home automation, smart factories and so on), including, of course, critical systems such as transportation and medical devices. Being able to formally model and certify CPS is thus a major challenge nowadays.

This work was supported by grant ANR-17-CE25-0005 (The DISCONT Project https://discont.loria.fr) from the Agence Nationale de la Recherche (ANR).

© Springer Nature Switzerland AG 2020
A. Raschke et al. (Eds.): ABZ 2020, LNCS 12071, pp. 169–185, 2020.
https://doi.org/10.1007/978-3-030-48077-6_12

The design of formal modelling approaches for CPS development and/or certification has been addressed in various ways. In [4], Alur defines the hybrid automata formalism to model hybrid systems. Hybrid model-checkers such as HyTech, d/dt, PHaVer or SpaceEx can then be used to establish properties such as reachability.

In terms of modelling techniques, [18] have proposed HybridCSP as a hybrid extension of CSP [17]. [7] proposes a continuous extension of Action System. In the same manner, [8] proposes an hybrid extension to the Event-B method.

Proof-based approaches have also been used to try and formally prove CPS. [9] use Coq to that extent, starting from an annotated C program. [21] uses a special formalism (hybrid programs) to model and to prove hybrid systems using KeYmaera. Event-B has been used for modelling similar systems in [23] and [10].

However, all these approaches still require formal modelling expertise, where the developer needs to establish correctness using complex proof systems involving discrete and continuous mathematics features and proof rules. As a consequence, the use of these methods on a large scale is hindered, in particular in formal system engineering. So, easing CPS formal modelling and verification activities in presence of both discrete and continuous behaviours is still a challenge.

To address this challenge, we propose a systematic correct-by-construction approach to design hybrid systems based on the definition of architecture patterns. Indeed, one commonly used method in formal system engineering is to provide formalised generic patterns where relevant generic properties are established. Furthermore, those patterns can be instantiated for specific systems. In such a setting, the system developer selects the most adapted pattern and instantiates it. Proof obligations, in particular regarding well-definedness, may need to be discharged in order to inherit all the properties of the generic pattern.

In our previous work [14], we used Event-B to design and formalise commonly used architecture patterns (AP) for centralised hybrid systems. We based those patterns on our generic approach [15], allowing to model both discrete and continuous behaviours. In this paper, we extend these architecture patterns to model distributed hybrid systems i.e. systems that manage multiple autonomous subsystems, linked together by a communication network. A case study is given as a possible instantiation of this pattern, involving independent liquid tanks enforcing a global invariant that expresses safety properties.

This paper is organised as follows: Sect. 2 gives an overview of Event-B and Sect. 3 presents hybrid modelling features needed for hybrid system development. Section 4 introduces the architecture patterns identified when modelling hybrid systems. Section 5 recalls our generic method for designing hybrid systems in Event-B. Section 6 introduces a case study to support our work, which is solved in Sect. 7. Finally, Sect. 8 provides an assessment of the approach, and Sect. 9 concludes the paper and discusses possible future research directions.

2   Modelling Hybrid Systems with Event-B

Event-B method [2] supports the development of *correct-by-construction* complex systems. First order logic and set theory are used as core modeling language.

The design process consists of a series of refinements of an abstract model leading to the final concrete model. Refinement progressively contributes to add design decisions to the system.

Event-B *machines* formalize models described as state-transitions systems and a set of proof obligations are automatically generated for each model.

Notation. We use the superscripts ^{A} and ^{C} to denote abstract and concrete features.

Table 1. Model structure

Context	Machine	Refinement
CONTEXT Ctx	**MACHINE** M^A	**MACHINE** M^C
SETS s	**SEES** Ctx	**REFINES** M^A
CONSTANTS c	**VARIABLES** x^A	**VARIABLES** x^C
AXIOMS A	**INVARIANTS** $I^A(x^A)$	**INVARIANTS** $J(x^A, x^C) \wedge I^C(x^C)$
THEOREMS T_{ctx}	**THEOREMS** $T_{mch}(x^A)$	...
END	**VARIANT** $V(x^A)$	**EVENTS**
	EVENTS	**EVENT** evt^C
	EVENT evt^A	**REFINES** evt^A
	ANY α^A	**ANY** α^C
	WHERE $G^A(x^A, \alpha^A)$	**WHERE** $G^C(x^C, \alpha^C)$
	THEN	**WITH**
	$x^A :\mid BAP^A(\alpha^A, x^A, x^{A'})$	$x^{A'}, \alpha^A\colon W(\alpha^A, \alpha^C, x^A, x^{A'}, x^C, x^{C'})$
	END	**THEN**
	...	$x^C :\mid BAP^C(\alpha^C, x^C, x^{C'})$
		END
		...
(a)	(b)	(c)

- *Event-B Contexts (Table 1.a). Contexts* are the static part of a model. They set up all the definitions (carrier sets s and constants c), axioms (A) and theorems (T_{ctx}) needed to describe the required concepts.
- *Event-B Machines (Table 1.b).* A *machine* describes the dynamic part of a model as a transition system. A set of guarded events modifying a set of variables (state) represents the core concepts of a machine. *Variables x, invariants $I(x)$, theorems $T_{mch}(x)$, variants $V(x)$* and *events evt* (possibly guarded by G and/or parameterized by α) are defined in a machine. *Invariants* and *theorems* formalize safety system properties while *variants* define convergence properties (reachability).
- *Event-B Refinements (Table 1.c).* A system is gradually designed by introducing properties (functionality, safety, reachability) at various abstraction levels. *Refinement* decomposes a *machine*, a state-transitions system, into a more concrete one, with more design decisions (refined states and events) while moving from an abstract level to a less abstract one. Abstract and

concrete variables are related by gluing invariants ensuring properties preservation between abstract and concrete models.

- *Proof Obligations (PO) and Property Verification.* To establish the correctness of an Event-B model, a set of POs are automatically generated from the calculus of substitutions. They need to be proved.
- *Extensions with mathematical theories.* In order to handle theories beyond set theory and first order logic, an Event-B extension to support externally defined mathematical theories has been proposed. It offers the capability to introduce new datatypes through the definition of new types, sets operators, theorems and associated rewrite and inference rules, in so-called *theories*.
- *Event-B and its IDE Rodin.* It offers resources for model edition, automatic PO generation, project management, refinement and proof, model checking, model animation and code generation. Several provers, like SMT solvers, are plugged to Rodin. In particular, a plug-in allows to define theories [11].

3   Hybrid Systems Modelling Features

Modelling hybrid systems requires handling of continuous behaviours. We thus need to access specific mathematical objects and properties, not natively available in Event-B. These concepts such as differential equations and their associated properties have been modelled through an intensive use of Event-B theories and have been used to model various case studies found in [13–15].

In order to deal with continuous objects, theories have been defined for continuous functions, (ordinary) differential equations as well as for their properties. They are used throughout the defined models. Their complete definitions are available at https://irit.fr/~Guillaume.Dupont/models.php. Some of those concepts as they are used in this paper are recalled below.

Time. A notion of time is needed to define continuous behaviors. We thus introduce dense time $t \in \mathbb{R}^+$, modeled as a continuously evolving variable.

System State. According to the architecture of hybrid systems, we have identified two types of states.

- **Discrete state** $x_s \in STATES$, variable that represents the controller's internal state. It evolves in a pointwise manner with instantaneous changes.
- **Continuous state** $x_p \in \mathbb{R}^+ \to S$ represents the plant's state and evolves continuously. It is modelled as a function of time with values in space S.

Hybrid Modeling Features. Modeling hybrid systems requires the introduction of multiple specific features which are defined below.

- **DE**(S) type for differential equations which solutions evolve over set S
- **ode**(f, η_0, t_0) represents the ODE[1] $\dot{\eta}(t) = f(\eta(t), t)$ with initial condition $\eta(t_0) = \eta_0$

---

[1] Ordinary Differential Equation.

```
THEORY
  TYPE PARAMETERS E, F
  DATA TYPES
    DE(F)
      CONSTRUCTORS
        ode(fun : ℙ(ℝ × F × F), initial : F, initialArg : ℝ)
  OPERATORS
    solutionOf <predicate> (D_R : ℙ(ℝ), η : ℝ⁺ ⇸ F, eq : DE(F))
    CauchyLipschitzCondition <predicate> (D_R : ℙ(ℝ), D_F : ℙ(F), eq : DE(F))
    Solvable <predicate> (D_R : ℙ(ℝ), eq : DE(F))
      direct definition
        ∃x · x ∈ (ℝ⁺ ⇸ F) ∧ solutionOf(D_R, x, eq)
    ...
END
```

Fig. 1. Differential equation theory snippet

- **solutionOf**$(D, \eta, \mathcal{E})$ is the predicate stating that function η is a solution of equation $\mathcal{E}$ on subset D
- **Solvable**$(D, \mathcal{E}, \mathcal{I})$ is the predicate stating that equation $\mathcal{E}$ has a solution defined on subset D so that the solution satisfies the constraint $\mathcal{I}$

These features have been encoded in a theory from which we show a snippet on Fig. 1 (the theory accumulates more than 150 operators and 350 properties).

Other, more specialised expressions and predicates are defined (*FlowEquation*, *FlowODE*) in additional theories. Note that all these definitions use *algebraic datatypes* together with axioms, theorems and proof rules.

In the following, we use x to denote the union of discrete and continuous state variables.

Continuous Assignment. Continuous variables are essentially functions of time and are at least defined on $[0, t]$ (where t is the current time). Updating such variables thus requires to 1) make the time progress from t to $t' > t$, and 2) append to the already existing function a new piece corresponding to its extended behavior (on $[t, t']$) while ensuring its "past" (i.e. whatever happened on $[0, t]$) remains unchanged.

Similarly to the classic Event-B's before-after predicate (*BAP*), we define a *continuous before-after predicate* (*CBAP*) operator, denoted $:|_{t \to t'}$, as follows:

$$x_p :|_{t \to t'} \mathcal{P}(x_s, x_p, x_p') \,\& \mathcal{I} \equiv [0, t] \lhd x' = [0, t] \lhd x \tag{PP}$$

$$\wedge \mathcal{P}(x_s, [t, t'] \lhd x_p, [t, t'] \lhd x_p') \tag{PR}$$

$$\wedge \forall t^* \in [t, t'], x_p'(t^*) \in \mathcal{I} \tag{LI}$$

We note $CBAP(x_s, x_p, x_p') \equiv PP(x_p, x_p') \wedge PR(x_s, x_p, x_p') \wedge LI(x_p, x_p')$. The operator consists of 3 parts: past preservation and coherence at assignment point (PP), before-after predicate on the added section (PR), and local invariant preservation (LI). The discrete state variables x_s do not change in the interval $[t, t']$ but the predicate $\mathcal{P}$ may use it for control purposes.

Note that this operator is well-defined if and only if $t' > t$, as otherwise the interval $[t, t']$ would not be well-defined.

From the above definition, shortcuts can be introduced for readability purposes:

- Continuous assignment: $x :=_{t \to t'} f \,\&\, \mathcal{I} \;\equiv\; x :|_{t \to t'} x' = f \,\&\, \mathcal{I}$
- Continuous evolution along a solvable differential equation $\mathcal{E} \in \mathbf{DE}(S)$:
  $x :\sim_{t \to t'} \mathcal{E} \,\&\, \mathcal{I} \;\equiv\; x :|_{t \to t'} \mathbf{solutionOf}([t, t'], x', \mathcal{E}) \,\&\, \mathcal{I}.$

4   Architecture Patterns for Modelling Hybrid Systems

One of the most common architectures found in CPS (see Fig. 2) is a discrete software controller, which interacts by some means (e.g. actuators) with a plant and its physical environment (continuous physical phenomenon) in a closed-loop schema. Input from sensors is processed and output is generated and communicated to actuators [12]. Commands from a user or another controller may also be addressed to the controller.

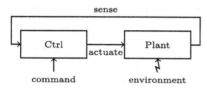

Fig. 2. Generic hybrid system representation

Controllers are characterised by discrete state variables and transitions corresponding to control decisions; as for plants, they are defined by continuous state variables whose evolution is generally described using differential equations. Sensors, user commands control decision and actuators modify these variables.

In this paper, we focus on the verification of the correctness of such discrete controllers, which require correct composition of discrete and continuous models. We claim that correctness should arise from a design process based on sound abstractions and models of the relevant laws of physics.

Hybrid systems may combine the behaviours of multiple separated components (plants or controllers), which can lead to very different control strategies, following the number of controllers and plants to be controlled. Therefore, the generic architecture given in Fig. 2 should be refined into three types of architecture patterns:

- **Single-to-Single** AP corresponds to hybrid systems with one controller and one plant. Examples of hybrid systems corresponding to this pattern addressed in the literature include the automatic car braking [15], the *signalised left-turn assist* [13], heating systems [16], etc.
- **Single-to-Many** AP describes hybrid systems with one controller and many plants (more than one). This pattern corresponds to centralised control. An example of hybrid system corresponding to this pattern is the control of a global volume distributed over several tanks formalised with hybrid automata and with Event-B in [14].
- **Many-to-Many** AP characterises hybrid systems with many controllers and many plants. This pattern refers to the case where several hybrid systems are integrated together to implement a given function. Examples of such systems are UAV or rover fleet control modelled in Event-B [22].

All the patterns defined above refine the one of Fig. 2. The controller and the plant components may be refined to one or many components. These refinements introduce specific properties and behaviours associated to each pattern. The *single-to-single* AP defines one discrete state for the controller and one continuous state for the plant. The *single-to-many* AP defines a controller with one discrete state able to build a global continuous state aggregating the many states of each and every plants.

Finally, the *many-to-many* AP allows to define distributed hybrid systems where each component has a partial view of all the other systems. Here, it is difficult to build a global state of the whole system. Therefore, an approximation of this global state is used by each system controller to take control decisions. Then, the correctness of this approximation shall be ensured to establish global invariants. In other words, local invariants associated to each hybrid system contribute to ensure the global invariant of the whole system composed of all the hybrid systems.

Note: it is worth noticing that the case of *many-to-many* AP may be defined as a set of hybrid systems corresponding to either *single-to-single* AP or *single-to-many* AP. In the last case, *single-to-many* is abstracted by a *single-to-single* system, providing modular verification.

5 Methodology for Hybrid System Design

The pattern presented in Sect. 4 served as a basis for setting up a methodology for hybrid system design. This methodology has been first presented in [15]. It consists of a generic Event-B model that abstracts hybrid systems following the pattern of Fig. 2. This model is then instantiated via refinement. Discrete models can be derived in the same manner [6].

Note that the generic model introduces typically continuous concepts such as differential equations and dense time. It therefore heavily relies on the theory extension of Event-B implemented as a plug-in (see Sect. 3).

5.1   A Generic Event-B Model for Hybrid Systems

```
VARIABLES t , x_s , x_p
INVARIANTS
    inv1 : t ∈ ℝ
    inv2 : x_s ∈ STATES
    inv3 : x_p ∈ ℝ+ ↠ S
    inv4 : [0, t] ⊆ dom(x_p)
```

Model State. The generic model deals with three variables. x_s represents the controller's discrete state that belongs to $STATES$ set consisting of the states of the system's mode automaton.

x_p is the system's continuous state. It is a function of time (inv3) valued in the (continuous) *state space* S, usually $\mathbb{R}^n$. It represents the physical quantities that are sensed and/or controlled. Last, we recall that variable t models the physical, dense time.

Model Behaviour. The defined model follows the control-command principle depicted on Fig. 2. Two categories of events are defined. Discrete events are instantaneous. They are associated with changes in the state of the mode automaton either internal (*Transition* event) or induced by the sensing of the plant's state (*Sense* event). Continuous events, on the contrary, are not instantaneous. They describe the Plant's behaviour, either following environmental changes (*behave* event) or caused by actuation (*actuate* event). Note that all these generic events will be refined later for developing particular hybrid systems.

Transition. Transition events (corresponding to *command* arrow and the *Ctrl* box of Fig. 2) model internal changes in the controller. They represent user commands, internal timers or non-deterministic choices that occur in the discrete part of the system (mode automata). It updates the state of the automaton (`act1`).

```
Transition
ANY s
WHERE
    grd1:  s ∈ ℙ1(STATES)
THEN
    act1:  x_s :∈ s
END
```

Sense. Sensing events (corresponding to *sense* arrow of Fig. 2) model changes in the controller induced by the reading of the plant's state, generally obtained from sensors. As they are fired according to the plant's state and to the mode automaton's state, they are guarded by a predicate over $x_p(t)$ and x_s (`grd3`). The purpose is to change state x_s in action `act1` of the mode automaton.

```
Sense
ANY s , p
WHERE
    grd1:  s ∈ ℙ1(STATES)
    grd2:  p ∈ ℙ(STATES
           ×ℝ × S)
    grd3:  (x_s ↦ t ↦ x_p(t)) ∈ p
THEN
    act1:  x_s :∈ s
END
```

Behave. Behave events (corresponding to the *environment* arrow of Fig. 2) represent changes in the plant due to the environment: rain, wind, etc. These events enforce, in action `act1`, the dynamics of the plant to comply with a differential equation under solvability condition (`gdr2`) but without any condition on the state of the mode automaton.

```
Behave
ANY eq , t'
WHERE
    grd1:  eq ∈ DE(S)
    grd2:  Solvable([t, t'], eq, ⊤)
THEN
    act1:  t, x_p :∼_{t→t'} eq & ⊤
END
```

Actuate. Actuation events (corresponding to the *actuate* arrow of Fig. 2) model changes in the plant induced by the controller (generally performed by actuators). These events enforce, in action `act1`, the dynamics of the plant to comply with a differential equation under solvability condition (`gdr2`) and a constraint H on the plant evolution domain (`gdr5` and `gdr6`). Moreover, unlike for *Behave*, since *Actuate* results from a change in the controller, it is guarded by a predicate on the mode automaton (`gdr4`).

```
Actuate
ANY eq , s , H , t'
WHERE
    grd1:  eq ∈ DE(S)
    grd2:  Solvable([t, t'], eq, H)
    grd3:  s ⊆ STATES
    grd4:  x_s ∈ s
    grd5:  H ⊆ S
    grd6:  x_p(t) ∈ H
THEN
    act1:  t, x_p :∼_{t→t'} eq & H
END
```

As mentioned above, both *Behave* and *Actuate* are continuous events. They rely on the continuous evolution operators defined in Sect. 3. Both events enforce plant behaviour by setting up a corresponding differential equation.

5.2   Semantics

The semantics of hybrid models we use is close to the one of Hybrid Event-B [8], hybrid programs in [21] or continuous action systems [7,20].

In classical Event-B semantics, each model is associated with a discrete state-transition system, in which transitions are the fired machine events and states consist of the machine's variables. A system is hence characterised by a set of licit traces i.e. a set of fired events that abide by the system's invariants.

In our approach, discrete events are timeless, while continuous ones have a duration. In order to properly handle the modelling of continuous behaviours, the semantics of Event-B is enhanced to handle modelling of continuous phenomena which are, in nature, different from discrete behaviours. We have identified two categories of events: discrete (instantaneous) events, which use discrete assignments operators such as $:|$ and $:=$ and continuous (not instantaneous) events that span over some duration and use continuous assignment operators, namely $:|_{t \to t'}$ and $:=_{t \to t'}$. Note that, if several (continuous or discrete) events guards are enabled, these enabled events are fired non deterministically.

A model is then defined as follows. After initialisation, continuous events (*Behave* and *Actuate* events) run continuously unless a discrete, instantaneous event is enabled (either a *Sense* or a *Transition* event). In this case, discrete events are preemptive. This protocol ensures that when the conditions (events' guards) are met, the controller is able to trigger control actions (*Sense* or *Transition*) that may or may not change the continuous behaviour of the plant (through triggering an *Actuate* event). Unlike *Actuate*, the *Behave* event neither requires control action to be triggered nor any plant evolution constraint H. Sensing actions using the *Sense* event will re-establish the correct plant behaviour via the control loop in order to further trigger an *Actuate* event.

5.3   The Generic Model in Rodin

The generic model is the entry point for the method. Specific hybrid system models are obtained by refining it, providing the various witnesses issued from event parameters and substituted variables. In itself, this model generates 13 proof obligations that are easily discharged. Among them there is an important obligation stating that if equation e is solvable then $x :\sim_{t \to t'} e$ is feasible.

This approach has been successfully applied to various case studies. [13, 15] show a class of systems with one controller and one plant while [14] demonstrates the possible use of the method for a system with one controller and several plants.

Models for the generic approach, including the above-mentioned case studies can be found at https://www.irit.fr/~Guillaume.Dupont/models.php.

6   Case Study: The Water Tank Problem

We now illustrate our approach for formally verifying hybrid systems patterns with a well-known control theory problem: keeping the volume of liquid inside a tank between specific bounds proposed by [5].

6.1   Abstract System

The problem is depicted on Fig. 3 and can be described as follows: one or more tanks are filled with a liquid and connected to an input and an output pump.

A controller can access the global volume V of all tanks and may control their pumps to start filling or emptying them. The goal of the controller is to keep the whole volume between V_{low} and V_{high}.

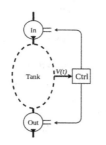

Fig. 3. Abstract tank

The following safety requirements are defined. Let V, be the volume of the tanks (continuous state being controlled). V is physically bounded by 0 and V_{max}, such that $V_{high} \leq V_{max}$, and it shall satisfy the following properties:

SAF1 The volume never overflows nor underflows: $V(t) \in [V_{low}, V_{high}]$

SAF2 The variation of the volume is bounded (to avoid excessive turmoil in the tank): $|\dot{V}(t)| < \Delta V_{max}$

At this level, it is not needed to know the specific characteristics of the tanks (i.e. their shapes, their number, the behaviour of the pumps, the way the controller accesses V and so on). They are simply abstracted away so as to keep this description as generic as possible. The system is later refined for specific tanks and using specific architecture patterns.

6.2   Architecture Patterns as Abstract System Refinements

The system formerly introduced can be refined to illustrate the three architecture patterns identified in Sect. 4 and depicted on Fig. 4.

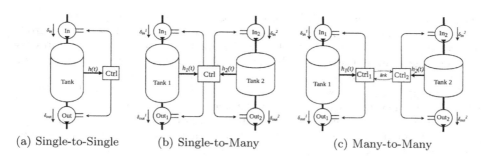

(a) Single-to-Single   (b) Single-to-Many   (c) Many-to-Many

Fig. 4. Three refinement patterns for the case study

Single-to-Single Architecture Pattern. Within a refinement, the abstract model of Fig. 3 is instantiated by a concrete system composed of one controller managing one cylinder-shaped tank (see Fig. 4a). The abstract plant's volume is refined using the gluing invariant $V = B \cdot h$, where B is the surface of the cylinder's base and h is the height of liquid in the tank (easier to measure than the direct volume). Constraints on h are strengthened by the well-definedness condition $V_{max} \leq B \cdot H_{max}$, ensuring that the cylinder can contain (at least) volume V_{max}.

As a matter of simplification, the pumps are associated with a fixed flow rate and are either open (full flow) or closed (no flow), with no intermediate state. Therefore, a differential equation for the system is $\dot{h} = in \cdot \delta_{in} + out \cdot \delta_{out}$, where *in, out* are the states of the pumps and δ_{in}, δ_{out} are their respective flows. This pattern has been previously instantiated in [13,15].

Single-to-Many Architecture Pattern. The same case study can be used to illustrate the second architecture pattern, which involves a single controller and many plants. In this case, we assume that the controller has a global view of the system. In other words, it *knows* all the plants' continuous state variables.

Figure 4b depicts a simplified case for two cylinder tanks, but it scales to any number of tanks of various shapes provided the differential equations that governs these plants are known. For two cylindrical tanks, the gluing invariant is $V = B_1 \cdot h_1 + B_2 \cdot h_2$ where B_1 and B_2 are the surface of the cylinders bases and h_1 and h_2 are the height of liquid in the tanks. The associated differential equations given as witnesses for instantiation are defined by a linear combination.

However, the interesting property in this instantiation relates to the feasibility of the refinement. Indeed, an additional well-definedness condition, expressed as an invariant, states that $V_{max} \leq B_1 \cdot H_{1,max} + B_2 \cdot H_{2,max}$ as to guarantee that the maximum abstract volume can be contained by the two cylinders representing the concrete plant.

This pattern has been thoroughly instantiated and studied in [14].

Note: All the Event-B models corresponding to the two architecture patterns discussed above are available at https://www.irit.fr/~Guillaume.Dupont/models.php. We did not discuss them in this paper due to space limitations. More details can be found in [14,15].

Section 7 below, focuses only on the Event-B models corresponding to the most complex architecture pattern: *many-to-many*.

7   Application of the Many-to-Many Architecture Pattern

In this section we present the last refinement chain corresponding to the *many-to-many* architecture pattern of Sect. 4. The details of the Event-B models are given below for the case study of the water tank following the instantiation of this specific architecture pattern from Fig. 4c.

Refinement Strategy. The refinement strategy is similar to the one used with the *single-to-single* and *single-to-many* patterns. It consists in instantiating the generic model of Sect. 5. Depending on the number of components (controllers/-plants), the generic parts for controller and plant are refined.

Note that the instantiation of the generic model is achieved by providing witnesses to the parameters of the generic events of the Event-B models, i.e. providing a witness for an existential proof obligation.

Two refinements leading to the final Event-B model are defined. First, an abstract tank model corresponding to the system presented in Sect. 6.1 is built as an instance of the generic model of Sect. 5. Then, the final instantiated architecture pattern of Fig. 4c is modelled as a refinement of this model, providing witnesses for generic parameters. The two refinements are summarised below in Sects. 7.1 and 7.2.

7.1 Abstract Tank Model

```
MACHINE WaterTank_base REFINES Generic
VARIABLES  t, V, x_s
INVARIANTS
    inv1:  V ∈ ℝ⁺ ⇸ S
    inv2:  [0, t] ⊆ dom(V)
    inv2:  V = x_p
    inv3:  V ∈ 𝒟¹([0, t], ℝ)∧
           ∀τ · τ ∈ [0, t] ⇒ |V̇(τ)| ≤ ΔV_max
    inv4:  ∀τ · τ ∈ [0, t] ⇒ V_low ≤ V(τ) ≤ V_high
```

Machine State. The controlled variable is the volume V. As mentioned in Sect. 6.1 (SAF1 and SAF2), this quantity shall remain between V_{low} and V_{high} and its derivative $(\dot{V})$ shall be bounded by the ΔV_{max} constant.

The system operates in 4 modes: *Emptying* (volume decreases), *Filling* (volume increases), *Normal* (volume varies in an arbitrary way between V_{low} and V_{high}) and *Stable* (volume does not vary) defining the set *STATES*.

Transition and Sense. When the volume reaches V_{low} (resp. V_{high}) the system moves to *Filling* (resp. *Emptying*) mode. Outside of these restrictions, the system may evolve arbitrarily from one mode to another, via *transition* events. Transition events are guarded by a stricter version of the safety invariant as to prevent the system from deliberately going into an unsafe mode.

```
ctrl_sense_too_high REFINES Sense
WHERE
    grd1:  V_high ≤ V(t)
WITH s:  s = {Emptying}
     p:  p = STATES × ℝ⁺ × {V* | V_high ≤ V*}
THEN
    act1:  x_s := Emptying
END

ctrl_transition_normal REFINES Transition
WHERE
    grd1:  V_low < V(t)
    grd2:  V(t) < V_high
WITH s:  s = {Normal}
THEN
    act1:  x_s := Normal
END
```

```
ctrl_actuate_pumps REFINES Actuate
ANY  e, ss, t'
WHERE
    grd1:  e ∈ DE(S)
    grd2:  Solvable([t, t'], e)
    grd3:  FlowEq(ss, [t, t'], e)
    grd4:  ss ∈ STATES
    grd5:  x_s = ss
    grd6:  V_low < V(t) < V_high
WITH x'_p:  x'_p = V'
     s:  s = {ss}
     H:  H = {V* | V_low < V* < V_high}
THEN
    act1:
        V :∼_{t→t'} e & {V* | V_low < V* < V_high}
END
```

Behave and Actuate. The system performs actuation on the pumps. At this level, the shape of the tank(s) and the behaviour of the pumps are not known yet. The only constraint the actuation shall enforce is that whenever the system is in a specific state, the provided differential equation for actuation is such that its solutions have the expected behaviour (e.g. decreasing solutions when in *Emptying* mode, increasing solutions when in *Filling* mode, etc.).

This constraint is captured by the $FlowEq(x_s, D, e)$ predicate of guard grd3 and is defined in a theory, where x_s is the controller's state, D is the domain on which the predicated behaviour is expected to be true and e is the given equation.

7.2   Many-to-many Model

The model presented below corresponds to the system depicted on Fig. 4c.

```
MACHINE WaterTank_2Ctrl_2Tanks REFINES WaterTank_base
VARIABLES t, V₁, V₂, V₁ˢⁱᵐ, V₂ˢⁱᵐ, x¹ₛ, x²ₛ, Δ₁ˢⁱᵐ, Δ₂ˢⁱᵐ
INVARIANTS
    inv11:  V₁ ∈ ℝ⁺ → S ∧ [0, t] ⊆ dom(V₁)
    inv12:  V₁ˢⁱᵐ ∈ ℝ⁺ → S ∧ [0, t] ⊆ dom(V₁ˢⁱᵐ)
    inv13:  x¹ₛ ∈ STATES
    inv14:  ∀τ · τ ∈ ℝ⁺ ⇒ V₁(τ) + V₂ˢⁱᵐ(τ) ≤ V_high − Δ₂ˢⁱᵐ
                        ∧ V_low + Δ₂ˢⁱᵐ ≤ V₁(τ) + V₂ˢⁱᵐ(τ)
    inv15:  Δ₁ˢⁱᵐ ∈ ℝ⁺
    inv16:  ∀τ · τ ∈ ℝ⁺ ⇒ |V₂(τ) − V₂ˢⁱᵐ(τ)| ≤ Δ₂ˢⁱᵐ
    inv21−26:  −− similar to inv11−16 with V₂
    inv01:  V = V₁ + V₂
    inv02:  xₛ = guess_gs(x¹ₛ ↦ x²ₛ)
```

Machine State. In this refinement, we want to control two tanks (although this could be extended to any number of tanks). Each tank has its own volume, V_1 and V_2 (see Fig. 4c) behaving as a global invariant. The abstract volume V is hence refined using the gluing invariant $V = V_1 + V_2$, and the safety invariant becomes $V_{low} \leq V_1 + V_2 \leq V_{high}$ (corresponding to $V_{low} \leq B_1 \cdot h_1 + B_2 \cdot h_2 \leq V_{high}$). Each volume V_i is bounded by $V_{i,max}$, and in order to have a coherent refinement we need to have $V_{max} \leq V_{1,max} + V_{2,max}$. The controller discrete abstract state x_s is glued (inv02) with the two discrete controllers states using the *guess_gs* operator.

Each tank is controlled by an independent controller. In a many-to-many pattern, a controller does not know exactly the state of the other controllers (i.e. what the other controllers are doing). However, the physics asserts that an estimation of this other state, and as such of the global state, can be built. To model this situation, two additional continuous variables, V_1^{sim} (resp. V_2^{sim}) are introduced. They allow the controller 2 (resp. 1) to simulate (i.e. estimate) V_1 (resp. V_2).

Because it is an estimation, V_i^{sim} is associated to a bound, Δ_i^{sim}, that represents the maximum error allowed for the controller to ensure a correct behaviour. We then need to have, at any time and for all i: $|V_i - V_i^{sim}| \leq \Delta_i^{sim}$, i.e.: V_i^{sim} is a *precise enough* approximation of V_i. Again, these properties are borrowed from the physics.

Transition and Sense. Controller 1 needs to enforce the (local) invariant $V_{low} + \Delta_2^{sim} \leq V_1 + V_2^{sim} \leq V_{high} - \Delta_2^{sim}$, and similarly for controller 2. This enforcement is used to prove the initially defined global invariant.

```
ctrl_sense_too_low_1 REFINES
        ctrl_sense_too_low
WHERE
    grd1:  V₁(t) + V₂ˢⁱᵐ(t) ≤ V_low + Δ₂ˢⁱᵐ
THEN
    act1:  x¹ₛ := Filling
END
```

Behave and Actuate. The system's actuation is established using continuous refinement as presented in [14]: the witness for e (abstract differential equation) is a predicate that links the solutions of e_1 and e_2 (concrete differential equations) such that the sum $V_1^* + V_2^*$ of any pair of solutions (V_1^*, V_2^*) of (e_1, e_2) is a solution of e, in addition to having the relevant general constraints (namely, a correct behaviour as per the system's current state). The witness for V' is given to establish the invariant after actuation.

```
ctrl_actuate_pumps REFINES ctrl_actuate_pumps
ANY   ss , e₁ , e₂ , ss₁ , ss₂ , V₁^{sim*} , V₂^{sim*} , t'
WHERE
   grd01−02:  ss ∈ STATES ∧ ss = guess_gs(ss₁ ↦ ss₂)
   grd11−13:  e₁ ∈ DE(S) ∧ Solvable([t, t'], e₁) ∧ FlowEq(ss₁, [t, t'], e₁)
   grd14−15:  ss₁ ∈ STATES ∧ x_s^1 = ss₁
   grd16:  V₁^{sim*} ∈ ℝ⁺ ↠ S ∧ [t, t'] ⊆ dom(V₁^{sim*}
   grd17:  ∀V₁* · V₁* ∈ ℝ⁺ ↠ S ∧ [t, t'] ⊆ dom(V₁*)∧
              solutionOf([t, t'], V₁*, e₁) ⇒ (∀t* · t* ∈ [t, t'] ⇒ |V₁*(t) − V₁^{sim*}(t)| ≤ Δ₁^{sim})
   grd21−27:  −− similar to grd11−17 with V₂
   grd30:  V_{low} < V₁(t) + V₂^{sim}(t) < V_{high}
   grd31:  V_{low} < V₁^{sim}(t) + V₂(t) < V_{high}
WITH
   V':  V' = V₁' + V₂'
   e :  e ∈ DE(S) ∧ Solvable([t, t'], e) ∧ FlowEq(guess_gs(ss₁ ↦ ss₂), [t, t'], e)∧
        (∀V₁*, V₂* · V₁* ∈ ℝ⁺ ↠ S ∧ [t, t'] ⊆ dom(V₁*) ∧ V₂* ∈ ℝ⁺ ↠ S ∧ [t, t'] ⊆ dom(V₂*)∧
        solutionOf([t, t'], V₁*, e₁) ∧ solutionOf([t, t'], V₂*, e₂)
        ⇒ solutionOf([t, t'], V₁* + V₂*, e))
THEN
   act1:  V₁, V₂, V₁^{sim}, V₂^{sim}:|_{t→t'}
        solutionOf([t, t'], V₁', e₁) ∧ solutionOf([t, t'], V₂', e₂)∧
        V₁^{sim'} = V₁^{sim*} ∧ V₂^{sim'} = V₂^{sim*}
        &{V_{low} < V₁(t) + V₂^{sim}(t) < V_{high} ∧ V_{low} < V₁^{sim}(t) + V₂(t) < V_{high}}
END
```

8   Assessment

The work presented in this paper showed that the generic model proposed in [15] applies to different architecture patterns of hybrid systems. Below, we provide an assessment of the approach with respect to the proof effort and set up methodology. The models presented in this paper have been developed on the Rodin platform and all the generated proof obligations were discharged.

Complete models can be found at https://irit.fr/~Guillaume.Dupont/models.php.

Proof Effort. The abstract tank model generated 107 proof obligations, most of which are invariant (about 40%) or well-definedness (about 21%) related. Well-definedness also appears often in proofs subgoals. These POs are usually easy to prove, at least on paper. Feasibility POs, related to solution existence, are those difficult to prove.

As for the *many-to-many* model, it yields 156 proof obligations, among which a good proportion (53%) consists of invariant POs alone. Again, most of them are not hard to discharge. The model also yields quite a few guard strengthening POs (around 15%) that ensure that the controllers behave properly despite the estimation it makes of the system. But the hardest POs to discharge are the one regarding refinement (witness well-definedness and feasibility, and simulation).

A great interest of the proposed methodology is there: the only complex proofs to carry on are related to *refinement*. Proofs for complicated invariants and so on have been realised at the *abstract* level and are done once and for all.

Tool Support. Because of our heavy use of the theory plug-in in Rodin, proof automation (including SMTs and external provers) is nearly nonexistent for discharging the generated POs. Proof is thus mostly interactive, and even simple steps such as basic well-definedness are to be done fully manually using the interactive prover. That being said, the possibility to define rewrite and inference rules greatly improves the prover's overall ergonomy.

Methodology. The use of patterns as methodological basis is not new in system engineering. The availability of architecture patterns offers a methodological guide to system designers, who simply need to identify which pattern matches the hybrid system under design and instantiate it with refinements and witnesses.

The generic model offers a framework that is formally proven once and for all. It corresponds to a *customisation* of Event-B to offer resources for modelling controllers, plants, sensing and actuation, integrating both discrete and continuous behaviours. Proofs are done once for all and the designer does not need to re-prove them. This generic model is used as a ground model for further designs.

Each defined architecture pattern is formalised as an instance of the generic model. The pattern to be chosen for instantiation depends on the number of controllers and plants required in the model. Instantiation is performed using Event-B refinement.

One of the interests of the Event-B method is the capability to check well-definedness and feasibility conditions, which is particularly useful during instantiation. In our developments, it has been extensively used to provide conditions about the soundness of the defined instantiations. For example, it has been used to state that the cylinders given as refinement are capable of storing an abstractly specified volume of liquid V_{max}.

9   Conclusion and Future Work

This paper presented a framework for modelling hybrid systems. It relies on a formal model of different hybrid systems architecture patterns formalised with the Event-B method using the Rodin platform. These patterns, commonly used when designing hybrid systems, are characterised by the number of controlled plants and by the kind of control strategy (centralised or distributed). Because this framework is formalised at a generic level, it offers a systematic methodology for hybrid systems development and verification.

The approach extensively uses the mathematical extensions capabilities offered by the theory plug-in of Event-B, allowing to enrich Event-B models with continuous behaviours. Data types for reals, continuous functions, differential equations and so on have been defined within a sound Event-B theory. The available axioms and theorems were used to prove the relevant safety properties of the developed systems expressed as machine invariants. The developed models are scalable (modulo proof efforts), as they can deal an arbitrary number of state variables. Witnesses for the sets $STATES$ and S are provided at instantiation using gluing invariants.

This work revealed several research perspectives. Below, we summarise the identified future research actions.

Need for Other Domain Theories. Although the definition of generic architecture patterns has reduced the number and the complexity of proof obligations and their proofs, the proof effort still needs to be reduced. Providing other sound domain theories contributes to such a reduction. One of the main extensions to our work consists in enriching the proposed framework with other theories. Two kinds of theories are expected: theories for other types of control and theories where the physics of considered plants is formalised. A library of such theories would help for such hybrid systems developments by making explicit knowledge in physics and in other related domains [3].

Methodology. From the method formalisation point of view, the major improvement is to leverage the formalisation of architecture patterns at a higher abstraction level to handle controllers and plants as first class mathematical objects.

Other patterns where the number of hybrid systems evolves dynamically could be considered. In this case, each system would have a partial knowledge of its environment. This kind of patterns may help to model autonomous aspects. However, defining safety properties remains a major challenge for such patterns.

Integration of Simulation Tools. To handle the traditional hybrid systems development processes where simulation is extensively used, coupling the developed models with simulation tools, like in [1], would help in animating these models.

Acknowledgment. We thank T. S. Hoang for his help with Rodin's Theory plug-in and R. Banach for the helpful discussions related to Event-B hybridation.

References

1. Project INTO-CPS: Integrated Tool Chain for Model-based Design of Cyber-Physical Systems. http://into-cps.au.dk/about-into-cps
2. Abrial, J.R.: Modeling in Event-B: System and Software Engineering. Cambridge University Press, Cambridge (2010)
3. Aït-Ameur, Y., Méry, D.: Making explicit domain knowledge in formal system development. Sci. Comput. Program. **121**(C), 100–127 (2016)
4. Alur, R., Courcoubetis, C., Henzinger, T.A., Ho, P.-H.: Hybrid automata: an algorithmic approach to the specification and verification of hybrid systems. In: Grossman, R.L., Nerode, A., Ravn, A.P., Rischel, H. (eds.) HS 1991-1992. LNCS, vol. 736, pp. 209–229. Springer, Heidelberg (1993). https://doi.org/10.1007/3-540-57318-6_30
5. Alur, R., Henzinger, T.A.: Modularity for timed and hybrid systems. In: Mazurkiewicz, A., Winkowski, J. (eds.) CONCUR 1997. LNCS, vol. 1243, pp. 74–88. Springer, Heidelberg (1997). https://doi.org/10.1007/3-540-63141-0_6
6. Babin, G., Aït-Ameur, Y., Singh, N.K., Pantel, M.: A system substitution mechanism for hybrid systems in Event-B. In: Ogata, K., Lawford, M., Liu, S. (eds.) ICFEM 2016. LNCS, vol. 10009, pp. 106–121. Springer, Cham (2016). https://doi.org/10.1007/978-3-319-47846-3_8
7. Back, R.J., Petre, L., Porres, I.: Continuous action systems as a model for hybrid systems. Nord. J. Comput. **8**(1), 2–21 (2001)
8. Banach, R., Butler, M., Qin, S., Verma, N., Zhu, H.: Core hybrid Event-B I: single hybrid Event-B machines. Sci. Comput. Program. **105**, 92–123 (2015)
9. Boldo, S., Clément, F., Filliâtre, J.C., Mayero, M., Melquiond, G., Weis, P.: Trusting computations: a mechanized proof from partial differential equations to actual program. Comput. Math. Appl. **68**(3), 325–352 (2014)
10. Butler, M., Abrial, J.R., Banach, R.: Modelling and refining hybrid systems in Event-B and Rodin. In: From Action Systems to Distributed Systems: The Refinement Approach. Computer and Information Science Series, pp. 29–42. Chapman and Hall/CRC (2016)

11. Butler, M., Maamria, I.: Practical theory extension in Event-B. In: Liu, Z., Woodcock, J., Zhu, H. (eds.) Theories of Programming and Formal Methods. LNCS, vol. 8051, pp. 67–81. Springer, Heidelberg (2013). https://doi.org/10.1007/978-3-642-39698-4_5

12. Cardenas, A.A., Amin, S., Sastry, S.: Secure control: towards survivable cyber-physical systems. System **1**(a2), a3 (2008)

13. Dupont, G., Aït-Ameur, Y., Pantel, M., Singh, N.K.: Hybrid systems and Event-B: a formal approach to signalised left-turn assist. In: Abdelwahed, E., et al. (eds.) MEDI 2018. CCIS, vol. 929, pp. 153–158. Springer, Cham (2018). https://doi.org/10.1007/978-3-030-02852-7_14

14. Dupont, G., Aït-Ameur, Y., Pantel, M., Singh, N.K.: Handling refinement of continuous behaviors: a refinement and proof based approach with Event-B. In: 13th International Symposium TASE, pp. 9–16. IEEE Computer Society Press (2019)

15. Dupont, G., Aït-Ameur, Y., Pantel, M., Singh, N.K.: Proof-based approach to hybrid systems development: dynamic logic and Event-B. In: Butler, M., Raschke, A., Hoang, T.S., Reichl, K. (eds.) ABZ 2018. LNCS, vol. 10817, pp. 155–170. Springer, Cham (2018). https://doi.org/10.1007/978-3-319-91271-4_11

16. Henzinger, T.A.: The theory of hybrid automata. In: Proceedings of 11th Annual IEEE Symposium on Logic in Computer Science, LICS, pp. 278–292. IEEE Computer Society (1996)

17. Hoare, C.A.R.: Communicating Sequential Processes. Prentice-Hall, Upper Saddle River (1985)

18. Jifeng, H.: From CSP to hybrid systems. In: Roscoe, A.W. (ed.) A Classical Mind, pp. 171–189. Prentice Hall International (UK) Ltd., Upper Saddle River (1994)

19. Lee, E.A., Seshia, S.A.: Introduction to Embedded Systems - A Cyber-Physical Systems Approach, 1.5 edn. LeeSeshia.org (2014). http://leeseshia.org/

20. Meinicke, L., Hayes, I.J.: Continuous action system refinement. In: Uustalu, T. (ed.) MPC 2006. LNCS, vol. 4014, pp. 316–337. Springer, Heidelberg (2006). https://doi.org/10.1007/11783596_19

21. Logical Foundations of Cyber-Physical Systems. Lecture Notes in Computer Science. Springer, Cham (2018). https://doi.org/10.1007/978-3-319-63588-0_21

22. Singh, N.K., Aït-Ameur, Y., Pantel, M., Dieumegard, A., Jenn, E.: Stepwise formal modeling and verification of self-adaptive systems with Event-B. The automatic rover protection case study. In: 21st International Conference on Engineering of Complex Computer Systems, ICECCS 2016, pp. 43–52 (2016)

23. Su, W., Abrial, J.R., Zhu, H.: Formalizing hybrid systems with Event-B and the Rodin platform. Sci. Comput. Program. **94**(Part 2), 164–202 (2014). abstract State Machines, Alloy, B, VDM, and Z

Integration of iUML-B and UPPAAL Timed Automata for Development of Real-Time Systems with Concurrent Processes

Fatima Shokri-Manninen[1]([⊠]), Leonidas Tsiopoulos[1,2], Jüri Vain[2]([⊠]), and Marina Waldén[1]

[1] Åbo Akademi University, Turku, Finland
{fatemeh.shokri,marina.walden}@abo.fi
[2] Tallinn University of Technology, Tallinn, Estonia
{leonidas.tsiopoulos,juri.vain}@taltech.ee

Abstract. Developing safety-critical systems requires to consider safety and real-time requirements in addition to functional requirements. Event-B is a formalism that is visualised by iUML-B and supports the development of functional aspects having rich verification and validation tools. However, it lacks well-established support for timing analysis. UPPAAL Timed Automata (UTA), on the other hand, address timing aspects of systems, and enable model checking reachability and timing properties. By integrating iUML-B and UTA, we combine the best verifying and validating practices from the two methods achieving a formal development of systems. We present the mapping for translating iUML-B constructs to UTA. The novel aspect is the use of a multi-process trigger-response pattern to address the modelling and verification of reachability properties of complex systems with concurrent processes. The approach is demonstrated on an airport control system, where timing, fairness, as well as liveness properties play a vital role in proving safety requirements.

Keywords: Verification · Model checking · Timed automata · Event-B · iUML-B · UPPAAL · Real-time systems · Trigger-response patterns

1 Introduction

Correct-by-Construction Design (CCD) [1] plays an important role in the development of safety critical-systems, since it guarantees their reliability and correctness with respect to the system requirements. This is vital in cases where human safety and large financial assets are at stake. Correctness by construction is gained by the use of formal methods, which are mathematical methods for deriving a system based on its requirements. The main reasons for applying formal modelling is to avoid ambiguity or misunderstanding of system requirements and to detect problems early in system development.

A. Raschke et al. (Eds.): ABZ 2020, LNCS 12071, pp. 186–202, 2020.
https://doi.org/10.1007/978-3-030-48077-6_13

One of the formal methods which supports CCD in the development process is Event-B [1]. Event-B is based on set theory and supports design by stepwise refinement. Event-B tool RODIN with its plug-ins provides a rich support for this CCD. However, in spite of these beneficial aspects, Event-B lacks sufficient support for timing analysis and refinement of timed specifications. UTA [3] address timing aspects of systems providing efficient data structures and algorithms for their representation and verification, but are less focusing on supporting the refinement-based development and verification.

The goal of this paper is to advocate a model-based design method, where Event-B and UTA are combined to mutually complement each other. The motivation for selecting Event-B as the base formalism is that Event-B provides support for verifying infinite-sized models with advanced data structures using first-order logic. Additionally, it allows for correct-by-construction development via stepwise refinement. For mapping, we opt for UTA as it supports verification of real-time properties which are required before the implementation phase of the model. The design method consists of stepwise refining the system using Event-B for proving the functional and safety properties in each step. Each development step is then translated to a UTA model that can be validated via model checking and checked for real-time properties without re-checking non-timing related properties.

For the work in this paper we extend the earlier work on the Event-B to UTA mapping by introducing an intermediate representation in iUML-B for the generation of the control structure that serves as skeleton also for UTA. In particular, we investigate how the integrated approach addresses the development of complex real-time trigger-response pattern-based systems with concurrent processes and discuss the benefits of the integrated method in contrast to that when only one of the two formalisms is applied for the development and verification of safety critical Cyber-Physical Systems (CPS) [12,13].

2   Related Work

Formal development of CPSs requires continuous timing properties to capture real behaviour of these systems. As discrete timing constraints cannot describe some of the substantial dynamic properties like Zeno behaviour, essential discontinuities, and other singularities of real-time systems in the physical environment, continuous time constraints become the inseparable part of modelling CPS.

Event-B is a formal method for modelling a safety critical-system that originally lacks the notion of real-time. Recent attempts [5,8,10] have been made to integrate discrete time to Event-B using patterns, such as *delay, expriry, deadline* and *interval*. Since invariants on discrete time introduce noise to the provers [5], it easily leads to cases that are difficult to prove. All these timing properties contain a trigger and response pattern, which are modelled as events in Event-B. To capture all these timing properties we focus on their underlying trigger-response pattern, where a trigger must always be followed by a response.

Due to the lack of concept of time continuity, the above discrete time properties cannot always be applied to CPSs to embrace their continuous behaviour.

Zhu et al. addressed this problem by extension of deadline constraints [12]. Moreover, authors defined discrete task- and scheduler-based timing properties of each process and of concurrent tasks between processes, respectively. They refined task-based timing into scheduler-based timing by either a FIFO queue scheduling policy or a deferrable priority-based scheduling policy with aging. For addressing intermediate events between trigger and response events, they propose in [13] the conditional convergent notion. In this approach, intermediate events can converge if there is no response event enabled, assuming weak-fairness of intermediate events and eventual execution of the response event. While [13] addresses a single-process trigger-response pattern, we can model multiple trigger-response relations in UTA by applying our integrated approach. It allows verifying that the interleavings between concurrent processes do not cause deadlock while proving reachability, liveness and non-Zenoness properties in the model.

Compared to earlier research on combining Event-B and UTA [4, 11], we extend the mapping by considering sequencing of Event-B events in iUML-B diagrams, and control structures representing trigger-response patterns. Specifically, the iUML-B graphical design provides us with an untimed control structure identical to that of UTA. This is further elaborated by incorporating timing analysis at each refinement/design step. The straight-forward mapping proposed in [4] leads to too large models. In a later approach [11] an event-level mapping was introduced where each event from the Event-B specification was translated to an UTA and then parallely composed to form the full model. Additional optimisations were needed to aggregate the automata which model mutually exclusive events, and thus, reduce interleaving of model events.

In this paper the mapping is still based on the events but the UTA model structure is extracted from the iUML-B state diagram. By decorating the extracted control structure with UTA specific attributes we can verify the system's timing correctness and provide feedback to the Event-B side of the development for the feasibility of system events. For capturing the behaviour of processes based on the trigger-response pattern, we show in the following sections how concurrency with multi-process intermediate events is modeled and verified using an airport control system as a case study.

3   Preliminaries

3.1   Event-B and iUML-B

Event-B [1] is a state-based formalism for the development of reactive and distributed systems. Event-B uses refinement [2], which enables the system to be created in a stepwise manner gradually adding details into the model proving that each refinement step preserves the correctness of the previous steps. A model in Event-B, a machine, can be interpreted as a transition system where the variable valuations constitute the states and the events represent the transitions. Machines can be refined either via *superposition refinement* [9], where new features are added to the machine, or by *data refinement*, where abstract features are replaced by more concrete ones. Event-B is well supported by the Rodin Platform [6], which is extendable with plugins facilitating the modelling and verification.

iUML-B is an integrated form of the classical UML-B graphical front-end for Event-B [7] that is an extension of the Rodin Platform. It allows modellers to build a model through a diagrammatic design in the form of state-machines and class diagrams. The translator then generates Event-B automatically facilitating the modelling process. Class diagrams provide a way to model data relationships, while state-machines show the states and transitions of an Event-B machine. The guards and actions of the Event-B events form the guards and actions of the transitions in the state-machine diagrams. The operational semantics of the events are, hence, visualised with the state-machines.

In a state-machine with an transition e_1 between states S_1 and S_2, transition e_1 can be fired if the state is S_1 and the guard of the transition $G(t, v)$ evaluates to *true*. When e_1 is fired it changes the state to S_2 and may also modify other variables of the state-machine via actions $S(t, v)$. This corresponds to event e_1 in Event-B:

$$e_1 = \textbf{any } t \textbf{ where } state = S_1 \wedge G(t, v) \textbf{ then } state := S_2 \, || \, S(t, v) \textbf{ end}$$

Invariants may also be given in the states. They correspond to invariants in an Event-B machine. The state-machines can be refined in a corresponding manner to the Event-B machines concerning variables and events. Additionally, states can be nested in state-machines (i.e. states in a state), which is also often used when refining a system to model the increased level of detail in the states.

3.2   UPPAAL Timed Automata

UTA [3] are defined as a closed network of extended timed automata that are called processes. The processes are combined into a single system by synchronous parallel composition like that in process algebra CCS. The nodes of the automata graph are called locations and directed lines between locations are called edges. For each edge, which is a transition between two locations, conditions or guards can be defined. Whenever the guard holds, the edge can be fired, which leads to a new location. Communication and synchronisation between different automata is taken care of by *send* and *receive* actions. An action *send* over a channel h is denoted by $h!$ and its co-action, *receive* is denoted by $h?$.

Formally, an UTA is defined as the tuple $(L, E, V, CL, Init, Inv, T_L)$, where:

- L is a finite set of locations,
- E is the set of edges defined by $E \subseteq L \times G(CL, V) \times Sync \times Act \times L$, where
  - $G(CL, V)$ is the set of constraints in guards,
  - $Sync$ is a set of synchronisation actions over channels and
  - Act is a set of sequences of assignment actions with integer and boolean expressions as well as with clock resets.

- V denotes the set of integer and boolean variables,
- CL denotes the set of real-valued clocks ($CL \cap V = \varnothing$),
- $Init \subseteq Act$ is a set of assignments that assigns the initial values to variables and clocks,
- $Inv : L \to I(CL, V)$ is a function that assigns an invariant to each location, $I(CL, V)$ being the set of invariants over clocks CL and variables V and
- $T_L : L \to \{ordinary, urgent, committed\}$ is the function that assigns the type to each location of the automaton.

In *urgent* locations an outgoing edge will be executed immediately when its guard holds. *Committed* locations are useful for creating atomic sequences of process actions since an outgoing edge must be executed immediatelly without time passing.

UTA Requirement Specification Language. The requirement specification language (in short, query language) of UTA, used to specify properties to be model checked, is a subset of Timed Computation Tree Logic (TCTL) [3]. The query language consists of path formulae and state formulae. State formulae describe individual states, whereas path formulae quantify over paths or traces of the model and can be classified into *reachability*, *safety* and *liveness* [3]. For example, safety properties are specified with path formula A□φ stating that state formula φ should be true in all reachable states. In the next section we describe in more detail the TCTL formulae we apply in the rest of this paper.

4   Mapping from Event-B and iUML-B Models to UTA

We base our work here on the previous work by Vain et al. [11]. We assume that the system is developed stepwise using Event-B and iUML-B and prove the safety properties in each step using the proof system of this formalism. The result of each development step is then translated to UTA in order to have a model that can be validated via model checking and specifically checked for real-time properties avoiding re-checking of functional/safety properties. Note that due to the locality of refinements only those model fragments that are introduced by Event-B refinements need to be mapped to the corresponding UTA fragments. The rest of the UTA model defined in earlier steps remains untouched by the current Event-B refinement step.

Plant and Controller. In Event-B and iUML-B the model represents a holistic view to the control systems where the controller and the plant events are all given in one machine. However, when mapping the model to UTA these different kinds of events have to be identified. The *plant events* in the iUML-B state-machines are mapped to UTA plant automaton with corresponding states and transitions. This leads to a sequential model of the control system in UPPAAL. The states and state transitions of the state-machines are given in Event-B as global, but when mapped to UTA the states and transitions are partitioned by automata that introduces modularity to models and to verification. The *controller events* are each translated as in [11] to simple self-loop automata. All these events emulating self-loop transitions are composed in parallel. The communication with the plant takes place via channels by trigger and response actions.

In control systems, there might be several plants (processes) for a controller. In Event-B the setup of the system is given in the context machine. Only one state-machine is created for the system, but the plants/processes are specified as instances of the machine. When mapping this scenario to UPPAAL, one UTA template is created for the process and instantiated for the multiple processes.

Mapping of Functions and Predicates. Variables of integer and enumerated types in Event-B become integers in UTA, while finite sets and relations in Event-B are mapped to (multidimensional) arrays in UTA. We can then implement the set and relational operators as C-functions in UTA.

Mapping of Events. Transitions in iUML-B are generally translated to state transitions in UTA [11]. In Fig. 1 we exemplify the translation with an iUML-B state machine and Event-B code to the left and a corresponding UTA model to the right. Let

$$e = \textbf{any } p \textbf{ where } G(p, v) \textbf{ then } S(p, v) \textbf{ end}$$

be an event of Event-B, then

(i) the parameter p will appear in the select label of the UTA edge, which contains a comma separated list of **p : int** expressions where p is a variable name and *int* is a defined type (see Fig. 1).

(ii) the event guard $G(p,v)$ is mapped to the guard $G(V)$ of an edge where V denotes UTA variables corresponding to variables v (**p > 5** in Fig. 1).

(iii) the event action $S(p,v)$ corresponds to assignment statements (updates) $V' = S(V)$ of the UTA edge (**num := num + p** in Fig. 1).

For plants consisting of many processes, the instance of a plant is identified by a unique parameter value. The template may have parameters of type integer. This allows modelling the ANY-construct of Event-B, where the choice is *finite*. The parameter of a template specified by its type defines the instances (processes) of the template, one for each value in the parameter type.

Timing of Events. When mapping the Event-B model to UTA, we need to add timing explicitly to the model to be able to consider timing aspects like time-bounded reachability. When adding explicit timing constraints to UTA events, it is assumed that the occurrence of an event is instantaneous. An event may occur within some time interval $[lb, ub]$, where lb stands for lower bound and ub stands for upper bound, provided it is enabled by guard G. For specifying these constraints, new variables, namely the set CL of clocks is introduced. We usually assume the continuous intervals are of shape $[lb, ub]$, where lb and $ub \geq 0$, and $ub \geq lb$. Note that having infimum inf of clocks domains $inf\, dom(cl) = 0$ and guards $cl \geq inf$ it may introduce Zeno computations if there exists a loop in the model where the maximum of lower bounds of occurrence interval of each edge is equal to the infimum.

In general, the timing specification of events introduces bounded intervals of occurence that are specified as location invariant $\mathrm{inv}(CL) \equiv \wedge_i cl_i \leq ub_i$ and the guard $G_i(CL) \equiv \wedge_j cl_j \geq lb_j$ of its self-loop edge e_i that models an event. A set of clock conditions $G_i(CL)$ and $\mathrm{inv}(CL)$ indicate time constraints when an event e_i should be fired, i.e. not later than time ub_i (deadline) and not before time lb_i (delay) (see $cl <= ub$ and $cl >= lb$ in the UTA in Fig. 1).

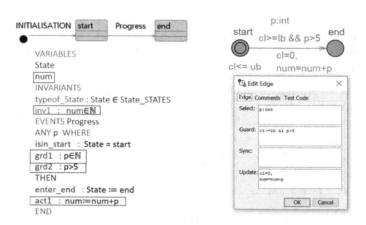

Fig. 1. Transition structure in iUML-B (left) and UTA (right)

Undelayed Reaction. In the context of multiple trigger-response patterns, some response should be fired immediately without delay in a critical situation. It brings the concept of priority based on timing. According to the timing constraint proposed in [8], the delay constraint is specified by $Delay(Trigger, Response, delay)$ that specifies the delay constraint where $delay = 0$. In UPPAAL, undelayed response is modelled with *urgent channel* (urgent chan) which is defined to be synchronizing executions of enabled edges without delay. Clock conditions on these transitions are not allowed. An alternative to model undelayed reaction encoded as a single (not synchronized) edge is to define its source location type as either *committed* or *urgent* or set this location invariant condition upper bound to 0.

Invariants. The invariants of Event-B are not directly translated into UPPAAL model-checking queries. However, these invariants can be specified and model checked as TCTL formulas of form $A\Box p$, where p is a first order state formula and the pair of modalities $A\Box$ requires p to be true in all reachable states of the model in UPPAAL. Formula p can involve predicates on model clocks to specify explicit timing constraints.

Time Bounded Reachability. For real-time applications, we consider time bounded reachability as one of the most fundamental properties. In UTA, the reachability of an event E (where E is specified in terms of after state of the event and/or valuation of state variables) from model initial state is expressible using TCTL formula pattern $A\Diamond E$ $\&\&$ $Clock \leq$ TB, for time bound TB.

Trigger-response properties are expressed as a special case of time bounded reachability where the reachability of a response event is always considered relative to its trigger events. Proving multiple trigger-response properties in Event-B presumes augmentation of the model with auxiliary "boolean property variables" which are set to true only when considered triggers receive a response (as $Landing\_permission(selfP) := TRUE$ in Fig. 5). Conjoining multiple trigger-response pairs of different processes is non-trivial and can easily cause misinterpretation. For instance, reactions to stimuli of different processes may occur in different states or even be mutually exclusive in some states. Therefore, the trigger-response properties for multiple concurrent processes need to be specified and checked separately.

The reachability of a response event (actually its post condition rp) from a trigger event tr (respectively its post condition tr) is then expressed in UTA and TCTL using *leads to* operator as $tr \dashrightarrow rp$. The multiple trigger-response properties can be specified and proved similarly. For instance, TCTL formula $tr_1\&\&...\&\&tr_n \dashrightarrow rp_1\&\&...rp_m$ where auxilliary boolean variables $tr_1,...,tr_n$ in the model register the occurrence of trigger 1 to trigger n and auxilliary variables $rp_1,...,rp_m$ register the occurrence of response events 1 to m.

In case of time bounded reachability of a response event, a property clock constraint should be conjoined to the right hand side of *leads to*. Here it should be granted that the property clock is reset in the model at the moment when the trigger (conjunction $tr_1\&\&...\&\&tr_n$) of considered trigger-response pair is set to true.

Liveness. In Event-B due to weak fairness, enabled processes will eventually be executed. If the system is deadlock free, there is always an event that can be executed. In UTA, the situation, where a transition is enabled but there is no finite interval specified in the location invariant (or not using location types *urgent*, *committed*), may result in an infinite waiting in that location. This provides behaviourally similar effect as deadlock. It means that regardless if one or more of the outgoing edges of that location are enabled, none of them will ever be executed because there is no upper bound that forces the edge to be executed in finite time. In that way, weak fairness is not sufficient to guarantee non-blocking in UTA and the progress must be granted by specifying the location type either *committed* or *urgent* or adding a time bound conjunct to each location invariant. Liveness can be proved by TCTL query $A\Box$ *not deadlock* provided all legal terminal locations are supplied with self-loop edges.

5   Overview of Case Study

In order to demonstrate the CCD methodology with integrated formal methods, we use an airport control system example. We propose this case study for presenting the verification of behavioral and timing properties by combining two complementing formalisms Event-B and UTA. We focus only on the landing control with one runway. Based on system requirements, we have two types of landing, namely, emergency landing and normal landing. The flight control is in charge of safe landing by giving airplanes permission to land at appropriate times. For normal landing planes may queue up to enter the landing runway. There are two queues with different priorities based on the planes' fuel level. An emergency landing has higher priority than both queues. In case of an emergency landing, no other plane can land and all landing requests are rejected. Only one emergency landing can take place at a time. As a safety requirement it is ensured that there is no plane in the runway before allowing another plane to use it.

The model of the airport system[1] consists of one abstract model with three refinement steps. The abstract model presents the general view of the airport system, with two landing modes. In the first refinement step, we introduce two queues with different priorities for normal landing. Next, we implement the FIFO policy of the queues. Finally, we introduce fuel level for each plane.

5.1 M0: An Abstract Model of Airport Control System

In the abstract model the behaviour of the system is depicted in the a state-machine diagram of iUML-B (in Fig. 2) where the different states of a plane to reach the final state (At_Gate) are modeled. In Event-B, we create a context that introduces the set $PLANES$ in addition to the generated implicit context which consists of the states in the state-machine. The state of a plane and the transitions between the states are generated automatically from the state-machine. We define a parameter $self\,P$ in the iUML-B diagram to represent the instances of $PLANES$, which is translated into corresponding Event-B events.

In the abstract Airport Control System in Fig. 2, a plane in state In_Air sends a landing request to the controller. If the response is positive, the plane can enter the landing queue. We define an invariant ($LQ\_Permission = TRUE$) in the $Landing\_Queue$ state which ensures that each plane in the landing queue has a landing queue permission.

The controller checks whether there is an ongoing emergency case or not via the boolean variable $Emergency\_Prog$. In case of an ongoing emergency, no new plane will get permission to land and will have to leave the airport.

We ensure safety on the critical section, the runway, using mutual exclusion with boolean variable $Runway\_Busy$ to ensure that no landing permission is admitted if there is a plane on the runway. When the plane is on the runway it will be given a gate by the controller and it moves to the final state At_Gate.

Emergency request can be sent in states $Landing\_Queue$ or In_Air. Only one emergency at a time can be handled at the airport. If there is already an emergency, emergency landing permission for that plane is rejected, and the plane has to leave the airport.

In order to be able to introduce timing properties to the airport system, we map the Event-B and iUML-B model into an UTA. In Event-B, the model of

---

[1] iUML-B and UTA models are found in: https://github.com/fshokri/FormalModels.git.

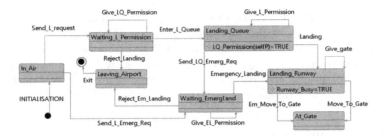

Fig. 2. The abstract model in iUML-B

the plane and the controller are integrated. However to model different timing behaviors of components in UPPAAL, we need to divide the model into plane and controller (Fig. 3) where simultaneous events are synchronised via channels.

For the plane template, we follow the same structure as in the iUML-B state-machine diagram, while for the controller we only consider events for giving permissions. The instantiation of the plane template for modelling the plane instances is modelled with template parameter *id*, while the handling of each plane by the controller is addressed with the *select* clause in UTA corresponding to the ANY event parameter in Event-B. This is described in Mapping of events in Sect. 4.

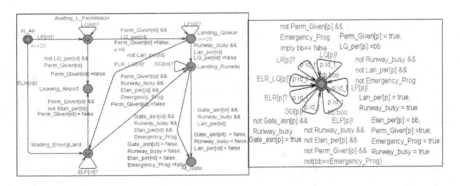

Fig. 3. The abstract model of plane (left) and controller (right) in UTA

We define clock constraints with upper bounds stated with invariants for locations waiting before triggering requests. We assign locations as *urgent* when waiting for responses from the controller. This way planes can progress immediately when a response is given. For the *Landing_Queue* location an upper bound is given allowing flexibility when later refining the behaviour for planes in the landing queues i.e., allowing time passing when queueing.

Comments on the Modelling. In Event-B, the controller in the abstract system gives landing queue entry permissions non-deterministically. Since there is no implicit timing in Event-B, this does not create a deadlock. However, in order to avoid deadlock in the corresponding state $Waiting\_L\_Permission$ of the UTA model, we define a variable ($Perm\_Given$) to indicate that permission has been given to make exit conditions from this state more deterministic. In this way each plane gets an answer for the landing request, either a positive one to move to the landing queue or a negative one to leave the airport.

5.2   M1: Introducing Two Queues

In the first refinement, we split the state $Landing\_Queue$ into two queues states, $High\_Priority\_Queue$ and $Low\_Priority\_Queue$. This is done by adding nested state-machines in iUML-B, while in UTA two separate states and transitions are created (Fig. 4).

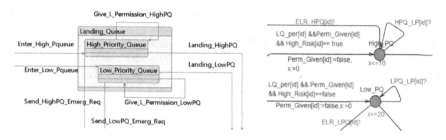

Fig. 4. The first level of refinement model excerpt in iUML-B (left) and UTA (right)

In M1, we introduce a new boolean variable $High\_Risk$ which states whether a plane has a high risk or not in the queueing situation. The plane with a high risk is eligible to move to $High\_Priority\_Queue$ with shorter waiting time, while in normal situation planes enter $Low\_Priority\_Queue$. If $High\_Priority\_Queue$ is not empty, a plane in $Low\_Priority\_Queue$ needs to wait for a number of planes (here at most three) in $High\_Priority\_Queue$ to land.

The nested state with queues in Event-B is translated to a separate state for each queue in UTA. The guards and actions of the events in Event-B are translated in a straightforward manner to guards and updates in the UTA model.

5.3   M2: Implementing FIFO Method for Each Queue

In the second refinement step, we implement the FIFO policy for each queue. Since the functionality of two queues is the same, we focus on the high priority queue in the Event-B model (Fig. 5).

The queues have positions (@inv1) and are of limited length (@inv4). Via event $Enter\_High\_Pqueue(selfP)$ plane $selfP$ can enter the high priority queue $High\_Pqueue$ provided that it is a high-risk plane that has been given landing permission and that the high priority queue is not full. The plane $selfP$ will be

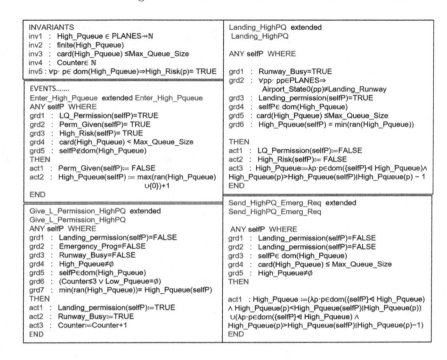

INVARIANTS inv1 : High_Pqueue ∈ PLANES↠ℕ inv2 : finite(High_Pqueue) inv3 : card(High_Pqueue) ≤Max_Queue_Size inv4 : Counter∈ ℕ inv5 : ∀p· p∈ dom(High_Pqueue)⇒High_Risk(p)= TRUE	Landing_HighPQ extended Landing_HighPQ ANY selfP WHERE grd1 : Runway_Busy=TRUE grd2 : ∀pp· pp∈PLANES⇒ 　　　　 Airport_State0(pp)≠Landing_Runway grd3 : Landing_permission(selfP)=TRUE grd4 : selfP∈ dom(High_Pqueue) grd5 : card(High_Pqueue) ≤Max_Queue_Size grd6 : High_Pqueue(selfP) = min(ran(High_Pqueue))
EVENTS....... Enter_High_Pqueue extended Enter_High_Pqueue ANY selfP WHERE grd1 : LQ_Permission(selfP)=TRUE grd2 : Perm_Given(selfP)= TRUE grd3 : High_Risk(selfP)= TRUE grd4 : card(High_Pqueue) < Max_Queue_Size grd5 : selfP∉dom(High_Pqueue) THEN act1 : Perm_Given(selfP):= FALSE act2 : High_Pqueue(selfP) := max(ran(High_Pqueue) 　　　　　　　　　　　　∪{0})+1 END	THEN act1 : LQ_Permission(selfP):=FALSE act2 : High_Risk(selfP):= FALSE act3 : High_Pqueue:=(λp·p∈dom({selfP}◁ High_Pqueue)∧ High_Pqueue(p)>High_Pqueue(selfP)|High_Pqueue(p) − 1 END
Give_L_Permission_HighPQ extended Give_L_Permission_HighPQ ANY selfP WHERE grd1 : Landing_permission(selfP)=FALSE grd2 : Emergency_Prog=FALSE grd3 : Runway_Busy=FALSE grd4 : High_Pqueue≠∅ grd5 : selfP∈dom(High_Pqueue) grd6 : (Counter≤3 ∨ Low_Pqueue=∅) grd7 : min(ran(High_Pqueue))= High_Pqueue(selfP) THEN act1 : Landing_permission(selfP):=TRUE act2 : Runway_Busy:=TRUE act3 : Counter:=Counter+1 END	Send_HighPQ_Emerg_Req extended Send_HighPQ_Emerg_Req ANY selfP WHERE grd1 : Landing_permission(selfP)=FALSE grd2 : Landing_permission(selfP)=FALSE grd3 : selfP∈ dom(High_Pqueue) grd4 : card(High_Pqueue) ≤ Max_Queue_Size grd5 : High_Pqueue≠∅ THEN act1 : High_Pqueue :=(λp·p∈dom({selfP} ◁ High_Pqueue) ∧ High_Pqueue(p)<High_Pqueue(selfP)|High_Pqueue(p)) ∪(λp·p∈dom({selfP}◁ High_Pqueue) ∧ High_Pqueue(p)>High_Pqueue(selfP)|High_Pqueue(p)−1) END

Fig. 5. The Event-B code for the second refinement with FIFO queue

inserted in the first free position of the queue which is position one if the queue is empty. In event *Give_L_Permission_HighPQ*, landing permission is given to the first plane in the queue provided that less than three planes from the high priority queue have landed in a row or low priority queue is empty. Event *Landing_High_Pqueue(selfP)* models plane *selfP* leaving the high priority queue and entering the landing runway. As a result of landing, queueing planes are shifted in the queue. If there is an emergency situation while the plane is in the queue the plane will leave the queue (event *Send_HighPQ_Emerg_Req*).

In the UTA model, we use functions for enqueuing and dequeuing, for a smooth implementation of the FIFO queue corresponding to the lambda expressions in Event-B. For example, action *act*1 of event *Send_HighPQ_Emerg_Req* in Fig. 5 is mapped to C-like functions in UTA as in Fig. 6. The left one (*Em_deHPqu_idx*) appears in the guard and the right one (*Em_deHPqueue*) in the update of the transition from location *High_PQ* to *Waiting_EmergLand* in Fig. 8.

5.4   M3: Introducing Fuel Consumption

In the third refinement step, we introduce variables *Plane_Fuel* and *Fuel_count* in our Event-B model to indicate fuel consumption. Variable *Plane_Fuel* gives the fuel level (*High*, *Medium* and *Low*) for each plane, while *Fuel_count* is a

//Finding position of plane in queue	//Removing the plane based on its position
id_t Em_deHPqu_idx(id_t plane) {	void Em_deHPqueue(id_t i){
int i=0;	--lenH;
while (HPqueue[i] != plane)	while (i < lenH) {
{ i++;}	HPqueue[i] = HPqueue[i + 1];
return i;}	i++;} HPqueue[i] = 0;}

Fig. 6. UTA C-like functions for the second refinement with FIFO queues

variant of type natural number to show that the superposed fuel consumption will not take over the behaviour of the system.

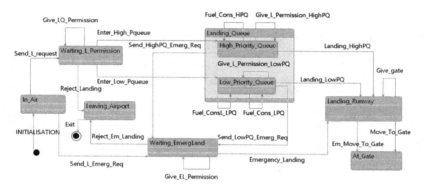

Fig. 7. The last level of refinement model in iUML-B

A plane with fuel level *Medium* enters *High_Pqueue*, while a plane with fuel level *High* is eligible for *Low_Pqueue*. Events *Fuel_Cons_HPQ*, *Fuel_Cons_LPQ* and *Fuel_ConsL_LPQ* will decrease plane fuel while waiting for permission to enter the landing runway (see Fig. 7). If the fuel level drops to *Low*, the plane will send an emergency request. The plane with the emergency request will reach the *At_Gate* location if there is no other emergency situation progressing.

The iUML-B state-machine is directly mapped to a UTA (see Fig. 8). However in the UTA model, the variable *Plane_Fuel* is mapped to a variable of type enumerated set which assigns a numerical value for the fuel level of each plane. The fuel consumption events are the events of the plane which are considered as intermediate events. The execution of these events depends on the delayed response from the controller for assigning landing permission. For avoiding delays on transitions triggering emergency cases, we defined the *ELP, ELR_HPQ, ELR_LPQ* channels in UTA to be *urgent*.

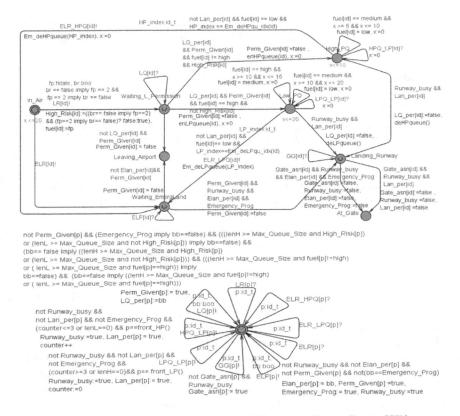

Fig. 8. The last level of refinement of Plane and Controller in UTA

5.5   Analysis Results

Proof Statistics: Machines M0 (abstract specification) and M1 (first refinement) in Event-B are automatically proved almost (100%) by the Rodin tools. For machine M2, where we introduce the FIFO mechanism for the queues, more complex proof obligations were generated of which 57% were automatically proved. In M3, where we define fuel consumption, 85% were automatically discharged. By triggering the interactive provers the rest of the proof obligations were discharged to get a fully proved model.

TCTL Queries: To verify the real-time behaviour of our UTA model based on multi-trigger and response pattern, we specify correctness properties in TCTL (Table 1). The properties which occur and need analysis in real systems include concurrency, deadlock freedom, non-Zenoness, liveness and reachability as well as the existence of intermediate events. Note that the mutual exclusion and fairness have already been proved using Event-B proof support. The timing related properties such as time bounded reachability and trigger-response timing correctness are verified in UPPAAL. Some of the most characteristic timing properties of the case study are exemplified by Queries 1 to 3 in Table 1.

Table 1. TCTL queries based on multi-trigger and response pattern

Id	Query	Result
1	M3_Planes(1).Waiting_L_Permission && High_Risk[1] && fuel[1]! =high && not Emergency_Prog && lenH < Max_Queue_Size && LQ_per[1] --→ M3_Planes(1).High_PQ && aux_clk <= tb1	Satisfied
2	M3_Planes(1).Waiting_EmergLand && Elan_per[1] && Runway_busy && Gate_asn[1] --→ M3_Planes(1).At_Gate && aux_clk <= tb2	Satisfied
3	A<> forall (i : int [0,4]) Planes(i).At_Gate || M3_Planes(i).Leaving_Airport & & Gclk <= 120	Satisfied

Query 1 exemplifies a simpler timed trigger-response property satisfied by our system for plane instance 1. It states that when the plane will trigger its landing permission request to the controller, including also information about risk and fuel level, the response by the controller, if there is no ongoing emergency and if the queue is not full, will lead to the plane reaching location $High\_PQ$ within time $tb1$. Note that aux_clk is an auxiliary clock used only for the verification of the query. Constant $tb1$ comes from system requirements expressing the upper time bound explicitly for this trigger-response property.

Query 2 exemplifies a more complex timed trigger-response property, including intermediate events, satisfied by our system for plane instance 1. It states that when the plane triggers an emergency landing case to the controller entering location $Waiting\_EmergLand$ this will lead to the plane finally reaching location At_Gate if the intermediate response events by the controller allow it, i.e., the controller giving the emergency landing permission, keeping the runway reserved and assigning a gate to this plane instance. Application and assumptions for aux_clk and time bound $tb2$ are as for Query 1.

Query 3 represents the full integral time-bounded reachability property satisfied by our system. It states that all plane instances reach eventually the legal terminal locations At_Gate or $Leaving\_Airport$ by the time the global clock of the system reaches 120 time bound. The upper bound for the system global clock for this query is based on the real-time constraints on resolving the landing situation by traffic situations given by model constraints.

6   Conclusion and Discussion

The novel aspect studied in this paper is the use of multi-process trigger-response pattern with intermediate events to address the modelling and verification of reachability properties of complex systems with concurrent processes. We first extended the Event-B to UTA mapping by incorporating iUML-B for the generation of the control structure that serves as the skeleton for UTA avoiding generation of too large UTA models. We then investigated how the integrated approach addresses the development of complex real-time trigger-response pattern-based systems with concurrent processes. We have shown that by using our integrated method we can address the development of complex real-time systems with concurrent processes without extending Event-B nor UTA standard features.

The co-use of Event-B and UTA and translation between them seems straight forward since we follow the control structure imposed by the iUML-B state-machine representation. Therefore, an automated translation from iUML-B is currently being investigated.

An essential observation is that introducing timing constraints by imposing them mechanically to Event-B or iUML-B model control structure often reveals modelling cases that are correct from untimed perspective, but may appear to be infeasible from the perspective of timing. For example, introducing non-zero durations to triggering conditions of events, may introduce some blocking conditions that results in violation of liveness properties proved on the initial Event-B model. Hence, the combination of the two approaches has proved to be beneficial to the development of coherent well-timed models.

Acknowledgments. This work has received funding from the Electronic Component Systems for European Leadership Joint Undertaking under grant agreement No 737494. This Joint Undertaking receives support from the European Union's Horizon 2020 research and innovation programme and Sweden, France, Spain, Italy, Finland, the Czech Republic. This was also supported by the ERDF funded centre of excellence project EXCITE (2014-2020.4.01.15-0018) and the Estonian Ministry of Education and Research institutional research grant no. IUT33-13.

References

1. Abrial, J.R.: Modeling in Event-B: System and Software Engineering. Cambridge University Press, Cambridge (2010)
2. Back, R.J., von Wright, J.: Refinement Calculus: A Systematic Introduction. Graduate Texts in Computer Science. Springer, Heidelberg (1998). https://doi.org/10.1007/978-1-4612-1674-2
3. Behrmann, G., David, A., Larsen, K.G.: A tutorial on UPPAAL 4.0. Department of computer science, Aalborg university (2006)
4. Berthing, J., Boström, P., Sere, K., Tsiopoulos, L., Vain, J.: Refinement-based development of timed systems. In: Derrick, J., Gnesi, S., Latella, D., Treharne, H. (eds.) IFM 2012. LNCS, vol. 7321, pp. 69–83. Springer, Heidelberg (2012). https://doi.org/10.1007/978-3-642-30729-4_6
5. Cansell, D., Méry, D., Rehm, J.: Time constraint patterns for event B development. In: Julliand, J., Kouchnarenko, O. (eds.) B 2007. LNCS, vol. 4355, pp. 140–154. Springer, Heidelberg (2006). https://doi.org/10.1007/11955757_13
6. Jastram, M., Butler, M.: Rodin User's Handbook: Covers Rodin v 2.8 (2014)
7. Said, M.Y., Butler, M., Snook, C.: A method of refinement in UML-B. Softw. Syst. Modeling **14**(4), 1557–1580 (2013). https://doi.org/10.1007/s10270-013-0391-z

8. Sarshogh, M.R., Butler, M.: Specification and refinement of discrete timing properties in Event-B. Electron. Commun. EASST **46**, 1–15 (2012)
9. Snook, C., Waldén, M.: Refinement of statemachines using Event B semantics. In: Julliand, J., Kouchnarenko, O. (eds.) B 2007. LNCS, vol. 4355, pp. 171–185. Springer, Heidelberg (2006). https://doi.org/10.1007/11955757_15
10. Sulskus, G., Poppleton, M., Rezazadeh, A.: An interval-based approach to modelling time in Event-B. In: Dastani, M., Sirjani, M. (eds.) FSEN 2015. LNCS, vol. 9392, pp. 292–307. Springer, Cham (2015). https://doi.org/10.1007/978-3-319-24644-4_20
11. Vain, J., Tsiopoulos, L., Boström, P.: Integrating refinement-based methods for developing timed systems. In: From Action Systems to Distributed Systems, pp. 199–214. Chapman and Hall/CRC (2016)
12. Zhu, C., Butler, M., Cirstea, C.: Refinement of timing constraints for concurrent tasks with scheduling. In: Butler, M., Raschke, A., Hoang, T.S., Reichl, K. (eds.) ABZ 2018. LNCS, vol. 10817, pp. 219–233. Springer, Cham (2018). https://doi.org/10.1007/978-3-319-91271-4_15
13. Zhu, C., Butler, M., Cirstea, C.: Semantics of real-time trigger-response properties in Event-B. In: 2018 International Symposium on Theoretical Aspects of Software Engineering (TASE), pp. 150–155. IEEE (2018)

Formal Distributed Protocol Development for Reservation of Railway Sections

Paulius Stankaitis[1](✉), Alexei Iliasov[1], Tsutomu Kobayashi[2],
Yamine Aït-Ameur[3], Fuyuki Ishikawa[2], and Alexander Romanovsky[1]

[1] School of Computing, Newcastle University, Newcastle upon Tyne, UK
p.stankaitis@newcastle.ac.uk
[2] National Institute of Informatics, Tokyo, Japan
[3] INPT–ENSEEIHT, 2 Rue Charles Camichel, Toulouse, France

Abstract. The decentralisation of railway signalling systems has the potential to increase railway network capacity, availability and reduce maintenance costs. Given the safety-critical nature of railway signalling and the complexity of novel distributed signalling solutions, their safety should be guaranteed by using thorough system validation methods. In this paper, we present a rigorous formal development and verification of a distributed protocol for reservation of railway sections, which we believe could deliver benefits of a decentralised signalling while ensuring safety and liveness properties. For the formal distributed protocol development and verification, we devised a multifaceted framework, which aims to reduce modelling and verification effort, while still providing complementary techniques to study protocol from all relevant perspectives.

Keywords: Formal verification · Distributed resource allocation · Performance analysis · Event-B · PRISM model checker · Railway signalling

1 Introduction

Railway signalling is a safety-critical system whose responsibility is to guarantee a safe and efficient operation of railway networks. In recent decades there have been proposals to utilize distributed system concepts (e.g. [13,24]) in railway signalling as a way to increase railway network capacity and reduce maintenance costs. These emerging distributed railway signalling concepts propose using a radio-based communication technology to decentralise contemporaneous signalling systems[1]. Because of their complex concurrent behaviour, distributed systems are notoriously difficult to validate and this could curtail the development and deployment of novel distributed signalling solutions.

In recent years there has been a push (e.g. [12,22]) by the industry with a strong focus on distributed systems to incorporate formal methods into their

---

[1] A single signalling computer may be responsible for controlling tens of routes (case studies [18,20]) whereas novel distributed systems would reduce that number.

© Springer Nature Switzerland AG 2020
A. Raschke et al. (Eds.): ABZ 2020, LNCS 12071, pp. 203–219, 2020.
https://doi.org/10.1007/978-3-030-48077-6_14

system development processes to improve system assurance and time-to-market. Yet, despite that for years the railway domain has proved to be a fruitful area for applying various formal methods [3,7], considerably less has been done in applying them for distributed railway systems by industry and academia. Therefore, the long-term aim of our research is to lower the effort the barriers to applying formal methods in developing correct-by-construction distributed signalling systems.

In order to manage the modelling and verification complexity of distributed protocols we are working towards an integrated multifaceted methodology, which is based on three concepts: stepwise renement, communication modelling patterns and validation through proofs. In spite of advancements in proof automation, it might be too onerous to mathematically prove the model in early development stages. Therefore, it is also desirable that the framework should support model animation and scenario validation. It is also paramount that the framework should support quantitative evaluation; as stated by Fantechi and Haxthausen [10], distributed signalling solutions will only be adopted in practice if system availability is demonstrated. The authors (as discussed in [10]) of related researches did not consider liveness and fairness properties, which directly affect system availability. In our proposed multifaceted methodology we integrate stochastic simulators for quantitative analysis.

In this paper, we present a research, which uses the proposed methodology to formally develop and verify a distributed railway signalling protocol, which would deliver decentralised signalling benefits, while meeting high safety requirements. The developed distributed signalling protocol is based on serialisability and is inspired by protocols used in transactions processing [4,8,11] in centralised and distributed database systems. The main objective of our protocol is to guarantee mutual exclusion of railway sections while ensuring systems liveness. In a nutshell, our key contributions are the formally proved distributed railway section allocation protocol inspired by past protocols for database systems and the formalisation of the multifaceted verification framework.

Related Work. In Fantechi and Haxthausen [10] the authors formalise the railway interlocking problem as a distributed mutual exclusion problem and discuss the related literature on distributed interlocking (e.g. [9,13,24]). In principle all railway models share similar high-level safety, liveness and fairness requirements, as summarised on page 2 in [10]. One difference between our work and the studies overviewed in [10] is the interlocking engineering concept and the system model (e.g. allowed message delays). Another difference is the formal consideration of liveness and fairness requirements. In our work we not only prove the safety properties of the protocol, but also ensure systems liveness, fairness and analyse performance.

A similar distributed signalling concept is presented as a case study in [1]. The authors verified their system design via a simulation approach and only considered scenarios with up to two trains. In our verification approach we prove the distributed signalling system mathematically and hence guarantee its safety for any number of trains. In the paper by Morley [21] the author formally proved a distributed protocol, which is used in the real-world railway signalling systems to reserve a route, which is jointly controlled by adjacent signalling systems. Even though, the distributed signalling concepts of our works are different, the effects of message delays to the safety were considered in both works.

The rest of the paper is organised as follows. Section 2 outlines the motivation for developing the protocol, semi-formally describes its functionality, elicits the requirements and introduces its specifications and the properties to be proved. Section 3 further discusses the integrated methodology we are proposing. The following section briefly discusses formal model development and also provides technical details on property verication and performance analysis. In the last section we summarise our work and discuss future work directions.

2   Distributed Resource Allocation Model and Protocol

The distributed railway signalling can increase networks capacity (as trains could run closer), improve systems agility to delays and possibly reduce repair costs. On the other hand, an increased system complexity and a safety-critical (SIL4) nature requires the highest level of safety assurance. In order to apply formal methods one must clearly state system requirements and specifications. In the following subsections we describe an abstract model of the distributed railway system and its requirements as well as the $stage_1$ of the distributed protocol, which guarantees the safety and liveness of the distributed system.

2.1   High-Level Distributed System Model and Requirements

We abstract the railway model and instead of trains, routes and switches our system model consists of agents and resources (resources controllers). The system model permits message exchanges only between agents and resources, and messages can be delayed. Each resource controller has an associated queue-like memory, where agents allocation order can be stored. A resource also has a promise (ppt) and read pointers (rpt), which respectively indicate the currently available slot in the queue and the reserved slot (with an associated agent) that currently uses the resource. An agent has an objective, which is a collection of resources an agent will attempt to reserve (all at the same time) before using and eventually releasing them.

SAF_1 | A resource will not be allocated to different agents at the same time.

SAF_2 | An agent will not use a resource until all requested resources are allocated.

LIV_1 | An agent must be eventually allocated requested set of resources.

LIV_2 | Resource allocation must be guaranteed in the presence of message delays.

Requirements 1: High-level systems safety and liveness requirements

The main objective of the protocol is to enable safe and deadlock-free distributed atomic reservation of collection of resources. Where by a safe resource reservation we mean that no two different agents have reserved the same resource at the same time. The protocol must also guarantee that each agent eventually gets all requested resources - partial request satisfaction is not permitted. The main high-level safety and liveness requirements of the distributed system are expressed in Requirements 1.

The following section attempts to justify the need for an adequate distribute protocol by discussing problematic distributed resource allocation scenarios.

2.2 Problematic Distributed Resource Allocation Scenarios

Let us consider Scenarios 1–2 (visualised in Fig. 1) to see how requirement LIV_1 could not be guaranteed (while ensuring SAF_2) without an adequate distributed resource allocation protocol.

Scenario 1. In this scenario, agents a_0 and a_1 are attempting to reserve the same set of resources $\{r_0, r_1\}$. Agents start by firstly sending request messages to both resources. Once a resource receives a request message, it replies with the current value of the promised pointer ($ppt(r_k)$) and then increments the $ppt(r_k)$. For instance, in this scenario, resource r_0 firstly received a request message from agent a_0 and thus replied with the value $ppt(r_0) = 0$, which was then followed by a message to a_1 with an incremented $ppt(r_0)$ value of 1. In Figure, we denote a_n^* as the $ppt(r_k)$ value sent to a_n. Request messages at resource r_1 have been received and replied in the opposite order.

In this preliminary protocol, after an agent receives promised pointer values from all requested resources, it sends messages to requested resources to lock them at the promised queue-slot. In this scenario, agent a_0 was promised queue-slots $\{(r_0, 0), (r_1, 1)\}$ while a_1 queue-slots $\{(r_0, 1), (r_1, 0)\}$. If agents would lock these exact queue-slots, resource r_0 would allow a_0 to *use* it first, while r_1 would concurrently allow a_1. The distributed system would deadlock and fail to satisfy LIV_2 requirement as both agents would wait for the second *use* message to ensure SAF_2.

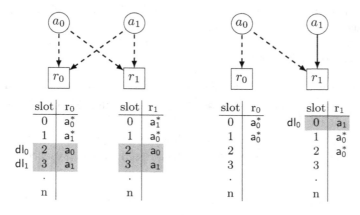

Fig. 1. Problematic scenarios: Scenario 1 (left) and Scenario 2 (right)

In order to prevent the cross-blocking type of deadlocks, an agent should repeatedly re-request the same set of resources (and not lock them) until all received promised queue slot values are the same. We define a process of an agent attempting to receive the same promised queue slots as an agent forming a distributed lane (dl).

A distributed lane of agent a_n is $dl(a_n) = \{(r_k, s), (r_{k+1}, s), \ldots, (r_{k+m}, s)\}$, where $\{r_k, r_{k+1}, \ldots, r_{k+m}\}$ are all resources requested by agent a_n and s is the queue slot value promised by all requested resources. Important to note, that this solution relies on the assumption, that there is a non-zero probability of distinct messages arriving at the same destination in different orders, even if they are simultaneously sent by different sources.

The modified situation is depicted in Scenario 1, where, after agents $\{a_0, a_1\}$ initially receiving $\{(r_0, 0), (r_1, 1)\}$ and $\{(r_0, 1), (r_1, 0)\}$ slots, mutually re-request resources again. This time they receive $\{(r_0, 2), (r_1, 2)\}$ and $\{(r_0, 3), (r_1, 3)\}$ slots, and are able to form distributed lanes $dl_0(a_0)$ and $dl_1(a_1)$.

Scenario 2. However, simply re-requesting the same resources might result in a different problem. In Scenario 2, agent a_1 has requested and has been allocated a single resource r_1 which in turn modified $ppt(r_1)$ to 1 while $ppt(r_0)$ remained 0. If another agent a_0 attempts to reserve resources $\{r_0, r_1\}$, it will never receive the same promised pointer values from both resources, and hence, will not be able to lock them.

To address the two issue described above, we developed a two-stage protocol, where the $stage_1$ of the distributed protocol specifies how an agent forms a distributed lane. $Stage_2$ of the protocol, which is out of this paper scope, addresses other deadlock scenarios, which can occur after agents form distributed lanes. In the following subsection we semi-formally describe the $stage_1$ of the protocol.

2.3   Semi-formal Description of the Stage$_1$

An agent, which intends to reserve a set of resources starts by sending request messages to resources. The messages are sent to those resources which are part of agents current objective. In the provided pseudocode excerpt, we first denote relations sent_requests and objective where they are mappings from agents to resource collections (ln. 1–3 Algorithm 1). The messages request are sent by an agent a_n to a resource r_k ($r_k \in$ objective$[a_n]$) until sent_requests$[a_n]$ = objective$[a_n]$ (images are equal). When a resource r_k receives a request message from an agent a_n it responds with a reply message which contains the current promised pointer value of resource $ppt(r_k)$ to that agent and increments the promised pointer (ln. 2–4 Algorithm 2). After sending all request messages an agent waits until reply messages are received from requested resources and then makes a decision.

Algorithm 1 Agent stage$_1$ communication algorithm

1: **while** sent_requests[a_n] $\neq$ objective[a_n] **do**
2:    request(a_n) $\rightarrow r_k$ sending request message from agent a_n to resource r_k
3: **end**
4: **wait until** received_replies[a_n] = objective[a_n]
5: **while** |replies[a_n]| $\neq$ 1 **do**   cardinality of agents received slot indices
6:        $m' = \max(\text{replies}[a_n]) + 1$
7:        sent_srequests[a_n]$' = \varnothing$ reset sent special request messages buffer
8:        received_replies[a_n]$' = \varnothing$ reset received reply messages buffer
9:            **while** sent_srequests[a_n] $\neq$ objective[a_n] **do**
10:                srequest(a_n, m) $\rightarrow r_k$
11:            **end**
12:        **wait until** received_replies[a_n] = objective[a_n]
13: **end**
14: **while** sent_write[a_n] $\neq$ objective[a_n] **do**
15:    $m' = \max(\text{replies}[a_n])$
16:    write(a_n, m) $\rightarrow r_k$
17: **end** The end of stage$_1$ of the protocol.

When all received promised pointer values are the same (a distributed lane can be formed) an agent completes the stage$_1$ by sending write, to all requested resources, messages which contain the negotiated index (ln. 14–17 Algorithm 1). But if one of the received promised pointer values is different an agent will start a renegotiation cycle (ln 5–13 Algorithm 1). By sending a srequest messages which contain a desired slot index to resources. A desired index is computed by taking the maximum of all received promised pointer values and adding a constant (one is sufficient) - ln. 6 Algorithm 1. A resource will reply to srequest message with the higher value of the current ppt(r_k) or received srequest message value and will update the promised pointer (ln. 5–7 Algorithm 2). After sending all srequest messages, an agent waits for reply messages and then restarts the loop if received slot indices are not the same.

Algorithm 2 Resource stage$_1$ communication algorithm

1: **switch** received_message **do**
2:    **case** request(a_n)
3:        reply(ppt(r_k), r_k) $\rightarrow a_n$
4:        ppt(r_k)$' =$ ppt(r_k) $+ 1$
5:    **case** srequest(a_n, n)
6:        reply($\max(\text{ppt}(r_k), n), r_k$) $\rightarrow a_n$
7:        ppt(r_k)$' = \max(\text{ppt}(r_k), n) + 1$

SAF$_3$ | An agent will not send write (form a distributed lane) messages until all receive promised pointer values are identical.

SAF$_4$ | Agents with overlapping resource objectives will negotiate distributed lanes with different index.

LIV$_3$ | An agent will eventually negotiate a distributed lane.

Requirements 2: Low-level protocol stage$_1$ safety and liveness requirements

It is important to note that the stage$_1$ protocol solution to the described deadlock scenarios has a stochastic nature and one needs to guarantee that a desirable state is probabilistically reachable. In Requirements 2 we summarise requirements for the stage$_1$ of the protocol.

After an agent completes stage$_1$ and thus negotiates a distributed lane it will start protocol stage$_2$ to prevent other deadlock scenarios. Predominantly because of papers verification focus towards properties from stage$_1$ (all complimentary verification/analysis techniques used) we provide protocol stage$_2$ description in the online appendix[2].

3   Multifaceted Modelling and Verification Framework

As stated before, the long-term objectives of our research are to reduce modelling and verification effort of distributed systems and to have a multifaceted framework to study protocols from all relevant perspectives. In the introduction, we defined key formal concepts the framework should rely on and in the following section we discussed protocol requirements we need to guarantee.

The following subsections proposes an engineering process with different formal techniques each of which is efficient to handle parts of above requirements and help to manage modelling and verification complexity.

3.1   Formalised Multifaceted Verification Framework

For any adequate formal system development, system requirements should be clearly stated, and so, this is the first step (Step 1 in Fig. 2) in the modelling process. Currently, we do not suggest or provide a specific structural approach for defining distributed system requirements. The next step (Step 2) in the process is developing and verifying a pivotal formal model. The purpose of formally modelling a distributed system is to have a formal artefact, which can be animated, analysed and formally verified.

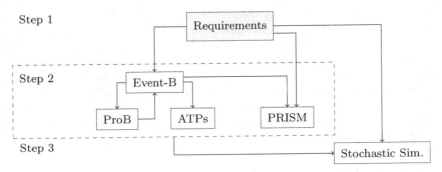

Fig. 2. Multifaceted modelling and verification framework

---

[2] A complete protocol description and formal models can be found at http://stankaitis. uk/2019/02/.

For the development and verification of pivotal functional system models we selected the Event-B [2] specification language, which has previously been successfully used for modelling and verification of various distributed protocols [5,15,16]. The Event-B method provides an expressive modelling language, flexible refinement mechanism and is also proof driven, meaning model correctness is demonstrated by generating and discharging proof obligations with available automated theorem provers [6,17]. The method is supported by tools such as ProB [19] which enable animating and model-checking a model. On the other hand, the Event-B method does not have an adequate probabilistic reasoning support, which, for example, was essential for verifying the distributed railway section reservation protocol. Therefore, it was decided to integrate the well-known PRISM [14] stochastic model checker into the framework, so stochastic system's properties can be verified.

The last step (Step 3) in the proposed engineering process is analysing a developed distributed system's performance. For that, we have implemented a high-fidelity protocol simulator which could help to evaluate protocols under normal or *stressed* conditions. Following subsections provide more detail on how each of the formal techniques would be used in the development and verification of a distributed protocol.

3.2   Step 2: Developing Functional Pivot Models in Event-B

A formal functional Event-B model can have a multitude of uses, but the main application is for formally proving properties about the distributed system. The completed distributed system's model in Step 2 should cover all requirements and specifications, and would be considered correct when all generated proof obligations are proved.

The model development approach we propose is a rather standard and starts with the abstract model which formally specifies the objective of the distributed protocol. In fact, distributed aspects of the system are ignored at this model level and the abstract model considers a centralised configuration. The abstract model is then iteratively refined by introducing more details about the distributed protocol, primarily by modelling communication aspects. To reduce modelling effort we previously developed communication modelling patterns and described a generic model refinement plan in [23]. A key aspect of our methodology is the scenario validation and analysis. Particularly, in early protocol development stages, it might be too onerous to verify a model only to discover design mistakes. To facilitate design exploration we apply animation and model-checking enabled by ProB. Nonetheless, the final (concrete) model should be proved by adding invariants to the model and proving generated proof obligations with available automated theorem provers.

3.3   Step 2: Proving Stochastic Properties with PRISM

As the distributed signalling protocol had a stochastic nature it was important to formally demonstrate that a satisfying state could be reached. Probabilistic or liveness properties are hard to formalise and prove in the Event-B method. Therefore, it was decided to prove progress of the protocol outside of Event-B by redeveloping part of the model ($stage_1$) in the PRISM model checker.

The drawback of using PRISM model checker, if a bounded problem abstraction cannot be found, the verification is limited to bounded models. As we could not find protocol's $stage_1$ abstraction, we created a skeleton model, which then could be instantiated to model specific scenarios of $stage_1$ with n agents, m resources and other initial conditions. Additionally, we developed a model generator, which can automatically instantiate the skeleton model to capture a random scenario and run probabilistic verification conditions.

3.4   Step 3: Analysing System's Performance

With Event-B and PRISM we aim to demonstrate that the protocol addresses the formulated requirements but it is necessary in our application domain to understand how the protocol is going to perform under various conditions if it were deployed in a real system. To conduct such a simulation we have implemented a high fidelity protocol simulator that can be populated with any number of resources and agents while realising any conceivable agents' goal formation and message delivery policies.

The simulator is parametrised with a function of probability of picking a certain message out of a pool of available messages. The probability function is itself parametrised by message source, destination, timestamp and type. The simulation would help to answer how fast, in terms of vital steps such as messages sent, a protocol's $stage_1$ can be completed and how the performance is affected by messages delays. With function D we can simulate slow agents and resources, fair, arbitrary and unfair delivery policies, agents that operate much faster than others and so on.

4   Formal Protocol Modelling, Verification and Analysis

In this section we present the application of previously introduced modeling and verification framework for developing distributed railway signalling protocol. In Sect. 2 we defined protocol's requirements (Step 1), thus following subsections focuses on formal methodology aspects.

4.1   Step 2. Formal Protocol Model Development in Event-B

We apply the Event-B formalism to develop a high-fidelity functional model and prove the protocol functional correctness requirements. We follow the modelling process presented in Sect. 3.2. Important to note that the protocol model was redeveloped multiple times as various deadlock scenarios were found with ProB animator and model-checker. Below, we overview the final (verified) model.

Modelling was started by creating an abstract model context which contains constants, given sets and uninterpreted functions. In the abstract context, we introduced three (finite) sets, to respectively represent agents (agt), resources (res) and objectives (obj). The context also contains an objective function which is a mapping from objectives to a collection of resources (ob $\in$ obj $\rightarrow$ $\mathbb{P}$(res)) and an enumerated set for agents status counter.

The dynamic protocol parts, such as messages exchanges, are modelled as variables and events computing next variable states and contained in a *machine*. According to the proposed model development process, the initial machine (abstract) should summarise the objective of protocol, which is an agent completing an objective (locking all necessary resources). To capture that, the abstract protocol machine contains two events, respectively modelling an agent locking and then releasing a free objective (ob $\in$ obj). The abstract model is refined by mostly modelling communication aspects of the distributed signalling protocol and for that we use a backward unfolding style where the next refinement step introduces preceding protocol step. Below, we overview the refinement chain and properties we proved at that modelling stage.

Refinement 1 (Abstract ext.). In this refinement we introduce resources into the model and now an agent tries to fulfill the objective by locking resources. Previous two events (lock/release) are now decomposed to two for each and capture iterative locking and releasing of resources.

Refinement 2. The abstract models are firstly refined with $stage_2$ part of the protocol. In the refinement, r_2, we introduced lock, response and release messages and associated events into the model. In this step we also demonstrated that the protocol $stage_2$ ensures safe distributed resource reservation by proving an invariant. The invariant states that no two agents will be both at resource consuming stage if both requested intersecting collections of resources.

Refinement 3. Model r_3, is the bridge between protocol stages $stage_1$ and $stage_2$ and introduces two new messages write and pready into the model.

Refinement 4. The final refinement step - r_4 - models $stage_1$ of the distributed protocol which is responsible for creating distributed lanes. Remaining messages request, reply, srequest and associated events are introduced together with the distributed lane data structure. In this refinement we prove that distributed lanes are correctly formed (req. $SAF_{3\text{-}4}$).

4.2   Step 2: Proving Functional Correctness Properties in Event-B

As shown in Sect. 2.2 (Scenarios 1 - 2) high-level system's requirements can only be met if an agent invariably and correctly forms a distributed lane. The probabilistic lane forming eventuality (LIV_3) is discussed separately while in the following paragraphs we focus on the proof regarding requirements $SAF_{3\text{-}4}$.

SAF_3 is required to ensure that agent's resource objectives are not satisfied or satisfied on full. The model addresses this via event *guards* restricting enabling states of the event that generates an outgoing write message. To cross-check this implementation we add an invariant that directly shows that SAF_3 is maintained in the model. For illustrative purposes we focus on details of verifying a slightly more interesting case of SAF_4 and assume that SAF_3 is proven.

Requirement SAF_4 addresses potential cross-blocking deadlocks or resource double locking due to distributed lane overriding. The strategy is to prove the requirement is to show that agents that are interested in at least one common resource (related) always form distributed lanes with differing indices. We start by assuming that agents only form distributed lanes if all received indices are the same (proved as SAF_3). Then, if a resource (or resources) shared between any two related agents send unique promised pointer values to these agents, these indices will be distributed lane *deciders* as all other indices from different resources must be the same to form a distributed lane. Hence, to prove SAF_4 it is enough to show that each resource replies to a request or special request message with a unique promised pointer value.

resource_reply_general $\widehat{=}$
ANY
    rq, rp
WHERE
    grd_1 rq $\in$ req take a sent request message
    grd_2 rp $\in$ REQ $\setminus$ rep create a new reply message
    grd_3 repd(rp) = reqs(rq) destination of reply message is source of request message
    grd_4 reps(rp) = reqd(rq) source of reply message is destination of request message
    grd_5 repn(rp) = ppt(reps(rp)) reply message contains promised pointer
THEN
    act_1 rep := rep $\cup$ {rp} add new message to reply channel
    act_2 req := req $\setminus$ {rq} remove request message from request channel
    act_3 ppt(res) := ppt(res) + 1 increment promised pointer
    act_4 $\mathbf{his_{ppt}(res)} := \mathbf{his_{ppt}(res)} \mathbin{⩤} \{(\mathbf{his_{wr}(res)}) \mapsto \mathbf{ppt(res)}\}$
    act_5 $\mathbf{his_{wr}(res)} := \mathbf{his_{wr}(res)} + 1$
END

Fig. 3. Event-B model excerpt of a resource sending a reply message (Color figure online)

To prove that all resources replies to a request or special request message with a unique promised pointer value, we firstly introduced a history variable his_{ppt} of type $his_{ppt} \in (res \to (\mathbb{N} \nrightarrow \mathbb{N}))$ into our model. The main idea behind the history variable was to chronologically store the promised pointer values sent by a resource. We also introduced a time-stamp variable his_{wr} of the type $his_{wr} \in res \to \mathbb{N}$ to chronologically order the promised pointer values stored in the history variable.

After introducing history variables, we modified events resource_reply_general and resource_reply_special, which in the protocol update the promised pointer variables, by adding two new actions (see Fig. 3). The first action act_4 updates the history variable with the promised pointer value (ppt(res)) that was sent to the agent at the time stamp $(\mathsf{his_w r(res)})$. The second action, $\mathbf{act_5}$, simply increments resource's res time-stamp $(\mathsf{his_w r(res)})$ variable.

inv_saf_4 $\forall r, n_1, n_2 \cdot r \in \mathsf{RES} \wedge n_1, n_2 \in \mathrm{dom}(\mathsf{his_{ppt}}(r)) \wedge n_1 < n_2 \Rightarrow$
$$\mathsf{his_{ppt}}(r)(n1) < \mathsf{his_{ppt}}(r)(n2)$$

Action $\mathbf{act_4}$ updates a history variable for a resource res with the current write stamp and promised pointer $(\mathsf{ppt(res)})$ value sent. The next action $\mathbf{act_5}$ simply updates the resource's write stamp. We can then add the main invariant to prove (**inv_saf_4**) which states that if we take any two entries n1, n2 of the history variable for the same resource where one is larger, then that larger entry should have larger promised pointer value.

inv_his_ppt $\forall \mathsf{res} \cdot$ $(\mathsf{his_{wr}(res)} = 0 \wedge \mathsf{his_{ppt}(res)} = \varnothing)$
$$\vee (\quad \mathrm{dom}(\mathsf{his_{ppt}(res)}) = 0 .. \mathsf{his_{wr}(res)} - 1$$
$$\wedge \mathsf{his_{ppt}(res)}(\mathsf{his_{wr}(res)} - 1) = \mathsf{ppt(res)} - 1)$$

To prove that resource_reply_{general, special} preserve **inv_saf_4**, the following properties play the key role: (1) the domain of $\mathsf{his_{ppt}}$ (i.e., 'indices' of $\mathsf{his_{ppt}}$) is $\{0, \ldots, \mathsf{his_{wr}} - 1\}$, (2) $\mathsf{his_{ppt}}(\mathsf{his_{wr}} - 1) < \mathsf{his_{ppt}}(\mathsf{his_{wr}})$. Property (2) holds because $\mathsf{his_{ppt}}(\mathsf{his_{wr}})$ is the maximum of promised pointer (ppt) and special request slot number and promised pointer is incremented as resource_reply_{general, special} occurs. We also specified these properties as an invariant (**inv_his_ppt**) and proved they are preserved by the events which helped to prove **inv_saf_4**.

Proof Statistics. In Table 1 we provide an overall proof statistics of the Event-B protocol model which may be used as a metric for models complexity. The majority of the generated proof obligations were automatically discharged with available solvers and even a large fraction of interactive proofs required minimum number of steps. We believe that a high proof automation was due to modelling patterns [23] use and SMT-based verification support [6,17].

Table 1. Event-B protocol model proof statistics

Model	No. of POs	Aut. discharged	Int. discharged
context c_0	0	0	0
context mes.	9	9	0
machine m_0	12	12	0
machine m_1	23	21	2
machine m_2	59	43	16
machine m_3	43	32	11
machine m_4	103	57	46
Total	249	174	75

4.3   Step 2: Proving Liveness (req. LIV₃) with PRISM

In this subsection, we discuss stochastic model checking results with which we intend to prove level that LIV_3 requirement is preserved. In particular, we focus on showing that LIV_3 requirement is ensured in Scenario 2 (Sect. 2.2).

In order to demonstrate that LIV_3 requirement holds in Scenario 2 (Sect. 2.2) we used stage₁ protocol's skeleton PRISM model to replicate Scenario 2. In this experiment we were interested in observing the effects a promised pointer offset has on an probability of agent forming a distributed lane while the upper limit of the promised pointer is increased[3] (n in Scenario 2). Early experiments showed that verification would not scale well (several hours for a single data-point) if we would increase the number of resources and agents above two resources and three agents (each agent trying to reserve both resources) so we kept these parameters constant.

For each scenario, we would run a quantitative property: $P = ?$ [F dist₀ > -1] which asks what is the probability of an agent negotiating a distributed lane until the upper promised pointer limit is reached. The three curves (red, green and violet) in Fig. 4 show the effect a promised pointer offset has on negotiation probability as queue depth is increased. Results suggest that increasing the offset reduces the probability of negotiating a distributed lane as queue depth is increased, but the probability still approaches one as the number of rounds is increased (Fig. 4).

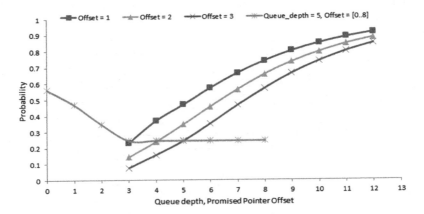

Fig. 4. Scenario 2 with varied resource promised pointer offset and queue depth.

To further see the effects of the offset, we considered a different experiment where the same quantitative property would be run when the number of possible renegotiations value is kept constant and offset is increased (light blue plot).

---

[3] Instead, of ppt upper limit we decided plot the probability against the queue depth, (*offset* - n) as it directly shows how many times an agent can renegotiate resources.

Results indicate that offset has only effect until a specific threshold and after that the probability of agent negotiating a distributed lane is not affected by the offset. These results suggest that the situation in Scenario 2 does not violate LIV_3 requirement as distributed lanes can be negotiated.

4.4   Step 3: Analysing Performance

The goal of this part is to study the protocol performance under various stress conditions and thus provide assurances of its applicability in real life situations. To build simulation, we simply capture protocol's $stage_1$ behaviour using a program. We are also able to obtain bounds on the number of messages required to form lanes in different setups. This can be directly translated into real-life time bounds on the basis of point to point transmission times.

Simulation Construction. Simulation is setup as a collection of actors of two types - agents and resources - and an orchestration component observing and recording message passing among the actors. A message is said to be in transit as soon as it is created by an actor. Every act of message receipt (and receipt only) advances the simulation (world) clock by one unit. Hence, any number of computations leading to message creation can occur in parallel but message delivery is sequential. To model delays we define a function that probabilistically picks a message to be delivered among all the messages currently in transit. A special message, called skip, is circulated to simulate idle passage of time. This message is resent immediately upon receipt by an implicit idle actor.

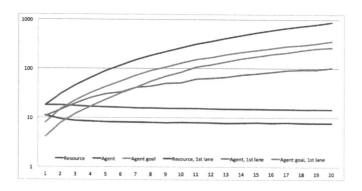

Fig. 5. Time to form all or first lanes, logarithmic scale.

Let M be set of all messages that can be generated by agents and resources. Also, let skip $\notin M$ denote the skip message and $M' = M \cup \{skip\}$. By its structure, set M' is countable (each message identified by unique integer) and one can define a measure space over M'. Let D signify the probability that some message $m \in M \subseteq M'$ from message pool M is selected for reception. We shall define D via the current message pool, the attributes of m such its source, destination, time stamp and protocol stage, and the world time: $D = D(M, m, t) = D(M, m.s, m.d, m.c, m.o, t)$. Here M is the set of available message, $m.s$ and $m.d$ are the message source and destination agent or resource, $m.c$ is the message type (e.g., WRITE), $m.o$ is the message timestamp (the point of its creation) and t is the world clock. Defining differing probabilities D we are able to address most scenarios of interest.

Uniform Distribution. With $D(M, m, t) = \text{card}(M)^{-1}$ the simulator picks a message from M using a uniform distribution. It is an artificial setting as the time in transit bears no influence over the probability of arrival. Counter-intuitively, the said probability may decrease with the passage of time when new messages are created quicker than they are delivered. The skip message has equal probability with the rest so the system "speeds up" when M is large. The plots in Fig. 5 shows how the protocol performance changes when the number of resources (Resource line), agents (Agent lines), and resources an agent attempts to acquire (Agent goal) increase. We plot separately time to form all lanes and any first lane. The values plotted are averaged over 10000 runs.

5 Conclusions and Future Work

In this paper we proposed a multifaceted framework with which we aim to reduce modelling and verification of distributed (railway signalling) systems. The framework was applied in the development of the novel distributed signalling protocol. Starting only with high-level system requirements we developed an early formal protocol prototype which with the help of ProB was refined as subtle deadlock scenarios were discovered. This in part is the advantage of a stepwise development supported by Event-B as complex distributed models can be *decomposed* into smaller problems and errors found earlier. The stepwise distributed protocol development as also shown before [5,15,16] together with adequate tools [6,17] helped to achieve fairly high verification automation. On the other hand, protocol verification was complicated by the need of stochastic reasoning and not adequate Event-B support for reasoning about probabilistic properties. The current solution relied on a model redevelopment in stochastic model checker PRISM which did not scale well for verification of larger scenarios. As a future direction it is essential to address this problem by most likely improving stochastic reasoning in Event-B. In the future we would also like to a much closer tool integration and support an automatic translation to PRISM and the stochastic simulator.

References

1. INTO-CPS Project. Case Studies 2, Deliverable D1.2. Technical report, November 2016. http://projects.au.dk/fileadmin/D1.2a_Case_Studies.pdf
2. Abrial, J.R.: Modeling in Event-B: System and Software Engineering. Cambridge University Press, New York (2013)

3. Behm, P., Benoit, P., Faivre, A., Meynadier, J.-M.: Météor: a successful application of B in a large project. In: Wing, J.M., Woodcock, J., Davies, J. (eds.) FM 1999. LNCS, vol. 1708, pp. 369–387. Springer, Heidelberg (1999). https://doi.org/10.1007/3-540-48119-2_22

4. Bernstein, P.A., Shipman, D.W., Rothnie Jr., J.B.: Concurrency control in a System For Distributed Databases (SDD-1). ACM Trans. Database Syst. 5(1), 18–51 (1980)

5. Cansell, D., Méry, D.: Formal and incremental construction of distributed algorithms: on the distributed reference counting algorithm. Theor. Comput. Sci. 364(3), 318–337 (2006)

6. Déharbe, D., Fontaine, P., Guyot, Y., Voisin, L.: Integrating SMT solvers in rodin. Sci. Comput. Program. 94(P2), 130–143 (2014)

7. Essamé, D., Dollé, D.: B in large-scale projects: the canarsie line CBTC experience. In: Julliand, J., Kouchnarenko, O. (eds.) B 2007. LNCS, vol. 4355, pp. 252–254. Springer, Heidelberg (2006). https://doi.org/10.1007/11955757_21

8. Eswaran, K.P., Gray, J., Lorie, R.A., Traiger, I.L.: The notions of consistency and predicate locks in a database system. Commun. ACM 19(11), 624–633 (1976)

9. Fantechi, A., Haxthausen, A.E., Nielsen, M.B.R.: Model checking geographically distributed interlocking systems using UMC. In: 25th Euromicro International Conference on Parallel, Distributed and Network-Based Processing (PDP), pp. 278–286, March 2017

10. Fantechi, A., Haxthausen, A.E.: Safety Interlocking as a distributed mutual exclusion problem. In: Howar, F., Barnat, J. (eds.) FMICS 2018. LNCS, vol. 11119, pp. 52–66. Springer, Cham (2018). https://doi.org/10.1007/978-3-030-00244-2_4

11. Gray, J., Reuter, A.: Transaction Processing: Concepts and Techniques, 1st edn. Morgan Kaufmann Publishers Inc., San Francisco (1992)

12. Hawblitzel, C., et al.: IronFleet: proving safety and liveness of practical distributed systems. Commun. ACM 60(7), 83–92 (2017)

13. Haxthausen, A.E., Peleska, J.: Formal development and verification of a distributed railway control system. IEEE Trans. Softw. Eng. 26(8), 687–701 (2000)

14. Hinton, A., Kwiatkowska, M., Norman, G., Parker, D.: PRISM: a tool for automatic verification of probabilistic systems. In: Hermanns, H., Palsberg, J. (eds.) TACAS 2006. LNCS, vol. 3920, pp. 441–444. Springer, Heidelberg (2006). https://doi.org/10.1007/11691372_29

15. Hoang, T.S., Kuruma, H., Basin, D., Abrial, J.R.: Developing topology discovery in event-B. Sci. Comput. Program. 74(11), 879–899 (2009)

16. Iliasov, A., Laibinis, L., Troubitsyna, E., Romanovsky, A.: Formal derivation of a distributed program in event B. In: Qin, S., Qiu, Z. (eds.) ICFEM 2011. LNCS, vol. 6991, pp. 420–436. Springer, Heidelberg (2011). https://doi.org/10.1007/978-3-642-24559-6_29

17. Iliasov, A., Stankaitis, P., Adjepon-Yamoah, D., Romanovsky, A.: Rodin platform why3 plug-in. In: Butler, M., Schewe, K.-D., Mashkoor, A., Biro, M. (eds.) ABZ 2016. LNCS, vol. 9675, pp. 275–281. Springer, Cham (2016). https://doi.org/10.1007/978-3-319-33600-8_21

18. Iliasov, A., Taylor, D., Laibinis, L., Romanovsky, A.: Formal verification of signalling programs with SafeCap. In: Gallina, B., Skavhaug, A., Bitsch, F. (eds.) SAFECOMP 2018. LNCS, vol. 11093, pp. 91–106. Springer, Cham (2018). https://doi.org/10.1007/978-3-319-99130-6_7

19. Leuschel, M., Butler, M.: ProB: a model checker for B. In: Araki, K., Gnesi, S., Mandrioli, D. (eds.) FME 2003. LNCS, vol. 2805, pp. 855–874. Springer, Heidelberg (2003). https://doi.org/10.1007/978-3-540-45236-2_46

20. Limbrée, C., Cappart, Q., Pecheur, C., Tonetta, S.: Verification of railway inter-locking - compositional approach with OCRA. In: Lecomte, T., Pinger, R., Romanovsky, A. (eds.) RSSRail 2016. LNCS, vol. 9707, pp. 134–149. Springer, Cham (2016). https://doi.org/10.1007/978-3-319-33951-1_10
21. Morley, M.: Safety assurance in interlocking design. Ph.D. thesis, University of Edinburgh, College of Science and Engineering, School of Informatics (1996)
22. Newcombe, C.: Why Amazon chose TLA$^+$. In: Ait Ameur, Y., Schewe, K.D. (eds.) ABZ 2014. LNCS, vol. 8477, pp. 25–39. Springer, Berlin (2014). https://doi.org/10.1007/978-3-662-43652-3_3
23. Stankaitis, P., Iliasov, A., Ait-Ameur, Y., Kobayashi, T., Ishikawa, F., Romanovsky, A.: A refinement based method for developing distributed protocols. In: IEEE 19th International Symposium on High Assurance Systems Engineering (HASE), pp. 90–97, January 2019
24. Whitwam, F., Kanner, A.: Control of automatic guided vehicles without wayside interlocking. Patent US 20120323411, A1 (2012)

Short Articles

Verifying SGAC Access Control Policies: A Comparison of ProB, Alloy and Z3

Diego de Azevedo Oliveira[(⊠)] and Marc Frappier[(⊠)]

Université de Sherbrooke, Québec, Canada
{dead1401,marc.frappier}@usherbrooke.ca

Abstract. This paper describes the formalisation of SGAC access control policies using Z3 and then we compare the performance with ProB and Alloy. SGAC is an attribute-based, fine-grain access control model that uses acyclic subject and resource graphs to provide rule inheritance and streamline policy specification. To ensure patient privacy and safety, four types of properties are checked: accessibility, availability, contextuality and rule effectiveness. Automatic translation of SGAC policies into each specification language has been defined. ProB offers the best verification performances, by two orders of magnitude. The performances of Alloy and Z3 are similar.

Keywords: Access control · Consent management · Verification · ProB · Formal model · Alloy · Z3

1 Introduction

SGAC (*Solution de Gestion Automatisée du Consentement/ Automated consent management solution*) [2] is a powerful, attribute-based, fine-grain access control model for EHR that uses acyclic subject and resource graphs to provide rule inheritance and streamline policy specification. To ensure patient privacy and safety, four types of properties are defined: *Accessibility, Availability, Contextuality,* and *Rule effectiveness*.

In [2], ProB [4] and Alloy [3] are investigated to verify these SGAC properties. ProB is mainly based on constraint logic programming using the CLP(FD) finite domain library of SICStus Prolog, while Alloy relies on Kodkod and SAT solvers. In this paper, we intend to complement this study by exploring a different technology, SMT solvers, using Z3 [1]. We present the translation of SGAC to SMT-LIB2 using the Python API for Z3. We then compare the performance of Z3 with that of ProB and Alloy using the translation described. We also improve this translation by fully taking into account rule conditions in contexts, instead of an abstraction as proposed in [2].

This work was supported in part by NSERC (Natural Sciences and Engineering Research Council of Canada).

A. Raschke et al. (Eds.): ABZ 2020, LNCS 12071, pp. 223–229, 2020.
https://doi.org/10.1007/978-3-030-48077-6_15

This paper is structured as follows. A brief overview of SGAC is presented in Sect. 2. Section 3 presents the formalisation of the SGAC model in Z3. Section 4 describes the formalisation of the properties to check. Section 5 brings the performance tests and compares each tool. We conclude this paper in Sect. 6.

2   Brief Introduction to SGAC

SGAC is an access control model with conflict resolution. Conflict resolution is based on a definition of precedence between rules; the rule with the highest precedence is chosen to determine the access decision. The precedence relation is not a total order. When there are several maximal elements, access is granted when all of them are permissions. The definitions provided in this section are taken from [2]. SGAC uses *directed acyclic graphs* (DAG). A sink of a DAG G is a vertex without any successor; $sink(G)$ denotes the set of all sinks of G.

An SGAC policy $P = (S, R, L)$ consists of a DAG S denoting subjects, a DAG R denoting resources, and a set of rules L. A rule $l \in L$ permits to specify who (subject) has access (action and modality) to what (resource) and when (priority and condition). A request is a demand the *subject* issues in order to execute an *action* on a *document*. A rule l applies to a request iff all of the following conditions are satisfied: the request subject is a descendant of the rule subject; the request resource is a descendant of the rule resource; the request action is the same as the rule action; the rule condition holds.

One strong point of SGAC is how it deals with conflict resolution. A *conflict* occurs when more than one rule apply to a request, and if they have different modalities. It is necessary to decide which rule has the highest precedence and determine the access decision. Let r_1, r_2 be two different applicable rules for a request:

1. If r_1 has a smaller priority than r_2, we say that r_1 has precedence over r_2.
2. If r_1 and r_2 have the same priority, and if the subject of r_1 is more specific than the subject of r_2 (i.e., the subject of r_1 is a descendant of the subject of r_2 in the subject graph), then r_1 has precedence over r_2.
3. If r_1 and r_2 have the same priority, and neither of their subjects is more specific than the other, then a prohibition has precedence over a permission.

Figure 1 provides a small example where a hospital has just one doctor, Edward, and he is part of the GP Physicians and the Psychologists groups. A patient was accepted to the hospital and the resources available are the exams: a blood test and an urine test. Four rules with the same priority are defined. In rule 1 there is a prohibition of access from the hospital to the exams. That way, just more specific groups may have access to content. In rule 2 the GP Physicians are permitted to access blood tests. In rule 3, the Psychologists are prohibited to access blood tests. In rule 4, Edward is allowed to access urine tests. Edward is only granted access to the urine test of the patient. Since rule 2 is overridden by rule 3, he is prohibited from accessing blood tests. This happens because the rules have the same priority, also neither rule2 is less specific than rule3 or vice-versa, and rule4 is more specific than rule1.

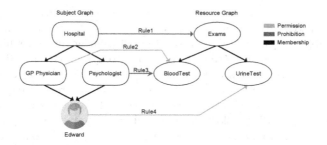

Fig. 1. SGAC graphs with rules

3   Formalisation of SGAC with Z3

Z3 [1] is a Satisfiability Modulo Theories (SMT) solver developed by Microsoft Research. It is specialized for solving background theories. Z3 supports arithmetic, fixed-size bit-vectors, extensional arrays, algebraic datatypes, uninterpreted functions and quantifiers. Several programming languages are available as front-end to interface with Z3, such as Ocaml, C++ and Python. Z3 uses combination theory and novel algorithms. It is composed of a congruence closure engine, a SAT solver-based and several default theory solvers or plug-ins.

Z3 does not natively support sets and relations. A set S that is a subset of a sort T can be represented by a boolean function $S_f \in T \to BOOL$. The predicate $s \in S$ is represented by $S_f(s)$. Similarly, an n-ary relation $r \subseteq T_1 \times \ldots \times T_n$ is represented by a function $r_f \in T_1 \times \ldots \times T_n \to BOOL$. A record $w \in W$, with $W = \mathsf{struct}(a_1 : T_1, \ldots, a_n : T_n)$, is represented by one function $a_{i,f}$ for each attribute a_i such that $a_{i,f} \in T_W \to T_i$, where T_W is a sort representing the set of all records. The value of an attribute a_i of w is given by $a_{i,f}(w)$.

The formalisation of SGAC in Z3 is highly inspired from the B specification of [2]. Z3 is not able to solve the SGAC specification in a single model, thus model staging is needed. Z3 does not natively support model staging. Thus, we use Python scripts to do model staging. In the first stage we calculate the graphs, their transitive closure and determine rule precedence. The second stage calculates the maximal applicable rules. The third stage verifies the SGAC properties. After solving a stage, we use Python to get the instance found and generate new constraints representing the values of the symbols solved in the next stage.

The sets of clause SETS of the B model are represented by sorts in our Z3 model. Although a sort in Z3 is infinite, it is possible to restrain its set of elements using constraints. For SGAC it is mandatory to use the elements that we nominated, and not let the solver choose others.

It is then possible to name the elements of the sort, using constants, and use them in the constraints. In our model, each element of subject, resource, rule, context and the two modalities is unique. A constraint must be added to state that each pair of constants are distinct from each other (*i.e.*, pairwise inequality).

To build the subject and resource graphs, we use a relation as previously described. We also compute the transitive closure of the subject and resource graphs externally in Python, taking advantage of their acyclicity, which is more efficient than the generic transitive closure operator provided in Z3Py.

A rule is represented by a structure as explained above. The set of requests is represented by a relation using the sinks of the subject and resource graphs, as in B. The next step is to specify conflict resolution and how the rules are ordered. We then define: *applicable*, takes the pair subject-resource, as a request, and decides if the rule is applicable to the pair, returning a boolean; *maxElem*, a function that was declared in the B definitions, responsible for giving the maximal rule elements for a given request; *isPrecedeBy*, that connects a subject, a resource, to two rules (*r1*, *r2*) and a boolean. The boolean only holds true when *r1* is less specific than *r2* and the two rules are part of the same request, represented as the subject and resource; *pseudoSink* (psdSink), returns all maximal applicable rules for a given request for a given context.

4   Properties Verification

Accessibility and Contextuality. Accessibility verifies whether a subject *sub* can access a resource *res* in a context *con*. Contextuality determine which contexts make a given request granted. Access is granted when the maximal applicable rules of each request (*sub*, *res*) under the context *con* are all permissions. We define the function *accessibility*(*sub*, *res*, *con*) that returns true when access is granted. Then, we add a constraint that holds if the request for the given context is accessible and we ask Z3 to solve it. In contrast to [2], where two formulas are used, we use a single formula to compute both.

$$accessibility(sub, res, con)$$
$$\Leftrightarrow \forall(rule).(psdSink(sub, res, con, rule) \Rightarrow r\_mod(rule, perm))$$
$$\wedge \exists(rule).(psdSink(sub, res, con, rule))$$

Availability. Finding hidden data allows one to warn the patient that within some conditions, their data may be out of reach. A document is defined hidden or unreachable under the context *con* if there is not a valid request under *con*.

The formalisation in Z3 checks if there is a document under the context that cannot be accessed by anyone. Z3 will return the context with hidden documents.

$$hiddenDataSet(con, res)$$
$$\Leftrightarrow res \notin dom(graph\_res) \wedge \forall(sub).(Request(sub, res)$$
$$\Rightarrow \neg(\forall(rule).(psdSink(sub, res, con, rule) \Rightarrow r\_mod(rule, perm))$$
$$\wedge \exists(rule).(psdSink(sub, res, con, rule))))$$

Rule Effectivity. A rule that can never be the determinant for the evaluation of a request is said ineffective. For instance, if we take two rules with different priorities, one of them has to be ineffective since one will always have precedence over the other. Effectivity of a rule r is formally defined in [2] as follows: Case r is a prohibition: there is at least one pair request-context where r is a maximal applicable rule, and r is the sole prohibition among the maximal rules for this pair; Case r is a permission: there is at least one pair request-context where r is the sole maximal rule.

$$
\begin{aligned}
&ineffectiveSet(rule1) \\
&\quad \Leftrightarrow \; \neg(\exists(sub, res, con). \\
&\qquad\qquad (Request(sub, res) \wedge conRule(con, rule1) \\
&\qquad\quad \wedge\, psdSink(sub, res, con, rule1) \\
&\qquad\quad \wedge\, (\neg(\exists(rule2).(psdSink(sub, res, con, rule2) \wedge\; rule1 \neq rule2))) \\
&\qquad\qquad \vee\; (r\_mod(rule1, proh) \\
&\qquad\qquad\quad \wedge\; \forall(rule2).(psdSink(sub, res, con, rule2) \\
&\qquad\qquad\qquad \wedge\; rule1 \neq rule2 \Rightarrow r\_mod(rule2, perm)))))
\end{aligned}
$$

5   Performance Test

In this section, we discuss the results of the performance tests we executed for the four checked properties. Tests were performed with randomly SGAC models. We vary the following parameters: the number of vertices in each graph (subject and resource), the number of rules, the number of contexts and the number of requests. We check all four SGAC properties by modifying only one parameter

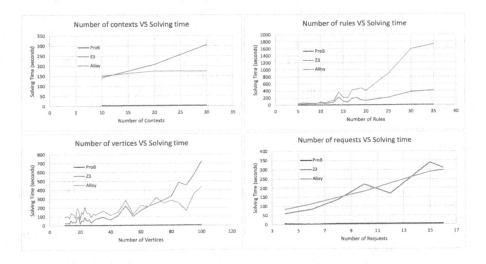

Fig. 2. SGAC performance tests.

at a time. For each defined value of the parameters, at least 6 randomly generated models are created and solved with Z3, PROB and ALLOY. The tests were performed on a Windows 10 64-bit OS, with 16 GB of RAM and Intel®Core™ i7-7700 3.60 GHz as CPU.

As shown in Fig. 2, PROB is faster than the other two solvers by two orders of magnitude in every occasion. Z3 is consistently better than ALLOY when varying the number of rules, while ALLOY outperforms Z3 when varying the the number of contexts. When varying the number of vertices, Z3 is slightly faster up to 75 vertices, after which ALLOY performs better than Z3. As detailed in [2], we use a staged model finding in PROB to solve the properties. The B model of SGAC uses constants to define the subject and resource graphs. The transitive closure of graphs are computed using the B closure operator, for which PROB provides an efficient implementation. B machine operations using set and relation operators are used to solve the four properties checked.

In our experiment, ALLOY is the only model that does not use staged model finding. We decided to investigate if staging could help in increasing its performance. We divided the ALLOY model into three smaller models, following the approach used in the B model. The instances found in one stage are used to build the next stage. This staged model finding cuts the computation time in half, but it is still outperformed by PROB.

6   Conclusion

In this paper we compared Z3 with the B and Alloy models of SGAC [2] for checking SGAC properties. Our experiment shows that PROB is still the most adequate of the three solvers for this task. It is quite easy to use staged model finding in B to increase performance, compared to Z3 and ALLOY. B operations can be easily used to compute the state variables needed to check the properties. During the development of the Z3 model, improvements were made to better take into account rule conditions. We were able to add constraints to the contexts, representing the formula of rule conditions. These modifications were also deployed on the B model. In future work, we plan to investigate the use of Z3 to further analyse rule conditions when a policy is constructed. Another approach would be to explore αRby [5], a deep embedding of ALLOY in Ruby.

References

1. De Moura, L., Bjørner, N.: Z3: an efficient SMT solver. In: Ramakrishnan, C.R., Rehof, J. (eds.) TACAS 2008. LNCS, vol. 4963, pp. 337–340. Springer, Heidelberg (2008). https://doi.org/10.1007/978-3-540-78800-3_24
2. Huynh, N., Frappier, M., Pooda, H., Mammar, A., Laleau, R.: SGAC: a multi-layered access control model with conflict resolution strategy. Comput. J. **62**(12), 1707–1733 (2018)
3. Jackson, D.: Software Abstractions: Logic, Language, and Analysis. MIT Press, Cambridge (2012)

4. Leuschel, M., Butler, M.: ProB: an automated analysis toolset for the B method. JSTTT **10**(2), 185–203 (2008)
5. Milicevic, A., Efrati, I., Jackson, D.: αRby - an embedding of alloy in ruby. In: Ameur, Y.A., Schewe, K. (eds.) ABZ 2014. LNCS, vol. 8477, pp. 56–71. Springer, Heidelberg (2014). https://doi.org/10.1007/978-3-662-43652-3_5

Account and Transaction Protocol
of the Open Banking Standard

Abdulaziz Almehrej[1], Leo Freitas[1], and Paolo Modesti[2(✉)]

[1] School of Computing, Newcastle University, Newcastle upon Tyne, UK
[2] Computer Science and Information Systems,
Teesside University, Middlesbrough, UK
p.modesti@tees.ac.uk

Abstract. To counteract the lack of competition and innovation in the financial services industry, the EU has issued the Second Payment Services Directive (PSD2) encouraging account servicing payment service providers to share data. The UK, similarly to other European countries, has promoted a standard API for data sharing: the Open Banking Standard. We present an overview of the results of a formal security analysis of the Account and Transaction API protocol.

1 Introduction

The lack of competition in the financial services industry has been one of the main factors that led the European Union to introduce the second version of the Payment Services Directive (PSD2) [14], which aims to improve competition by enabling and encouraging bank account holders to share, in a controlled and secure way, their account data. To provide a standard API for the sharing of customer data across different banks, the UK, similarly to other European countries, introduced the Open Banking Standard [13]. The regulation encompasses several API specifications suitable for different *Third Party Providers* (TPPs) who aim to service consumers that consent to sharing their data. The adoption of a standardised interface allows interoperability and simplifies the implementation of systems for sharing data between banks and TPPs.

Contribution. In this paper, we present an overview of a formal security analysis of the Open Banking Standard APIs, focusing on the verification of the correctness of the Account and Transaction API protocol. The work relies on a previously proposed methodology [5] which provided a practical approach to protocol modelling and verification. The methodology utilises the Alice and Bob notation (AnB) [9] to specify a formal model of the protocol that can be formally verified with the OFMC model checker [2]. We formalised and verified a number of security goals that are implicit in the requirements. Although most goals were satisfied in our analysis, the lack of rigourous definition of security properties in the standard can be a source of ambiguity, potentially leading to different interpretations of the security requirements in the implementation. To the best of

A. Raschke et al. (Eds.): ABZ 2020, LNCS 12071, pp. 230–236, 2020.
https://doi.org/10.1007/978-3-030-48077-6_16

our knowledge, our model, fully presented in [1], is the first attempt to formally analyse Open Banking protocols. Recently, other authors [7] made an evaluation of the integration of a web application with the Danish Nordea's Open Banking APIs considering the security threats of the underlying technology, in light of OWASP Top 10 Web Application Security Risks list. However, they did not analyse the security of Open Banking itself considering and assessing security goals as we did. Therefore, we believe this formal analysis can be valuable for stakeholders considering the adoption of a standard that can have a significant and long impact on the efficiency and security of the financial sector.

2   Open Banking Standard

The Open Banking Standard [13] aims at two key outcomes. The first one is an open API for sharing data regarding the services offered by *Account Servicing Payment Service Providers* (ASPSPs), *e.g.* banks. The other one is an open API for sharing the account data of *Payment Service Users* (PSUs) provided by ASPSPs. Open Banking is not only concerned about the API endpoints (e.g. location of resources accessible by third parties, such as developers, to build banking and financial applications), but also about data and security standards. The *data standard* provides data models to the API data format. The *API standard* covers the API's operational requirements. The *security standard* covers API security requirements. An *Account Information Service Provider* (AISP) is a regulated entity allowed by ASPSPs to access a PSU's account data if the PSU provides their consent. This type of access is read-only as the AISPs are not expected to directly affect the payment accounts they are allowed access to. An AISP can then provide different services having the PSU's account and transaction data, including applications that provide a user-friendly view of the states of the different payment accounts held by the PSU, budgeting advice, price comparisons and product recommendations.

Account and Transaction Protocol. The protocol is initiated with the PSU asking for information regarding their payment account(s) from an AISP (Step 1). The AISP then attempts to create an account access consent with the corresponding ASPSP, based on the access permissions agreed upon with the PSU. First, the AISP authenticates itself to the ASPSP through a client credential grant, which is an approach for machine-to-machine authentication. The ASPSP then provides the AISP with an access token used to request the creation of the consent resource (Step 2). At this point, the created account access consent has to be authorised to be used by the AISP to access the PSU's account data. This requires the PSUs to authenticate themselves to the ASPSP, followed by authorising the consent. During this phase, the PSU has to select the payment account(s) for which the chosen permissions should apply. The AISP then obtains an access token to the account data (Step 3). With this token, the AISP has to first retrieve the accessible accounts, including their unique IDs, through the accounts endpoint. The IDs can later be used to request the data of specific accounts (Step 4). To retrieve specific PSU account data (e.g. balances, transactions, direct debits, beneficiaries, *etc.*) the AISP will have to request the data via the appropriate link using the correct endpoint and method from the ASPSP.

3   Methodology and Security Goals

The formal verification of Open Banking API presented in this work is based on a protocol verification methodology proposed in [5]. The methodology utilises the Alice and Bob notation (AnB) [9] to specify a formal model of the protocol that can be formally verified through information flow (secrecy and authenticity) goals. Such notation abstracts from implementation details, but allows formal representation and analysis of the security-relevant characteristics of protocols.

An AnB specification comprises of several sections. The *Types* section declares the different identifiers used in the protocol. This includes the agents, constant and variable (random) numbers and transparent functions. Transparent functions are user-defined through their signature, thereby abstracting from their implementation details (*i.e.* they are uninterpreted). The *Knowledge* section describes the initial data each agent has before running the protocol. Fresh values are initialised at runtime. The information flow is described in the *Actions* section, where details about messages exchanged by agents are specified. Furthermore, the model can be used to verify specific security properties, such as (weak and strong) authentication and secrecy goals:

- **A weakly authenticates B on M**: agent A has evidence that the message M has been endorsed by agent B with the intention to send it to A (i.e. non-injective agreement [8]);
- **A authenticates B on M**: weak authentication plus evidence of the freshness of the message M (i.e. injective agreement [8]);
- **M secret between A, B**: message M is kept secret among listed agents.

The formal model captures the protocol requirements [12]. While the Open Banking API describes in details the information-flow, it lacks definitions of security goals that the exchanges between agents are meant to convey. Therefore, part of our work consisted in identifying suitable goals for the protocol model.

For the verification, we used the Open-Source Fixed-Point Model-Checker (OFMC) [10], a symbolic model-checker supporting the AnB notation. Moreover, the AnBx Compiler and Code Generator [11] was used to pre-process the model to benefit from a stricter type system and support the extension to AnB that allows named expression abstractions (*Definitions* section).

The goals we identified (and verified) are based on our understanding of the protocol and on its dependencies. For example, OAuth 2.0 security considerations [6, pp. 52–60], protocol use cases in [13, pp. 20–23] and our expectations of the protocol.

We identified eight goals: four on message secrecy, and four on authentication.

```
fAISPSecret(AISP) secret between AISP,aspspA              #G1
fPSUSecret(PSU) secret between PSU,aspspA                 #G2
ClientToken, AuthToken secret between AISP,aspspA,aspspR  #G3
# FAILED initially + Fixed #A2.3
PSU authenticates aspspR on fGetIntent(Intent)           #G4
aspspR authenticates PSU on SelectedAccounts             #G5
PSU weakly authenticates AISP on ASPSPAuthPSUEndP,AISPEndP #G6

# FAILED + Fixed #A4.1  #A4.2
Accounts secret between AISP, aspspR                      #G7
AISP authenticates aspspR on Accounts                    #G8
```

Two goals (G1 and G2) are obvious: the exchanged secrets/credentials between the AISP and PSU and the authorisation server remain secret whilst requesting for a client token (Action 2.1 in the specification) and acquiring consent authorisation (A3.1.2 and A3.3.3). That is because if the AISP credentials are leaked (A2.1), many attacks would be possible (for instance [6, Sect. 10.2] discusses client impersonation). Another secrecy goal (G3) states that various exchanged tokens (A2.2–A2.3 and A3.3.4–A4.1) remain secret between the AISP and the authorisation and resource servers. As tokens are AISP bound, a compromised token cannot be directly used. However, [6, Sect. 10.3] requires tokens to be confidential, to prevent attacks involving valid token injection [6, Sect. 10.12]. These goals clearly indicate the inherited potential vulnerabilities of the Account and Transaction Protocol (ATP) dependencies. The final secrecy goal (G7) is about the resource server message to the AISP (A4.2) and is obvious: account information must remain secret.

The authentication goals relate to the PSU authenticating the consent resource to authorise (G4), the resource server authenticating the PSU's selected accounts information (G5) and the AISP authenticating the PSU's account information from the resource server (G8). This last goal between the AISP and the resource server is crucial in verifying the integrity of the account data sent to the AISP by the resource server. In addition to direct data modification, it is important to verify that old data cannot be replayed. For instance, in the case of affordability check, if the PSU was an intruder and modified the data, they could trick an AISP into providing a product they are not eligible for. This goal also enforces fraud detection: if the transactional data can be modified by an intruder to hide fraudulent activity. Given the redirections from the PSU to the authorisation server and AISP (A3.1.1 and A3.3.1–A3.3.2), we weakly authenticate that those endpoints cannot be modified by an intruder to help avoid redirected URI manipulation [6, Sect. 10.6] and phishing attacks [6, Sect. 10.11] (G6).

Model Development. The Open Banking ATP is complex and with multiple dependencies. The AnB model aims to provide an abstract and accurate view of its essential aspects and to verify key properties. The initial AnB model was overly detailed with unnecessary data exchanges. To reach the right level of abstraction, we then decided to first determine the protocol goals prior to abstracting. Even after such endevour, verification was unwieldy: it ran for over two days without response. As is common within model checking problems, state explosion must be tackled beyond abstracting details, abstract on irrelevant data.

Restricting the role of the PSU, where it had to be different from the AISP and servers, considerably reduced the state space. This led to termination with goal verification to be reduced to about seven hours. This enabled us to identify further steps to abstract related to data, which reduced the verification time to about six minutes. A final abstraction, related to the various TLS-related steps, was to abstract them using AnB bullet channels, used to model channels providing authentication and/or secrecy properties at the end-points. The internal efficiency of OFMC dealing with such channels led the final version to verify within eight seconds. This exponential efficiency (up to 5 orders of magnitude) increase is not uncommon in model checking problems, so long the right abstractions are taken alongside expert knowledge of the tool's implementation.

Model Correctness. We used the OFMC model checker [2] to verify the eight goals described above. At first, three goals ($G4, G7, G8$) about PSU intent authentication and account information secrecy and integrity failed. This led us to check these goals independently in order to study their reason for failure quickly. The witness for the PSU authentication failure ($G4$) relates to the resource server authenticating with an unknown agent rather than the PSU. This was fixed by having the resource server being aware of the PSU's identity early on when setting up the access consent with the AISP (A3.2). Thus, this failure identifies a previously undocumented vulnerability, which our modification fixes.

The account information goals fail due to a limitation of bullet channels: they do not protect against replay attacks, hence their use here allowed breaking both secrecy ($G7$) and integrity ($G8$). The intruder could respond to the AISP's request for account data by replaying a previous message. This breaks integrity as the response received by the AISP, and perceived to be the account data, has been modified. As the data replayed is known to the intruder, it also breaks secrecy of the account data. However, the TLS protocol does protect against message replay [4, pp. 93–94]. To deal with this limitation and ensure that freshness would resolve the issue, we modified the model to include a nonce generated and sent by the AISP when requesting for the PSU account data and is expected to be part of the response.

These modifications enable checking all goals for one session. Multiple sessions verification is important as there could be attacks relying on multiple concurrent protocol runs. Due to increased state space and limited hardware, we were unable to fully verify the model for two parallel sessions. As customary in under such conditions (e.g. [3] for iKP and SET), we were able to obtain partial results by increasing the search space up to the available resource limits (search space depth: 15 plies, 14.5 GB RAM, 50 h to run) without being able to reach any attack state.

4   Conclusion

The novel Open Banking Account and Transaction protocol is a security-critical protocol, which is being enforced on the largest banks in Europe. Given the protocol's significance and expected wider use, verifying its correctness is crucial. Our findings were disseminated as part of a presentation on PSD2 at a UK Finance event, with representatives from Visa and MasterCard, as well as several banks. Some of the identified goals were known, others not. The audience was particularly keen on the time/cost analysis. Our future work will focus on the modelling of the protocol's state and transparent functions specification in VDM-SL: this is aimed at discovering underlying vulnerabilities related to the myriad of dependant technologies (*e.g.* OAuth2, TLS, *etc.*).

References

1. Almehrej, A., Freitas, L., Modesti, P.: Security analysis of the Open Banking Account and Transaction API Protocol. arXiv:2003.12776 (2020)
2. Basin, D., Mödersheim, S., Viganò, L.: OFMC: a symbolic model checker for security protocols. Int. J. Inf. Secur. **4**(3), 181–208 (2005). https://doi.org/10.1007/s10207-004-0055-7
3. Bugliesi, M., Calzavara, S., Mödersheim, S., Modesti, P.: Security protocol specification and verification with AnBx. J. Inf. Secur. Appl. **30**, 46–63 (2016). https://doi.org/10.1016/j.jisa.2016.05.004
4. Dierks, T., Rescorla, E.: The Transport Layer Security (TLS) Protocol Version 1.2, August 2008. https://tools.ietf.org/html/rfc5246. Accessed 22 Aug 2019
5. Freitas, L., Modesti, P., Emms, M.: A methodology for protocol verification applied to EMV® 1. In: Massoni, T., Mousavi, M.R. (eds.) SBMF 2018. LNCS, vol. 11254, pp. 180–197. Springer, Cham (2018). https://doi.org/10.1007/978-3-030-03044-5_12
6. Hardt, D.: The OAuth 2.0 Authorization Framework, October 2012. https://tools.ietf.org/html/rfc6749. Accessed 22 Aug 2019
7. Kellezi, D., Boegelund, C., Meng, W.: Towards secure open banking architecture: an evaluation with OWASP. In: Liu, J.K., Huang, X. (eds.) NSS 2019. LNCS, vol. 11928, pp. 185–198. Springer, Cham (2019). https://doi.org/10.1007/978-3-030-36938-5_11
8. Lowe, G.: A hierarchy of authentication specifications. In: CSFW 1997, pp. 31–43. IEEE Computer Society Press (1997)
9. Mödersheim, S.: Algebraic properties in Alice and Bob notation. In: International Conference on Availability, Reliability and Security (ARES 2009), pp. 433–440 (2009). https://doi.org/10.1109/ARES.2009.95
10. Mödersheim, S., Viganò, L.: The open-source fixed-point model checker for symbolic analysis of security protocols. In: Aldini, A., Barthe, G., Gorrieri, R. (eds.) FOSAD 2007-2009. LNCS, vol. 5705, pp. 166–194. Springer, Heidelberg (2009). https://doi.org/10.1007/978-3-642-03829-7_6
11. Modesti, P.: AnBx: automatic generation and verification of security protocols implementations. In: Garcia-Alfaro, J., Kranakis, E., Bonfante, G. (eds.) FPS 2015. LNCS, vol. 9482, pp. 156–173. Springer, Cham (2016). https://doi.org/10.1007/978-3-319-30303-1_10

12. Open Banking Limited: Account and Transaction API Specification - v3.1.1. https://tinyurl.com/qs643hq. Accessed 22 Aug 2019
13. Open Banking Working Group: The Open Banking Standard, February 2016
14. The European Parliament and the Council of the European Union: DIRECTIVE (EU) 2015/2366. Official Journal of the European Union, November 2015

Structuring the State and Behavior of ASMs: Introducing a Trait-Based Construct for Abstract State Machine Languages

Philipp Paulweber[1]([✉]), Emmanuel Pescosta[2], and Uwe Zdun[1]

[1] Faculty of Computer Science, Research Group Software Architecture,
University of Vienna, Währingerstraße 29, 1090 Vienna, Austria
{philipp.paulweber,uwe.zdun}@univie.ac.at
[2] Vienna, Austria

Abstract. Abstract State Machine (ASM) theory is a well-known state-based formal method to analyze, verify, and specify software and hardware systems. Nowadays, as in other state-based formal methods, the proposed specification languages for ASMs still lack easy-to-comprehend language constructs for type abstractions to describe reusable and maintainable specifications. Almost all built-in behaviors are implicitly defined inside a concrete ASM language implementation and thus, the behavior is hidden from the language user. In this paper, we present a new ASM syntax extension based on traits, which allows the specifier (language user) to define new type abstractions in the form of structure and behavior definitions to reuse, maintain, structure, and extend the functionality in ASM specifications. We describe the proposed language construct by defining its syntax and semantics. The decision to use a trait-based syntax extension over other object-oriented language constructs like interfaces or mixins was motivated and driven by the results of previously conducted empirical studies. Moreover, we outline details about the implementation of the trait-based syntax extension in our Corinthian Abstract State Machine (CASM) language implementation.

Keywords: Abstract State Machine · Trait · Structure · Modularization · CASM

1 Introduction

In 1993, Gurevich [1] introduced the ASM theory, which is a well-known state-based formal method consisting of transition rules and algebraic functions. It has been used extensively by scientists for a broad research field ranging from software and hardware to system engineering perspectives in order to specify, analyze, and verify systems in a formal way. ASMs are used to formally describe the

---

E. Pescosta—Member of CASM organization.

A. Raschke et al. (Eds.): ABZ 2020, LNCS 12071, pp. 237–243, 2020.
https://doi.org/10.1007/978-3-030-48077-6_17

evolution of function states in a step-by-step manner[1] and are used to describe sequential, parallel, concurrent, reflective, and even quantum algorithms. Based on the ASM theory by Gurevich [1], several theory improvements and ASM-based language implementations were developed, which were summarized by Börger and Stärk [2] and Börger and Raschke [3].

Prominent ASM languages and tools are AsmetaL [4], CASM [5], and Core-ASM [6]. Today, a common thread in the various ASM languages and tools, as well as in most other state-based formal methods, is that the proposed specification languages lack easy-to-comprehend abstractions to describe reusable and maintainable type specifications. While very few have embraced basic object-oriented abstractions such as classes and inheritance, more advanced type abstractions are usually missing. Therefore, in this paper we propose a new language construct for ASM specification languages to express type abstractions in the form of traits [7] to modularize specifications into structural state and behavioral parts.

2   Motivation

Modern object-oriented languages offer a variety of advanced type abstractions, and most offer either interfaces [8], mixins [9], or traits [7] in addition to classes and inheritance concepts. Interfaces establish a protocol and define method signatures to which a type has to conform [8]. They are often compared to a contract. Mixins define reusable behavior and structure that can be used to combine and form new types [9,10]. Traits are similar to interfaces except that they can define stateless behavior which depends on the trait itself [11]. There is a heated debate in the object-oriented community[2], which of these abstractions is best suited to promote reusable and maintainable type specifications, and many implementations combine different language constructs to define type abstractions. A notable example would be the programming language Scala [12], which offers a trait syntax that is similar to the Java 8 [13] interface syntax and offers mixins type abstractions through the class-based implementation and extension syntax. Another example of mixed type abstraction concepts, namely interfaces and traits, can be found in the programming language Rust [14], where the language user has to express every interface definition through traits, and the types have to conform to specified traits and implement all required functionalities.

In the world of ASMs, only AsmL [15] has introduced an object model in the language through classes and interfaces to represent type abstractions, and to achieve structuring of the ASM specifications. Only the ASM implementation and language XASM by [16] has introduced a sub-ASM construct to achieve a component-based modularization approach. A more generic concept called *ambient* ASMs [3] introduces the possibility to achieve hierarchical state partitioning through nesting of context-sensitive (sub)program environments. Based on this

---

[1] ASM theory was formerly called *Evolving Algebra*.

[2] See, e.g.: https://stackoverflow.com/questions/925609.

state of the art, we started to investigate the introduction of a new type abstraction language construct in ASMs. But which language construct is suitable for ASMs to represent such type abstractions?

Basically every language construct for forming type abstractions is suitable for ASMs, but it influences the understandability of the language considerably. For such an ASM extension, we consider the following properties important: (1) reuse and embed existing specifications; (2) describe built-in behavior of a language itself in the language; and (3) allow encapsulation of ASM states and corresponding behavior through modularization. Driven by the properties and questions raised, we conducted empirical studies to determine, which language construct – interfaces, mixins, or traits – is most understandable to ASM language users for expressing type abstractions [17]. The result of the experiments showed that the participants with strong object-oriented backgrounds (highly familiar with interfaces, not familiar with traits at all) had a similar to equal understanding of an interface and traits language construct in the experimental ASM syntax variants. Mixins, on the other hand, had a significantly lower understandability compared to traits and interfaces. Since the interface and traits type abstraction language constructs offer a similar to equal understandability, and novice language users seem to understand traits without even knowing the concept of traits, we investigated introducing traits into ASMs.

Moreover, the object-oriented communities often discuss traits more favorably than interfaces[3] and even point out that "Traits are Interfaces"[4] just with code-level reuse functionality. To gain a better understanding of how specifiers (language users) comprehend such trait-based specifications, we performed an eye-tracking experiment [17], where we observed the participants' gaze patterns. The results of this experiment showed that the participants could easily distinguish between behavioral and non-behavioral aspects of a given specification, when we applied our trait-based language construct to form state/behavior type abstractions.

3   A Trait-Based Construct for ASMs

This section proposes our trait-based language construct to extend the syntax of ASM specification languages. The syntax rules are defined and expressed in BNF (see Listing 1.1). The semantics of the proposed trait-based syntax extension is defined by lowering and transforming the new syntax elements to appropriate Turbo ASM [2] equivalent definitions (see example trait-based ASM Listing 1.2 and the transformed Turbo ASM Listing 1.3). The ASM specifications presented use the syntax of the CASM specification language[5]. The trait-based syntax extension is divided into three parts, namely *structural types, basic type behavior,* and *extended type behavior.*

---

[3] See, e.g.: https://stackoverflow.com/questions/9205083.

[4] See, e.g.: https://blog.rust-lang.org/2015/05/11/traits.html.

[5] For the CASM syntax description, see: https://casm-lang.org/syntax.

In order to modularize the states (functions not classified as derived) in ASM, we introduce a *structural type* construct (see Listing 1.1, Line 2–4), which allows a language user to group one or multiple functions together (similar to members of an object-oriented class) to form a new *structure* type (see **StructureDefinition** grammar rule). Each structure type defines a trait type through the defined state functions. The access to these functions is only allowed inside a proper basic behavior definition to clearly specify the access to an instantiated structure's state over dedicated behaviors (data encapsulation).

```
1  // Structural Types
2  StructureDefinition  ::= 'structure' Identifier '=' '{' ( FunctionDefinition )+ '}'.
3  StructureLiteral     ::= [Type] '{' [Identifier ':' Term (',' Identifier ':' Term)*] '}'.
4  Literal              ::= StructureLiteral | /* other literals */.
5  // Basic Type Behavior
6  ImplementDefinition      ::= 'implement' Identifier '=' '{'
7                               ( ObjectRuleDefinition | ObjectDerivedDefinition )+ '}'.
8  ObjectRuleDefinition     ::= 'rule' Identifier '(' 'this'
9                               ( ',' Identifier ':' Type )* ')' [ '->' Type ] '=' Rule.
10 ObjectDerivedDefinition  ::= 'derived' Identifier '(' 'this'
11                               ( ',' Identifier ':' Type )* ')' '->' Type '=' Term.
12 MethodCall               ::= Term '.' Identifier ['(' Term (',' Term)* ')'].
13 CallRule                 ::= MethodCall | ( Identifier [ '(' Term (',' Term)* ')' ] ).
14 Term                     ::= MethodCall | 'this'
15 // Extended Type Behavior
16 BehaviorDefinition       ::= 'behavior' Identifier '=' '{'
17                               ( ObjectRuleDeclaration | ObjectDerivedDeclaration
18                               | ObjectRuleDefinition | ObjectDerivedDefinition )+ '}'.
19 ImplementForDefinition   ::= 'implement' Identifier 'for' Identifier '=' '{'
20                               ( ObjectRuleDefinition | ObjectDerivedDefinition )+ '}'.
21 ObjectRuleDeclaration    ::= 'rule'    Identifier ':' 'Object' ('*' Type)* '->' Type.
22 ObjectDerivedDeclaration ::= 'derived' Identifier ':' 'Object' ('*' Type)* '->' Type.
```

Listing 1.1: Trait-Based ASM Syntax Extension

```
1  structure X = {
2    function f1 : -> Integer
3    function f2 : Integer -> Boolean
4  }
5
6
7
8
9
10
11 rule R1 =
12   let v1 = X{ f1: 1,
13     f2: ( 2 ) -> false } in skip
14 implement X = {
15   derived d1( this ) -> Integer =
16     this.f1
17
18   rule R2( this, a1 : Integer ) =
19     if a1 > -5 and this.d1 < 5 then
20       this.f2( a1 ) := true
21 }
22 behavior Y = {
23   derived d2 : Object -> Integer
24
25   derived d3( this ) -> Boolean
26     = this.d2 * this.d2 > 100
27 }
28 implement Y for X = {
29   derived d2(this) -> Integer = this.f1
30 }
31 // ...
```

```
1  domain X
2  function X_f1 : X -> Integer
3  function X_f2 : X * Integer -> Boolean
4  rule X_instantiate( a1 : Integer
5  , a2 : Integer -> Boolean ) -> X =
6    let object = new X in {
7      X_f1( object ) := a1
8      X_f2( object ) := a2
9      result := object
10   }
11 rule R1 =
12   let v1 = X_instantiate( 1,
13     { ( 2 ) -> false } ) in skip
14
15 derived X_d1( this : X ) -> Integer =
16   X_f1( this )
17
18 rule X_R2( this : X, a1 : Integer ) =
19   if a1 > -5 and X_d1(this) < 5 then
20     X_f2( this, a1 ) := true
21
22
23
24
25 derived X_d3( this : X ) -> Boolean
26   = X_d2( this ) * X_d2( this ) > 100
27 }
28
29 derived X_d2(this:X) -> Integer = X_f1(this)
30
31 // ...
```

Listing 1.2: Trait-Based ASM   Listing 1.3: Turbo ASM Equivalent

A *basic type behavior* (see Listing 1.1, Line 6–14) defines a set of rules and derived functions, which are associated with a certain domain type. We introduce a new `ImplementDefinition` to define a basic behavior consisting of one or more object-based derived function and/or rule definitions. The syntax for `ObjectRuleDefinition` and `ObjectDerivedDefinition` introduce a new keyword `this` as the first argument for all object-based rule and/or derived function definitions. The type of the argument variable `this` equals the type of the `ImplementDefinition` and it enables the access to the domain's or structure's behavior. The access happens through a `MethodCall` syntax, which uses a dot operator between a term, a target name, and a non-negative arity of arguments. The target name can be a function name or a rule name.

An *extended type behavior* (see Listing 1.1, Line 16–22) defines a set of rules and derived functions, and forms a new type in the type system. If a domain and/or structural type wants to use the functionality, it has to implement the extended behavior. The `BehaviorDefinition` defines an explicit trait with type name consisting of zero or more `ObjectRuleDeclaration` rule names and/or `ObjectDerivedDeclaration` derived function names. Please note that for all object-based declarations we introduced a generic `Object` argument type at the first position. The `Object` type gets checked against the domain or structural type which is implementing this declared behavior. A specifier can use the `Object` type for any other argument or target type in a declaration. Additionally, a trait can define a default behavior through zero or more `ObjectRuleDefinition` rule names and/or `ObjectDerivedDefinition` derived function names, which depends only on the functionality of the trait itself. Each domain and/or structural type that wants to support a certain behavior has to specify an `ImplementForDefinition` and provide the missing definitions of the trait declarations. If the trait defines a default behavior, the domain and/or structural type inherits this definition. This enables code reuse capabilities.

Listing 1.2 depicts an example trait-based ASM specification using all new syntax grammar rules and Listing 1.3 depicts the equivalent semantics-preserving Turbo ASM specification. The proposed trait-based syntax extension is realized in our CASM language implementation[6]. In order to provide a clean solution, we updated our CASM language front-end implementation and introduced two new internal AST representations before the specification gets transformed to the CASM-IR [5].

By introducing the proposed trait-based construct, we were able to explicitly specify the behavior of the CASM language itself in CASM in the form of a *prelude* (See footnote 6) specification, which gets automatically loaded (imported) for every parsed CASM specification. Each functionality of the CASM language (e.g. operators) is mapped to a behavior (trait) in the prelude specification. The language user can explore and extend the behaviors of CASM in CASM. Moreover, the prelude specification reduced the complexity of the CASM implementation.

---

[6] For sources, see: https://github.com/casm-lang/libcasm-fe/pull/205.

4   Conclusion

In this paper, we present a trait-based construct for ASM languages. It allows to specify composable models through the usage of domain and structural type objects, where the behavior can be defined and implemented in a reusable manner. The modularization and composing of object-oriented models is achieved by specifying structural states along with their behaviors clearly separated through traits. Novel about this contribution is that ASM language users can directly define the semantics of operations over domain (structure) types through this trait-based construct in the ASM language itself. To clearly separate structure and behavior, we only allow the definition of modifications to structural objects through a proper behavior definition. Based on previously conducted empirical studies, the current state of the art, and our current proposed trait-based construct, we believe that this is the first step towards clearer and more understandable ASM specifications by separating the structural (state) and behavioral elements through dedicated definitions.

References

1. Gurevich, Y.: Evolving Algebras 1993: Lipari Guide - Specification and Validation Methods, pp. 9–36. Oxford University Press Inc., New York (1995)
2. Börger, E., Stärk, R.: Abstract State Machines: A Method for High-Level System Design and Analysis. Springer, Heidelberg (2003). https://doi.org/10.1007/978-3-642-18216-7
3. Börger, E., Raschke, A.: Control state diagrams (meta model). Modeling Companion for Software Practitioners, pp. 297–315. Springer, Heidelberg (2018). https://doi.org/10.1007/978-3-662-56641-1_9
4. Gargantini, A., Riccobene, E., Scandurra, P.: A metamodel-based language and a simulation engine for abstract state machines. J. Univ. Comput. Sci. **14**(12), 1949–1983 (2008)
5. Paulweber, P., Pescosta, E., Zdun, U.: CASM-IR: uniform ASM-based intermediate representation for model specification, execution, and transformation. In: Butler, M., Raschke, A., Hoang, T.S., Reichl, K. (eds.) ABZ 2018. LNCS, vol. 10817, pp. 39–54. Springer, Cham (2018). https://doi.org/10.1007/978-3-319-91271-4_4
6. Farahbod, R., Gervasi, V., Glässer, U.: CoreASM: an extensible ASM execution engine. Fundam. Informaticae **77**(1–2), 71–104 (2007)
7. Curry, G., Baer, L., Lipkie, D., Lee, B.: Traits: an approach to multiple-inheritance subclassing. In: Proceedings of the SIGOA Conference on Office Information Systems, New York, NY, USA, pp. 1–9. ACM (1982)
8. Canning, P.S., Cook, W.R., Hill, W.L., Olthoff, W.G.: Interfaces for strongly-typed object-oriented programming. In: OOPSLA, pp. 457–467. ACM (1989)
9. Flatt, M., Krishnamurthi, S., Felleisen, M.: Classes and mixins. In: ACM SIGPLAN-SIGACT POPL, New York, NY, USA, pp. 171–183. ACM (1998)
10. Bracha, G., Cook, W.: Mixin-based inheritance. ACM Sigplan Not. **25**(10), 303–311 (1990)
11. Schärli, N., Ducasse, S., Nierstrasz, O., Black, A.P.: Traits: composable units of behaviour. In: Cardelli, L. (ed.) ECOOP 2003. LNCS, vol. 2743, pp. 248–274. Springer, Heidelberg (2003). https://doi.org/10.1007/978-3-540-45070-2_12

12. Odersky, M., Spoon, L., Venners, B.: Programming in Scala. Artima Inc., Walnut Creek (2008)
13. Potts, A., Friedel, D.H.: Java Programming Language Handbook. Coriolis Group Books, Scottsdale (2018)
14. Matsakis, N.D., Klock II, F.S.: The rust language. ACM SIGAda Ada Lett. **34**, 103–104 (2014)
15. Gurevich, Y., Rossman, B., Schulte, W.: Semantic essence of AsmL: extended abstract. In: de Boer, F.S., Bonsangue, M.M., Graf, S., de Roever, W.-P. (eds.) FMCO 2003. LNCS, vol. 3188, pp. 240–259. Springer, Heidelberg (2004). https://doi.org/10.1007/978-3-540-30101-1_11
16. Anlauff, M.: XASM- an extensible, component-based abstract state machines language. In: Gurevich, Y., Kutter, P.W., Odersky, M., Thiele, L. (eds.) ASM 2000. LNCS, vol. 1912, pp. 69–90. Springer, Heidelberg (2000). https://doi.org/10.1007/3-540-44518-8_6
17. Simhandl, G., Paulweber, P., Zdun, U.: Design of an executable specification language using eye tracking. In: EMIP 2019 (at ICSE 2019), May 2019

Exploring the Concept of Abstract State Machines for System Runtime Enforcement

Elvinia Riccobene[1] and Patrizia Scandurra[2]

[1] Dipartimento di Informatica, Università degli Studi di Milano, Milan, Italy
`elvinia.riccobene@unimi.it`
[2] Department of Economics and Technology Management, Information Technology and Production, Università degli Studi di Bergamo, Bergamo, Italy
`patrizia.scandurra@unibg.it`

Abstract. Modern intelligent software systems are rapidly growing in complexity and scale, and many real usage scenarios might be impossible to reproduce and validate at design-time. As envisioned by the Models@run.time research community, the use of formal models at runtime are fundamental to address this challenge. In this paper, we explore the concept of *ASM@run.time* and put this definition into the context of the *runtime enforcement* technique to address the runtime assurance of software systems. This is a work-in-progress research line.

1 Introduction

Modern intelligent software systems, such as those employed in smart infrastructures using big data, AI and IoT technologies, are rapidly growing in complexity and scale, and many real usage scenarios might be impossible to reproduce and validate at design-time. To address this challenge, the Models@run.time research community [4] has identified a reference architecture to equip a software system with a model running in tandem with the system to address software runtime assurance. Similar ideas have been proliferating in other contexts, such as *Digital twins* in the manufacturing domain [11], and *Living models* [9] in the field of Computer Automated Multi-Paradigm Modelling.

Among the different approaches and techniques proposed in literature that exploit the concept of model@runtime, *runtime enforcement* [6] is a runtime verification method that focuses on steering system executions with the goal of preventing and reacting to misbehaviours and failures. Runtime enforcement techniques enforce the software system to run according to its specification, for example the specification of safety assertions that describe situations (states) or actions that must be avoided (e.g., a train must not open its doors when moving). When a new (input) event occurs that may change the state of the software system, the model, if available, is used to evaluate safety assertions and prevent the system change if it violates an assertion on the runtime model of the system. This enforcement mechanism can be, therefore, used for *input sanitisation* [6] to

© Springer Nature Switzerland AG 2020
A. Raschke et al. (Eds.): ABZ 2020, LNCS 12071, pp. 244–247, 2020.
https://doi.org/10.1007/978-3-030-48077-6_18

protect the system from its (untrusted) environment. All inputs to the system shall enter first the enforcement mechanism which filters out those that could harm the system or ensure that all the necessary inputs are provided to the system. While classical runtime verification approaches (like [2,5] and [7] to name a few) generally focus on the *oracle problem*, namely assigning verdicts to a system execution, runtime enforcement focuses on ensuring the correctness of the sequence of events by possibly modifying or preventing the system execution [6].

In this paper we present some preliminary results of our work-in-progress investigation on the use of *Abstract State Machines* as models@run.time for runtime enforcement. In particular, we present the architecture of a runtime enforcement tool we have been developing within the ASMETA framework[1] – a set of tools for the ASM formal method – to check safety assertions of software systems at runtime. This mechanism exploits the concept of executable ASM models and it is based on a new component, the `AsmetaS@run.time`, that simulates the ASM models in tandem with the real software systems. We also envision some real scenarios in the context of safety-critical systems where we are applying the ASM@run.time enforcement approach.

2   Runtime Enforcement with AsmetaS@run.time

We here present a conceptual view of a runtime enforcement for *input sanitisation* [6] to protect the system from its (untrusted) environment. The proposed mechanism exploits the runtime simulator for ASMs, namely `AsmetaS@run.time`. This last tool was recently developed as part of the ASMETA toolset to allow the use of ASM models as runtime models. It supports simulation *as-a-service* features including model roll-back to the initial state after a failure of the model execution (e.g., invariant violations, inconsistent updates, ill-formed inputs, etc.) while processing an input event.

The intent of the proposed runtime enforcement mechanism is to evaluate safety assertions when there is a new (input) event that may change the state of the system and prevent the change if it violates an assertion on the ASM runtime model of the system. As shown in Fig. 1, every attempt (or only those considered critical) to change the system state is mediated by a process (the *enforcer*) that decides whether the change is safe. To make this decision (one per each observed event), the enforcer process evaluates the effect of the event on an ASM model of the behavior of the system (or a subpart of it dealing with the most critical requirements) that runs on-board the system as runtime model. If we are in a safe state (both the system and its runtime ASM model) and there is a new (input) event that may change this state, the enforcer performs first the state transfer on the ASM model (by feeding the input event to the ASM in terms of a monitored function value) and makes sure that the transition (that may take several machine steps – an ASM run) will take the ASM to a state without violating an invariant or generating an inconsistent update. If the ASM will produce a (safe) change of state, the enforcer confirms the state transfers also to

---

[1] http://asmeta.sourceforge.net/.

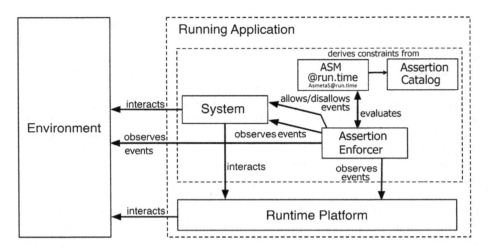

Fig. 1. Runtime safety assertion enforcement by `AsmetaS@run.time`

the system, otherwise (the change is considered unsafe) it prevents the system to react to the event and a state rollback of the ASM model is performed to move it back to its previous (safe) state before the input event was processed. Therefore, the system is allowed to react to an event only when the model successfully checked the safety constraints for the event.

We assume that there exists a *catalog of safety assertions* (expressed in any suitable language) describing all possible situations that may produce a violation of safety, and that these assertions have been expressed in the ASM runtime model in terms of ASM invariants. The catalog may be dynamically updated at runtime in case dangerous situations have not been foreseen at design time or because of unanticipated changes in the requirements when such changed requirements are added at runtime as effect of unanticipated adaptation (such as service-based applications plugging in new services and components discovered at runtime to improve quality of service). We assume that the invariants corresponding to the assertions added to the catalog dynamically are added to the ASM runtime model dynamically, as well.

The runtime enforcement technique could be useful to prevent the execution of unsafe commands in cyber physical systems where the environment is only partially observable [8], and, in general, in any safety-critical system, where the effects of not enforcing the safety assertions would lead to human hazards, as it happens for medical software [1]. We do not target hard real-time systems since these systems require dedicated solutions (e.g., real-time operating systems) and pose specific challenges.

3   Conclusion

In this paper, we have presented our long-term vision of using the ASM executable models as formal support to the runtime enforcement technique to assure safe execution of a software system. Our short-term plan is to complete the implementation of the other components (the `Assertion Catalog` and the `Assertion Enforcer`) of the proposed runtime enforcement mechanism based on the new `AsmetaS@run.time` component. We also want to test its effective operation in the area of safety-critical systems, as for example those in the medical software domain.

In the future, we plan to extend the ASM@run.time enforcement approach in the context of self-adaptive systems [3,10]. Our long term goal is to develop a complete framework able to deal with requirements changes also affecting the model behavior, and therefore providing model adaptation features at runtime.

References

1. Alemzadeh, H., Kalbarczyk, Z., Iyer, R., Raman, J.: Analysis of safety-critical computer failures in medical devices. IEEE Secur. Priv. **11**(4), 14–26 (2013). https://doi.org/10.1109/MSP.2013.49
2. Arcaini, P., Gargantini, A., Riccobene, E.: CoMA: conformance monitoring of Java programs by abstract state machines. In: Khurshid, S., Sen, K. (eds.) RV 2011. LNCS, vol. 7186, pp. 223–238. Springer, Heidelberg (2012). https://doi.org/10.1007/978-3-642-29860-8_17
3. Arcaini, P., Riccobene, E., Scandurra, P.: Formal design and verification of self-adaptive systems with decentralized control. ACM Trans. Auton. Adapt. Syst. **11**(4), 25:1–25:35 (2017)
4. Bencomo, N., Götz, S., Song, H.: Models@run.time: a guided tour of the state of the art and research challenges. Softw. Syst. Model. **18**(5), 3049–3082 (2019). https://doi.org/10.1007/s10270-018-00712-x
5. Calinescu, R., Kikuchi, S.: Formal methods @ runtime. In: Calinescu, R., Jackson, E. (eds.) Monterey Workshop 2010. LNCS, vol. 6662, pp. 122–135. Springer, Heidelberg (2011). https://doi.org/10.1007/978-3-642-21292-5_7
6. Falcone, Y., Mariani, L., Rollet, A., Saha, S.: Runtime failure prevention and reaction. In: Bartocci, E., Falcone, Y. (eds.) Lectures on Runtime Verification. LNCS, vol. 10457, pp. 103–134. Springer, Cham (2018). https://doi.org/10.1007/978-3-319-75632-5_4
7. Liang, H., Dong, J.S., Sun, J., Wong, W.E.: Software monitoring through formal specification animation. ISSE **5**(4), 231–241 (2009). https://doi.org/10.1007/s11334-009-0096-1
8. Pinisetty, S., Roop, P.S., Smyth, S., Allen, N., Tripakis, S., von Hanxleden, R.: Runtime enforcement of cyber-physical systems. ACM Trans. Embed. Comput. Syst. **16**(5s), 178:1–178:25 (2017). https://doi.org/10.1145/3126500
9. Tendeloo, Y.V., Mierlo, S.V., Vangheluwe, H.: A multi-paradigm modelling approach to live modelling. Softw. Syst. Model. **18**(5), 2821–2842 (2019). https://doi.org/10.1007/s10270-018-0700-7
10. Weyns, D., Iftikhar, M.U.: ActivFORMS: a model-based approach to engineer self-adaptive systems. CoRR abs/1908.11179 (2019). http://arxiv.org/abs/1908.11179
11. Zhuang, C., Liu, J., Xiong, H.: Digital twin-based smart production management and control framework for the complex product assembly shop-floor. Int. J. Adv. Manuf. Technol. **96**(1), 1149–1163 (2018)

ProB and Jupyter for Logic, Set Theory, Theoretical Computer Science and Formal Methods

David Geleßus and Michael Leuschel[(✉)] [iD]

Institut für Informatik, Universität Düsseldorf, Universitätsstr. 1,
40225 Düsseldorf, Germany
{dagel101,michael.leuschel}@hhu.de

Abstract. We present a tool for using the B language in computational notebooks, based on the Jupyter Notebook interface and the PROB tool. Applications of B notebooks include executable documentation of formal models, interactive manuals, validation reports but also teaching of formal methods, logic, set theory and theoretical computer science. In addition to B and Event-B, the tool supports Z, TLA$^+$ and Alloy.

1 Introduction and Motivation

The computational notebook concept has recently become popular in teaching and research, as it allows mixing executable code with rich text descriptions and graphical visualizations. We present a tool which enables B and other formal methods to be used in computational notebooks. Such notebooks have many applications, from teaching formal methods to documenting formal models or generating executable reference documents. Given the foundations of B in set theory and logic, and given that the Unicode syntax of B is identical to or very close to standard mathematical notation, our tool can also be used to produce notebooks for teaching mathematical foundations in general or theoretical computer science in particular. Given that our tool is based on the PROB tool, the notebooks also provide convenient access to its constraint solver.

2 Jupyter Kernel for B

Architecture. Jupyter Notebook [4] is a cross-platform computational notebook interface implemented in Python with a web-based frontend. Originally it was developed under the name IPython Notebook and only supported Python-based notebooks, but it has since been extended to allow using languages other than Python. Support for each language is provided by a Jupyter *kernel*: a language-specific backend that receives input from Jupyter Notebook, processes it using the target language, and returns the results to Jupyter. Jupyter communicates with kernels using a language-agnostic protocol, which allows implementing kernels in languages other than Python. In the case of PROB, the kernel was implemented in Java, as PROB provides a high-level Java API [1], and there is an

© Springer Nature Switzerland AG 2020
A. Raschke et al. (Eds.): ABZ 2020, LNCS 12071, pp. 248–254, 2020.
https://doi.org/10.1007/978-3-030-48077-6_19

existing Java implementation of the Jupyter kernel protocol by the jupyter-jvm-basekernel project [7].

The Jupyter Notebook web interface can also be extended using JavaScript-based plugins. This capability was used to implement syntax highlighting for B.

Interacting with B. At its core, the PROB Jupyter kernel is a simple REPL. It accepts standalone B expressions and predicates as input, which are evaluated or solved using PROB. The results are output as LATEX formulas and rendered by Jupyter Notebook, as shown in the following screenshot:

In [5]:

> 1   f = λx.(x ∈ Z|x * x) ∧ f(y) = 100

Out[5]:

TRUE

Solution:

- $f = \lambda x \cdot (x \in INTEGER \mid x * x)$
- $y = 10$

Markup cells can be used to provide documentation for the evaluation cells:

> The function $\delta s(x, \omega)$ computes the possible states after processing a word ω starting from the states x. For example, after processing the word 111 starting from the initial states S our automaton can be in the following states:

In [7]:   1   δs(S,[1,1,1])

Out[7]: $\{z0, z1, z2, z3\}$

Many PROB features can be accessed using notebook *commands*. For example, prefixing a B expression with the command :`table` displays the result as a table, which is useful for viewing complex values, such as sets of tuples. Additional commands include :`prettyprint` to pretty-print a predicate without evaluating it, :`type` to display an expression's static type, and :`solve` to solve a predicate using PROB's various solver backends (such as Kodkod or Z3).

To load a B machine, the B code can be input directly into a notebook cell, which allows for quick testing and prototyping of short machines. When written this way, the entire B machine needs to be placed in a single notebook cell (it is currently not possible to insert text cells in the middle of the machine), and it cannot refine, extend, or otherwise reference other machines. It is also possible to load external machine files using the :`load` command, which is more convenient for larger machines, and also supports loading machines that reference other machine files.

The loaded machine can be animated, using the :`exec` command to execute operations or events. While a machine is loaded, the input is evaluated in the current state of the animator, meaning that the loaded machine's constants and variables can be used in expressions and predicates. Additional commands such as `:check` and `:browse` are provided to examine the current state of invariants, assertions and operations. It is also possible to exercise PROB's model checker.

PROB's state visualisation features can be called using the `:show` and `:dot` commands, which can for example be used to visualise the current machine state or the animator's state space. The visualisation results are displayed directly in the notebook as raster or SVG images.

Jupyter Notebook's advanced code editing features, such as syntax highlighting and code completion, are also supported by the PROB kernel. Both regular B syntax and custom commands are highlighted, and completion is provided for B keywords, variable names, command names and parameters, etc. The "inspect" feature (accessed using Shift+Tab) provides quick access to command help directly inside the notebook interface.

Working with Notebooks. The ProB Jupyter kernel only handles the actual evaluation of the code in the notebook. Jupyter Notebook provides all other parts of the system: including the web frontend responsible for editing notebooks and rendering the kernel's outputs, and the file format used when saving notebooks.

Using the **nbconvert** tool provided by Jupyter, B notebook files can be converted to a variety of standard formats, including HTML, LaTeX, and PDF. This allows distributing notebooks in a format that can be viewed without Jupyter Notebook, although the resulting files cannot be edited and re-executed like the original notebook.

3   Applications

Industrial. As B notebooks can load and animate external machine files, they can be used to document the behavior of existing models. This is conceptually similar to a trace file, with the advantage that notebooks can include not just operation execution steps, but also explanatory text and evaluation/visualisation calls to demonstrate specific aspects of the machine's state.

Some of PROB's own documentation is currently being converted from static documentation pages to B notebooks. The *Modelling Examples* section, e.g., contains pages which start with an introductory text, usually describing a short logic puzzle or part of a real-world use case of PROB, followed by B code fragments modelling the problem in B and explanations of how PROB can be used to visualize, verify or solve the model. The notebook format is well-suited for this kind of documentation: the code in notebooks can be directly executed by the user and the respective visualisation features can be called directly from the notebook. Below is part of the documentation of PROB's external functions. This documentation is automatically up-to-date and users can experiment themselves with the various external functions before integrating them into their models:

This external function converts a string to lower-case letters. It currently converts also diacritical marks (this behaviour may in future be controlled by an additional flag or option).

Type: $STRING \rightarrow STRING$.

```
In [2]:    1   STRING_TO_LOWER("az-AZ-09-ääöü-ÄÖ")
```

Out[2]: "az-az-09-aaou-ao"

Teaching. In the context of teaching the B language as well as theoretical computer science in general, B notebooks can be used as a format for writing lecture notes and worksheets.

Lecture notes involving B expressions or machines can be written and distributed as notebooks, allowing students to execute the code for themselves and experiment with modifications. Due to its foundations in set theory, the B language can also be used to express many general theoretical computer science concepts, such as finite automata. These concepts can be demonstrated using B notebooks, taking advantage of the LaTeX output and graph visualisation capabilities to display the results in a format familiar to students.

B notebooks can also be used as a format for exercise sheets. Students are provided with a notebook that contains the exercise text and possibly some initial code. They can solve the exercises directly in the notebook and turn in the finished notebook file with their solutions. The nbgrader [9] extension for Jupyter provides support for writing exercise sheet notebooks: it allows marking cells with exercise text as read-only, and cells with solutions so they are removed when the exercises are distributed to students. The extension also assists with grading and also enables automated verification of solutions.

4   Conclusion, Related and Future Work

A formal model is usually derived from a natural language requirements document. A big issue is that of keeping the formal model and natural language in sync. A related issue is that of traceability, tracing natural language requirements to the formal model. In that setting the idea of literate programming [5], mixing the natural language documentation with the program, is appealing. The Z language [8] has always allowed literate programming by interleaving LaTeX commands with Z constructs. A similar capability for the B language is provided by PROB's LaTeX mode [6]. In comparison, the PROB Jupyter kernel focuses more on interactivity. Individual cells of a B notebook can be quickly edited and re-rendered/evaluated, whereas the PROB LaTeX mode can only render the entire document at once. However, the ability to write LaTeX code directly offers more flexibility in terms of formatting and layout, compared to a B notebook converted to LaTeX or PDF using `nbconvert`.

An open-source Jupyter kernel for TLA$^+$ [3] is available. Its feature set is similar to the basic features of the PROB Jupyter kernel: it supports evaluation of standalone TLA$^+$ expressions, as well as loading and checking of TLA$^+$ models using the TLC model checker.

A previous attempt at implementing a notebook-like interface for B was the PROB worksheet interface [2]. Its design and goals were very similar to our work, but the implementation provided its own custom web UI, server, and file format, mainly because extensible notebook implementations like Jupyter were not available at the time (2012–2013). In comparison, using Jupyter as a base significantly reduces the required implementation and maintenance work, and allows B notebooks to benefit from existing tooling for Jupyter, such as `nbconvert` and `nbgrader`.

In summary, this new tool provides a notebook interface to a variety of state-based formal methods. Along with some extensions of PROB itself, such as allowing Greek letters or subscripts in identifiers, it is also of use for applications in teaching of discrete mathematics or theoretical computer science.

Our tool is available for download at:

https://gitlab.cs.uni-duesseldorf.de/general/stups/prob2-jupyter-kernel

A   Appendix

Below we show two partial screenshots of a notebook for theoretical computer science. Observe that mathematical Unicode symbols, subscripts and Greek letters can be used in the B formulas and machines.

The function $\delta s(x, \omega)$ computes the possible states after processing a word ω starting from the states x. For example, after processing the word 111 starting from the initial states S our automaton can be in the following states:

```
In [7]:   1  δs(S,[1,1,1])
```
Out[7]: $\{z0, z1, z2, z3\}$

The automaton accepts the word 111 but not the word 101, because we have:

```
In [8]:   1  δs(S,[1,1,1]) ∩ F
```
Out[8]: $\{z2\}$

```
In [9]:   1  δs(S,[1,0,1]) ∩ F
```
Out[9]: $\emptyset$

These are all words of length 3 which are accepted by our automaton:

```
In [10]:   1  :table {x,y,z| [x,y,z]∈L}
```
Out[10]:

x	y	z
0	1	0
0	1	1
1	1	0
1	1	1

cluster_Z

Alternatively, we can view it as graph:

```
In [6]:   1  :dot expr_as_graph ("0",{x,y| x∈Z & y:δ(x,0)},
          2                       "1",{x,y| x∈S & y∈δ(x,1)})
```
Out[6]:

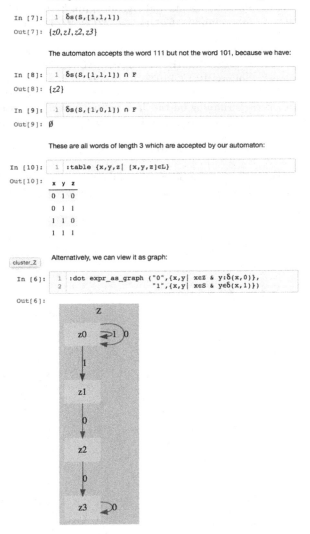

References

1. Bendisposto, J., Clark, J.: ProB Handbook. ProB 2.0. https://www3.hhu.de/stups/handbook/prob2/prob_handbook.html#prob2.0. Assessed 30 Jan 2020
2. Goebbels, R.: Worksheets für die Integration mit ProB. Bachelor's thesis, Heinrich-Heine-Universität Düsseldorf, 18 March 2013
3. Kelvich, S.: kelvich/tlaplus_jupyter: Jupyter kernel for TLA+, 9 December 2019. https://github.com/kelvich/tlaplus_jupyter/. Accessed 17 December 2019

4. Kluyver, T., et al.: Jupyter notebooks – a publishing format for reproducible computational workflows. In: Loizides, F., Schmidt, B. (eds.) Positioning and Power in Academic Publishing: Players, Agents and Agendas, pp. 87–90. IOS Press (2016)
5. Knuth, D.E.: Literate programming. Comput. J. **27**(2), 97–111 (1984). https://doi.org/10.1093/comjnl/27.2.97
6. Leuschel, M.: Formal model-based constraint solving and document generation. In: Ribeiro, L., Lecomte, T. (eds.) SBMF 2016. LNCS, vol. 10090, pp. 3–20. Springer, Cham (2016). https://doi.org/10.1007/978-3-319-49815-7_1
7. Spencer Park. SpencerPark/jupyter-jvm-basekernel: an abstract kernel implementation for Jupyter kernels running on the Java virtual machine. Revision ccfb7bb1, 14 May 2018. https://github.com/SpencerPark/jupyter-jvm-basekernel/. Accessed 02 Aug 2018
8. Spivey, J.M.: The Z Notation: A Reference Manual. Prentice-Hall, Upper Saddle River (1992)
9. Jupyter Development Team. nbgrader—nbgrader 0.5.4 documentation. Revision 808caf33, 18 July 2017. https://nbgrader.readthedocs.io/en/stable/. Accessed 20 Aug 2018

Existence Proof Obligations
for Constraints, Properties and Invariants
in Atelier B

Héctor Ruíz Barradas, Lilian Burdy, and David Déharbe[✉]

CLEARSY Systems Engineering, Aix-en-Provence, France
david.deharbe@clearsy.com

Abstract. Proof obligations of the B method and of Event B use predicates in the Constraints, Sets, Properties and Invariant clauses as hypotheses in proof obligations. A contradiction in these predicates results in trivially valid proof obligations and essentially voids the development. A textbook on the B method [3] presents three "existence proof obligations" to show the satisfiability of the Constraints, Properties and Invariant clauses as soon as they are stated in a component. Together with new existence proof obligations for refinement, this prevents the introduction of such contradictions in the refinement chain. This paper presents a detailed formalization of these existence proof obligations, specifying their implementation in Atelier B.

1 Introduction

The vaunted rigour of formal methods, such as B and Event-B, not only come from the use of a formal notation, but also from the generation and subsequent verification of proof obligations (POs). For instance, in Event-B [2], the model of a system is considered sound only when all POs have been demonstrated. In the B method [1], they guarantee that the refinement-based construction results in implementations faithful to their specification.

Typically, POs are generated at key steps of the design process. Invalid POs reveal errors in the source artefact. By inspecting these proof obligations, the user then identifies, possibly, remaining errors and fixes the source artefact. The process is repeated until all POs are discharged. To conduct the demonstrations, these methods demand that they are conducted with tools. In practice, this is accomplished by a mix of automatic proof and interactive proof. POs are thus the cornerstone of every such formal development.

A PO has the form $H \vdash G$, with H a set of hypotheses, and G the goal. Its validity may stem from a contradiction in H, i.e. have nothing to do with the goal. In the context of B and Event-B, a component with contradictory hypotheses in its POs will be (trivially) correct. In large developments, a contradiction may stay undetected. B addresses this issue with POs associated at the implementation level, i.e. at the very end of the development. At that point, this requires fixing the refinement chain up to the source of the contradiction, which

© Springer Nature Switzerland AG 2020
A. Raschke et al. (Eds.): ABZ 2020, LNCS 12071, pp. 255–259, 2020.
https://doi.org/10.1007/978-3-030-48077-6_20

is costly. Also, components in a B project that do not have an implementation (e.g., foreign interfaces) are not protected. Event-B does not fully address this issue.

Such situations can be easily avoided by adding so called "existence" POs whenever a contradiction may be introduced. An existence PO has the form $\Gamma \Rightarrow \exists V \cdot (\varphi)$, where Γ is the context predicate, φ the predicate that shall not be contradictory, and V a list of identifiers. A textbook on B [3] presents these POs, but without considering component visibility, inclusion and refinement. Existing tools for B and Event-B do not generate these, and we decided to add it to Atelier B. We present the formalization of the POs for the specification (Sect. 2) and the refinement (Sect. 3) levels. We discuss the case of standalone components, and generalize to components with dependencies.

2   Existence Proof Obligations in Specifications

Existence for Parameters. In B, specification components may have sets and scalar parameters. The CONSTRAINTS clause can be used to constrain these parameters. When the machine is instantiated, a PO asks to prove the establishment of the CONSTRAINTS clause, thus guaranteeing the absence of contradictions. If the parametrized machine is not instantiated, the CONSTRAINTS clause can contain undetected contradictions because no PO exists to detect them. Let p denote the parameters, C the predicate in the constraint clause, the existence PO given by [3] for parameters is $\exists p \cdot C$. It has been implemented as such in Atelier B.

Existence for Sets and Constants. The PROPERTIES clause state constraints on sets and constants declared respectively in the SETS and CONSTANTS clauses. Enumerated sets have a single possible valuation, and abstract sets must satisfy the implicit constraint that they are finite non-empty sets of integers. In this way, in order to prove the absence of contradictions in the predicate P of the PROPERTIES clause of a single machine, with no seen or included components, we define the following PO: $e\_sets \Rightarrow \exists(c, s) \cdot (P \wedge a\_sets)$, where e_sets is the conjunction of declarations of enumerated sets in the SETS clause, c is the list of abstract and concrete constants, s is the list of abstract sets, and a_sets is the conjunction of predicates $t \in FIN_1(INTEGER)$ for each variable t in s. Notice that the visibility rules of the language prohibit parameters in the predicate P, so it is useless to have predicate C as an antecedent.

If there are *seen* components in the machine, the predicates in the PROPERTIES clause from the seen components and their included components are in the antecedent of the PO. Moreover, for each abstract set u declared in the seen machine or declared in a machine included by the seen machine, the antecedent of the PO contains a predicate $u \in FIN_1(INTEGER)$. The definition of each enumerated set w declared in these machines is also in the antecedent.

If the machine *includes* components, the definition of their enumerated sets are in the antecedent of the PO, their abstract and concrete constants and the identifiers of their abstract sets are existentially quantified in the consequent and the predicates of their PROPERTIES clauses, together with the corresponding *a_sets* predicates, are in the body of the existential quantifier.

Following is an example of the existence PO for the SETS, CONSTANTS and PROPERTIES clauses for a standalone component:

SETS
$S1; S2 = \{UM, DOIS, TRES\};$
$S3 = \{UN, DEUX\}$
CONSTANTS
$c1, e1, e2, e3$
PROPERTIES
$c1 \in NAT \wedge e1 \in INT \wedge$
$e2 \in S2 \wedge e3 \in S1 \wedge$
$(e2 = UM \Rightarrow e1 = 1))$

PO:

$S2 = \{UM, DOIS, TRES\} \wedge$
$S3 = \{UN, DEUX\}$
$\Rightarrow$
$\exists(c1, e1, e2, S1) \cdot ($
$S1 \in FIN(INTEGER) - \{\{\}\} \wedge$
$c1 \in NAT \wedge e1 \in INT \wedge$
$e2 \in S2 \wedge e3 \in S1 \wedge$
$(e2 = UM \Rightarrow e1 = 1))$

Existence for State Variables. The predicate invariant may also contain contradictions. To prevent this, the existence PO of the INVARIANT clause for a standalone machine is $C \wedge P \wedge all\_sets \Rightarrow \exists(v) \cdot (I)$. The antecedent of this PO contains the predicates C and P from the CONSTRAINTS and PROPERTIES clauses. The predicate *all_sets* is the conjunction of *e_sets* and *a_sets* seen above. The quantifed variable v denotes the list of abstract and concrete variables of the machine.

If there are seen or included components, the antecedent is strengthened with the conjunction of their properties, assertions, invariants and their *all_sets* predicates. In this conjunction, we also consider the clauses of the components possibly included by the seen machines. Moreover, for the included components, the consequent of the PO quantifies over their variables and invariants.

3   Existence Proofs in Refinements

Refinement in B or Event B is used for stepwise development. Refinement POs are designed to be monotonic: If a component S is refined by a component T, these POs guarantee that the invariant of S is also preserved by operations in T. However, existence POs in a refinement are not monotonic in that sense. When an abstract constant or variable is refined by a concrete one, we still need to prove that the properties or invariants specified in the abstraction hold in the refinement.

Existence for Sets and Constants. For a refinement with no seen or included components and no seen or included components in any of its abstractions, the existence PO is intended to avoid contradictions in the predicate P of the PROPERTIES clause of the refinement and all properties of the previous refinements, denoted by the following predicate:

$$e\_sets \wedge abs\_e\_sets \Rightarrow \exists (c, c_a, s, s_a) \cdot (P \wedge a\_sets \wedge abs\_P \wedge abs\_a\_sets)$$

The predicates e_sets and a_sets are defined as before, abs_e_sets denotes the conjunction of declarations of enumerated sets, and abs_a_sets denotes the conjunction of $t \in FIN_1(INTEGER)$, for abstract sets t in previous refinements. Predicate abs_P is the conjunction of the properties predicates in the previous refinements. The variable lists c and s contain the constants of the refinement and its abstract sets. Finally, the lists c_a and s_a denote all constants and abstract sets in previous refinements. If the refinement or any of its abstractions contains seen or included components, the antecedent and the consequent are strengthened with the clauses of these components as it was done in the corresponding PO of the specification.

Existence for State Variables. The corresponding PO defined for specification components guarantees the absence of contradictions in the invariant. Also, the PO of the establishment of the invariant by the initialization $Init_a$ guarantees the existence of values of the abstract variables v_a satisfying the abstract invariant $I(v_a)$. The PO of the refinement of $Init_a$ by the initialization of a refined component $Init_c$ is not sufficient to guarantee the absence of contradictions in the refined invariant $J(v_c, v_a)$. Therefore, in order to prove the absence of contradictions in the invariant $J(v_c, v_a)$ we need to show that the assignment of some concrete values v to the concrete variables v_c is a refinement of $Init_a$. Formally this refinement is stated by $\exists v \cdot ([v_c := v]\neg[Init_a]\neg J$ which must be proved under the context of the refinement. After simplification, the existence PO for a standalone refinement and only standalone components in its abstractions is defined as follows:

$$C \wedge P \wedge all\_sets \wedge abs\_all\_sets \wedge abs\_P \Rightarrow \exists (v_c) \cdot (\neg[Init_a]\neg J)$$

where abs_all_sets is the conjunction of predicates all_sets of previous refinements, v_c is the list of abstract and concrete variables of the refinement and J is its invariant.

If there are seen or included components, the antecedent and consequent of the PO are strengthened with the corresponding clauses of these components.

4   Conclusion

This paper presents details of the generation of existence POs for the formal methods B and Event-B. These POs detect inconsistencies that would make trivial, but useless, the correctness of the components, as soon as they are introduced in the development. Their generation has been implemented and will be available in a future release of Atelier B.

References

1. Abrial, J.-R.: The B-Book, Assigning Programs to Meanings. Cambridge University Press, Cambridge (1996)
2. Abrial, J.-R.: Modelling in Event-B, System and Software Engineering. Cambridge University Press, Cambridge (2010)
3. Schneider, S.: The B-Method. Macmillan International, New York (2001)

VisB: A Lightweight Tool to Visualize Formal Models with SVG Graphics

Michelle Werth and Michael Leuschel[(✉)]

Institut für Informatik, Universität Düsseldorf, Universitätsstr. 1,
40225 Düsseldorf, Germany
{michelle.werth,michael.leuschel}@hhu.de

Abstract. Visualization is important to present formal models to domain experts and to spot issues which are hard to formalise or have not been formalised yet. VISB is a visualization plugin for the ProB animator and model checker. VISB enables the user to create simple visualizations for formal models. An important design criterion was to re-use scalable vector graphics (SVG) generated by off-the-shelf graphic editors using a lightweight and easy-to-use annotation mechanism. The visualizations can be used to formal models in B, Event-B, Z, TLA+ and Alloy.

1 Introduction and Background

The animator and model checker PROB [3] supports both classical B and Event-B, as well as several other formalisms (Z, Alloy and TLA+) which are translated to B. Animation allows the user to experiment with a model, inspecting states, and interactively choose events or operations to execute. Animation is very useful to validate functional behaviour of a model, but also to uncover unexpected behaviour related to issues or requirements the modeller has not yet thought about. Here *graphical visualization* of the current state of a formal model is often essential so that a human can more quickly validate the behaviour or spot unexpected behaviour. To cite Bryan Cantrill:[1] *"The visual cortex is unparalleled at detecting patterns."* and *"The value of visualization is not merely providing answers but especially provoking new questions."*

There are several visualization tools for formal models such as PVSio-Web [7] for PVS, various co-simulation tools for VDM such as [6], and JEB [8], AnimB[2] or Brama [5] for Event-B. There have been several visualization based on PROB in the past, such as the animation functions of [4], BMotionStudio [2] or BMotionWeb [1]. The animation function feature is based on declaring a set of images and writing a B expression which generates a matrix of image numbers. It is still available in current versions of PROB, but it is hard to generate larger, visually appealing visualizations. BMotionStudio still exists within Rodin for Event-B, but is not available for other formalisms and it can be cumbersome to generate

---

[1] https://www.slideshare.net/bcantrill/visualizing-systems-with-statemaps.
[2] Available at http://wiki.event-b.org/index.php/AnimB.

© Springer Nature Switzerland AG 2020
A. Raschke et al. (Eds.): ABZ 2020, LNCS 12071, pp. 260–265, 2020.
https://doi.org/10.1007/978-3-030-48077-6_21

complex visualizations using its editor. BMotionWeb is based on web technologies, and allows to generate very refined visualizations. However, its learning curve is quite steep, and due to its heavy use of web technology and associated frameworks can no longer be maintained by the PROB team. This situation was the starting point for the development of the present VISB technology: it should be both easy to use and maintain, it should not be bound to an editor but allow a user to generate the images using off-the-shelf applications or even to re-use existing images.

2   VisB Principles and Architecture

The core idea of VISB is to use SVG files as the basis of the visualization. An SVG file is shown in Listing 1.1. Such files can be produced by most off-the-shelf editors and their textual XML representation can be programmatically generated.

```
1  <svg height="200" width="200">
2    <circle id="button" cx="100" cy="100" r="80"
3      stroke="black" stroke-width="3" fill="green" />
4  </svg>
```

Listing 1.1. Small SVG file (`button.svg`)

Moreover, SVG files can contain object identifiers (such as `button` for the circle in Listing 1.1) and it is possible (e.g. using jQuery and JavaScript) to load an SVG file and programmatically find objects from an identifier and set attributes of the found objects, and immediately display the changes. This is the basis of VISB, whose core is written in Java, JavaFX and JavaScript, and whose architecture is shown in Fig. 1.

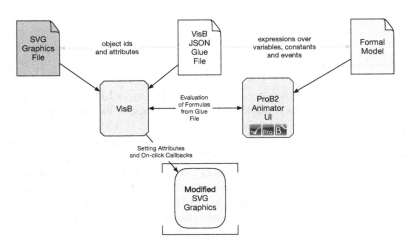

Fig. 1. VISB architecture

This architecture makes VISB easy to maintain because it allows the PROB team to mostly use Java and JavaFX in development, while cutting down the interactions with web languages (such as JavaScript) to a bare minimum.

The link to the formal model is provided by a lightweight glue file (see Listing 1.2), that provides two lists. VISB-items consist of SVG object identifiers, attributes, and expressions that provide the value the attribute should take depending on the state of the formal model. VISB-events link formal model events (aka operations or actions, depending on the formalism) to object identifiers. These events are executed when the object is clicked by the user.

The motivation was to keep the foundation of VISB simple, and not to require the user to learn any new programming language (e.g., JavaScript, Flash, ...). The user just has to know relevant expressions or variables from the formal model and corresponding object identifiers in the SVG graphics file. Moreover, VISB works for all of PROB's supported state-based formalisms (B, Event-B, Z, TLA+, Alloy) in an identical fashion.

```
1  {
2    "svg":"button.svg",
3    "items":[
4      {
5        "id":"button",
6        "attr":"fill",
7        "value":"IF button=TRUE THEN \"green\" ELSE \"red\" END"
8      }
9    ],
10   "events":[
11     {
12       "id":"button",
13       "event":"press_button",
14       "predicates":[
15         "status=TRUE"
16       ]
17     }
18   ]
19 }
```

Listing 1.2. Minimal example for VISB file

3   VisB Examples

One of the simpler examples of a VISB file is shown in Listing 1.2. The corresponding machine contains a bool variable and an operation called press_button that changes the status of this variable. We use the fact that PROB allows IF-THEN-ELSE and LET for expressions (to simplify the syntax of the VISB file). In Listing 1.2 the fill attribute of the SVG object with the identifier "button" is changed to green whenever the button variable "button" in the corresponding machine is set to true. This is realized with the IF-THEN-ELSE expression in the value attribute. For the visualization this means that the circle's color is changed from red to green, when the operation press_button is executed. Thanks to the VISB-events, the user can also execute press_button directly by clicking on the SVG object with the identifier "button".

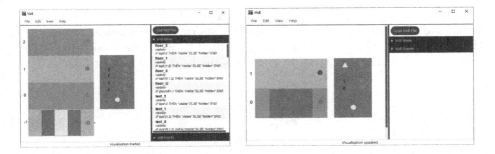

Fig. 2. Example of VisB visualization of lift model

The first visualisation created with VisB can be seen in Fig. 2. In the formal model, the state of a lift is represented by three variables: the current floor, an integer value between the ground floor and the top floor (`topf`), the current direction of the lift and a boolean variable indicating whether the door is open or not. In addition, the lift controller maintains the status of calling buttons inside the lift and on each floor. To cater for different number of floors, represented by the constant `topf`, we have made use of the SVG "visibility" attribute to hide unused floors (see right of Fig. 2). Note that each floor is represented by five graphical objects. To avoid having to hide each object of a given floor individually, we have grouped the objects for each floor together. VisB can then be used to hide or show all objects of a floor in one go, as shown in Listing 1.3.

```
1   ... {
2        "id":"gFloor_2",
3        "attr":"visibility",
4        "value":"IF topf>=2 THEN \"visible\" ELSE \"hidden\" END"
5   }, ...
```

Listing 1.3. VisB item with grouping of SVG elements

```
1   ... {
2        "id":"lift",
3        "attr":"y",
4        "value":"IF cur_floor=2 THEN \"3.207\" ELSIF cur_floor=1 THEN \"76.
         974\" ELSIF cur_floor=0 THEN \"150.474\" ELSE \"224.574\" END"
5   },
6   {
7        "id":"door_right",
8        "attr":"y",
9        "value":"IF cur_floor=2 THEN \"3.207\" ELSIF cur_floor=1 THEN \"76.
         974\" ELSIF cur_floor=0 THEN \"150.474\" ELSE \"224.574\" END"
10  },
11  {
12       "id":"door_left",
13       "attr":"y",
14       "value":"IF cur_floor=2 THEN \"3.207\" ELSIF cur_floor=1 THEN \"76.
         974\" ELSIF cur_floor=0 THEN \"150.474\" ELSE \"224.574\" END"
15  }, ...
```

Listing 1.4. Change "y" Attribute of the lift

Unfortunately, not all attributes can be changed for groups of SVG objects in this way. For example, the x and y coordinates cannot be changed for groups. Hence, to achieve the vertical movement of the lift cabin, we need three VISB items, each changing the attribute y to the same value (see Listing 1.4).

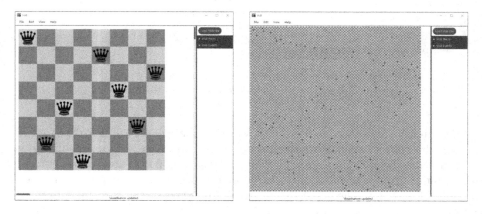

Fig. 3. Example of VISB visualization of N-queens problem

A solution to this drawback is to use embedded SVGs (i.e., nested SVG graphics embedded in the master SVG file) where it is possible to change the coordinates of those embedded SVGs. We have used this for the VISB visualization of the n-queens problem, partially shown in Listing 1.5, where the VISB items needed for the visualization of one given queen is shown. (Note, that the value of the second VISB item is not complete.) Additionally, each chess tile has a VISB event which triggers a B event to place a queen on that tile.

```
 1  ... {
 2        "id": "svgQueen1",
 3        "attr": "visibility",
 4        "value" : "IF 1:dom(queens) THEN \"visible\" ELSE \"hidden\" END"
 5  },
 6  {
 7        "id": "svgQueen1",
 8        "attr": "y",
 9        "value" :"IF  1|->2:queens THEN \"45\" ELSIF 1|->3:queens THEN \"90
             \" ELSIF 1|->4:queens THEN \"135\" [...] ELSIF 1|->20:queens
             THEN \"855\" ELSE \"0\" END"
10  },
11  {
12        "id": "svgQueen1",
13        "attr": "fill",
14        "value" : "IF is_attacked(1) & 1:dom(queens) THEN \"red\" ELSE \"
             black\" END"
15  }, ...
```

Listing 1.5. Example of VISB items for one queen in n-queens problem

For the n-queens problem, we programatically created the VISB file for the chess field and queens, which enabled us to visualize bigger chess fields (120 × 120), as you can see on the right in Fig. 3.

A more complex example can be found in our ABZ 2020 case study article in the present proceedings, where SVGs were received from coordinators of the case study and used to visualize various classical B and Event-B models.

In conclusion, thus far we seem to have met our goals of developing lightweight, easy-to-use and easy-to-maintain visualization technology, which nonetheless is flexible enough for creating simple academic visualizations up to complex, full-fledged industrial applications. VIsB is available for download at:

https://www3.hhu.de/stups/prob/index.php/VisB

References

1. Ladenberger, L.: Rapid creation of interactive formal prototypes for validating safety-critical systems. Ph.D. thesis (2016)
2. Ladenberger, L., Bendisposto, J., Leuschel, M.: Visualising event-B models with B-motion studio. In: Alpuente, M., Cook, B., Joubert, C. (eds.) FMICS 2009. LNCS, vol. 5825, pp. 202–204. Springer, Heidelberg (2009). https://doi.org/10.1007/978-3-642-04570-7_17
3. Leuschel, M., Butler, M.J.: ProB: an automated analysis toolset for the B method. STTT **10**(2), 185–203 (2008). https://doi.org/10.1007/s10009-007-0063-9
4. Leuschel, M., Samia, M., Bendisposto, J., Luo, L.: Easy graphical animation and formula viewing for teaching B. In: The B Method: From Research to Teaching, pp. 17–32 (2008)
5. Servat, T.: BRAMA: a new graphic animation tool for B models. In: Julliand, J., Kouchnarenko, O. (eds.) B 2007. LNCS, vol. 4355, pp. 274–276. Springer, Heidelberg (2006). https://doi.org/10.1007/11955757_28
6. Thule, C., Lausdahl, K., Gomes, C., Meisl, G., Larsen, P.G.: Maestro: the INTO-CPS co-simulation framework. Simul. Model. Pract. Theory **92**, 45–61 (2019)
7. Watson, N., Reeves, S., Masci, P.: Integrating user design and formal models within PVSio-web. In: Masci, P., Monahan, R., Prevosto, V. (eds.) Proceedings Workshop Formal Integrated Development Environment. EPTCS, vol. 284, pp. 95–104 (2018)
8. Yang, F., Jacquot, J., Souquières, J.: JeB: safe simulation of event-B models in JavaScript. In: Muenchaisri, P., Rothermel, G. (eds.) Proceedings APSEC 2013, pp. 571–576. IEEE Computer Society (2013)

Towards a Shared Specification Repository

Philipp Körner[(✉)](ID), Michael Leuschel[(✉)](ID), and Jannik Dunkelau(ID)

Institut für Informatik, Universität Düsseldorf,
Universitätsstr. 1, 40225 Düsseldorf, Germany
{p.koerner,michael.leuschel,jannik.dunkelau}@uni-duesseldorf.de

Abstract. Many formal methods research communities lack a shared set of benchmarks. As a result, many research articles in the past have evaluated new techniques on specifications that are specifically tailored to the problem or not publicly available. While this is great for proving the concept in question, it does not offer any insights on how it performs on real-world examples. Additionally, with machine learning techniques gaining more popularity, a larger set of public specifications is required. In this paper, we present our public set of B machines and urge contribution. As we think this to be an issue in other communities in scope of the ABZ as well, we are also interested in specifications expressed in other formalisms, for example Alloy, TLA$^+$ or Z.

1   Introduction and Motivation

Our group in Düsseldorf has collected since 2003 thousands of B and Event-B machines: our PROB repository contains around 13 000 machines, of which more than 3500 are publicly available. The examples are used for PROB's regression, performance and feature tests. Those public examples contain some duplicates, as they are compiled from different sources: e.g., from tickets in our bug tracker, teaching, literature, case studies, or student projects.

Naturally, not all machines are relevant to all research questions: infinite state spaces might be interesting in order to evaluate symbolic model checking techniques [11], whereas large yet finite state spaces are the important class for distributed model checking [10]. Other use cases, such as data validation [7] work by executing a model along one particular, linear path, while others, like constraint solving problems, sometimes work on machines without variables, consisting of a single state. Most recently, machine learning (ML) techniques are applied to model checking or synthesis as well, and require a large number of specifications, e.g., in order to extract and re-combine predicates [6]. Even with access to numerous machines, it is time-consuming and cumbersome to identify machines to use for benchmarking, especially since only a small amount of data can be presented in a typical research article. Without any doubt, other research groups have their individual set of B machines they use for testing and evaluation. Thus, we propose that individual sets of benchmarks from different

© Springer Nature Switzerland AG 2020
A. Raschke et al. (Eds.): ABZ 2020, LNCS 12071, pp. 266–271, 2020.
https://doi.org/10.1007/978-3-030-48077-6_22

parts of the community are combined into a global, shared repository. With this paper, we start this endeavour, and create an index of our specifications as described in Sect. 2. Benefits include:

– Benchmarks are publicly available and experiments can be replicated easily.
– Performance comparisons of several tools in different versions can be drawn.
– Suitable benchmarks can be quickly identified.
– Examples for translations between formalisms or ML are available.
– Particularly successful examples can be shared for teaching.

While we are most involved in the B and Event-B community, we think that similar issues are present in other communities which make up the ABZ conference. Thus, we explicitly want to invite everyone to contribute specifications written in other formalisms as well. The repository is located at:

https://github.com/hhu-stups/specifications

2 Proposed Index

Since our initial set of models is rather large, it is vital that a sufficient amount of meta-information is attached to the models. For this, we suggest usage of edn[1], a serialisation format with parsers available in most mainstream programming languages. For each specification, some basic information should be offered:

– Which formalism is this specification written in?
– A SHA-256 hash code to identify duplicates, and to ensure reproducibility of experiments regarding the specification.
– Number of deferred sets, enumerated sets, constants, state variables and operations/events, number of included machines, etc.
– Number of states and state transitions in the machine (if known).[2]
– Presence of invariant violations, deadlocks, etc. (if the property is known).
– Optional link to another (previous) model (e.g., a correction or evolution).

The information above is known to never change, but can be extended once further properties are considered. Additional information depending on the tool, its configuration or the use case altogether can be included as well, such as temporal properties (e.g., expressed in LTL or CTL) which are expected to hold or to be violated, tool name and version/revision which is able to parse or execute the specification, or settings, walltime and memory usage required for application of a technique such as model checking.

---

[1] Extensible Data Notation, see: https://github.com/edn-format/edn.
[2] Note that different tools count the number of transitions and states slightly differently. it might be necessary to keep track of the number of initial states and, e.g., the virtual constants setup states of prob. then, one can derive the expected statistics for other validation tools. some settings can also influence the number of states, e.g., the default scope for deferred sets or maximum number of transitions per operations. in that case, it is preferable not to specify a number of states, but rather include that number in a specific run of the tool (see below), that also includes the settings needed for replication.

Optional Fields. Naturally, this data must also be extensible via optional fields. For instance, additional information due to a new use case can be gathered, e.g., the amount of states when using state space reduction techniques. As runtime might depend on the hardware it was ran on, relevant data should be included as well. They also allow extension of the information, e.g., for further tools such as Atelier-B [4] or handling of entirely different file formats, e.g., Rodin [1] archives. In order to select suitable set of specifications, one can simply apply a filter predicate testing the formalism or dialect of it. Furthermore, optinal fields enable links between different machines (e.g., due to refinement or different parameter instantiation) and to external information, such as references to articles describing the model, descriptions of the models as well as the author(s) and their contact information. Finally, certain metrics do not make sense for specific use cases of a formalism, or cannot be applied to other formalisms at all. Thus, such data must not be a mandatory field (but may be mandatory for a given formalism)[3].

Filtering Specification. As previously mentioned, we use edn for the meta information because this format can easily be processed. A short example written in Clojure is given in Listing 1.1. There, all files containing meta-information in the directory are located (ll. 1–5). Then, they are read in and filtered (ll. 7–15). The expression starting in l. 9 returns a list of all file names of specifications written in the B formalism that are known to have a state space of at least 100 000 states. At the time of writing, there are 45 such machines. This example shows that finding specifications based on certain criteria is fairly easy and necessary for verification tool maintainers.

```
1   (def META-INF-DIR (java.io.File. "../meta-information"))
2
3   ;; get a sequence of all meta-information files in the directory
4   (def meta-files (remove (fn [file] (.isDirectory file))
5                            (file-seq META-INF-DIR)))
6
7   (defn read-meta-file [f] (read-string (slurp f)))
8
9   (->> meta-files
10      (map read-meta-file)
11      (filter (fn [data]
12                  (and (= (:formalism data) :b)
13                       (number? (:number-of-states data))
14                       (> (:number-of-states data) 100000)))))
15      (map :file))
```

Listing 1.1: Finding Specifications Based on Their Information

---

[3] It would be sensible to define different standard formats for different formalisms. These can be automatically enforced in a CI pipeline, e.g., by Clojure Spec [5], before pull requests are accepted.

Table 1. Overview of available machine meta data with a timeout of 30 min.

Errors on Load			310
Formalism		763 Event-B	2886 Classical B
Deadlock found	1080 yes	1576 no	683 timeout
Invariant violated	255 yes	2498 no	586 timeout
	max	*avg*	*usage in # machines*
States	1 000 002	8743	2624
Transitions	5 570 544	53 296	2624
Included Machines	13	1.18	3339
State Variables	10 000	7.49	2282
Operations	2000	6.00	2497
Deferred Sets	50	0.44	669
Enumerated Sets	19	0.79	1310
Invariants	10 000	9.39	1958
Constants	10 000	8.63	2090
Properties	12 015	17.51	2094
Static Assertions	188	1.46	646
Dynamic Assertions	54	0.20	200
Definitions	374	2.75	1265

Table 1 provides an overview of the information of B machines currently present in the repository, compiled after running each machine with a timeout of 30 min in the PROB model checker.

On Updating Versions. We strongly argue that the published version of a specification *must not be replaced*. Once they are online, they may be used by any researcher. Even though git clearly documents the history of a file, it would be unclear which version was used as a benchmark or presented in an article. If mistakes were spotted, new versions can be submitted *as a modified copy*.

3   Conclusions, Related and Future Work

We firmly believe that a shared repository of specifications will benefit all communities coming together at ABZ. Aside from making benchmarks available for replication, it can assist courses teaching the formal methods. Furthermore, it builds the foundation for exciting new research that relies on such a dataset.

Similar issues have been found in other communities. This led to the creation of central benchmarking sets, e.g., BEEM for models written in DVE [13], or the PRISM benchmark suite [12] for models written in PRISM. Yet, to our knowledge, it is not possible to contribute to these databases. This has led to criticism that, e.g., not many models that are large enough are featured. Also, a fixed set of benchmarks is not a viable approach in the B community, that creatively uses the B language in order to solve very different types of problems.

In other communities, such as SMT and SAT solving, shared benchmark sets are established for many years [3,8]. They both grow via community contributions and are the foundation for solver competitions [2,9]. SMT-LIB in particular is a success story, containing more than 100 000 benchmarks. There are many other examples for competitions and problem collections, e.g., SV-COMP[4], TPLP[5] [15], which we cannot exhaustively list here due to page limitations.

An interesting question we could not answer in this paper is to what extent our examples match the reality of (confidential) industrial specifications. An answer requires to take a closer look at the data that is available to us. When considering state space size, number of variables and operations as well as idioms used, e.g., usage of program counters or certain data structures, it might be possible to label some public machines accordingly.

Furthermore, research papers often contain links to download pages not only for benchmarks, but also tools themselves. Some tools presented years ago are hard or near impossible to find now. Some conferences, e.g., POPL, established artifact evaluation committees, yet making artifacts permanently available often is optional. ACM conferences offer different badges[6] depending on availability, replicability, etc. A similar, *mandatory* repository containing at least one binary version or even the source code of tools presented at conferences might prove useful to the research community as well. Worth mentioning here is the StarExec platform [14], that allows storage and execution of tools and benchmark problems, which may serve this effort to a satisfactory extent already.

In order for the presented endeavour to be successful, the effort of the entire community is required and their contributions to this repository will be appreciated.

Acknowledgement. Computational support and infrastructure was provided by the "Centre for Information and Media Technology" (ZIM) at the University of Düsseldorf (Germany). We thank the many persons who contributed to the repository (a list is available at the project's website).

References

1. Abrial, J.-R., Butler, M., Hallerstede, S., Voisin, L.: An open extensible tool environment for Event-B. In: Liu, Z., He, J. (eds.) ICFEM 2006. LNCS, vol. 4260, pp. 588–605. Springer, Heidelberg (2006). https://doi.org/10.1007/11901433_32
2. Barrett, C., de Moura, L., Stump, A.: SMT-COMP: satisfiability modulo theories competition. In: Etessami, K., Rajamani, S.K. (eds.) CAV 2005. LNCS, vol. 3576, pp. 20–23. Springer, Heidelberg (2005). https://doi.org/10.1007/11513988_4

---

[4] https://sv-comp.sosy-lab.org/2020/.

[5] Which inspired the second author to generate another library, Dozens of Problems for Partial Deduction https://github.com/leuschel/DPPD.

[6] Cf. https://www.acm.org/publications/policies/artifact-review-badging.

3. Barrett, C., Stump, A., Tinelli, C.: The SMT-LIB standard - version 2.0. In: Proceedings SMT 2010, July 2010
4. ClearSy. Atelier B, User and Reference Manuals. Aix-en-Provence, France (2016). http://www.atelierb.eu/
5. Clojure Spec Guide. https://clojure.org/guides/spec. Accessed 12 Mar 2020
6. Dunkelau, J., Krings, S., Schmidt, J.: Automated backend selection for PROB using deep learning. In: Badger, J.M., Rozier, K.Y. (eds.) NFM 2019. LNCS, vol. 11460, pp. 130–147. Springer, Cham (2019). https://doi.org/10.1007/978-3-030-20652-9_9
7. Hansen, D., Schneider, D., Leuschel, M.: Using B and ProB for data validation projects. In: Butler, M., Schewe, K.-D., Mashkoor, A., Biro, M. (eds.) ABZ 2016. LNCS, vol. 9675, pp. 167–182. Springer, Cham (2016). https://doi.org/10.1007/978-3-319-33600-8_10
8. Hoos, H.H., Stützle, T.: SATLIB: an online resource for research on SAT. In: SAT 2000, pp. 283–292. IOS Press (2000)
9. Järvisalo, M., Le Berre, D., Roussel, O., Simon, L.: The international SAT solver competitions. Ai Mag. **33**(1), 89–92 (2012)
10. Körner, P., Bendisposto, J.: Distributed model checking using ProB. In: Dutle, A., Muñoz, C., Narkawicz, A. (eds.) NFM 2018. LNCS. Springer, Cham (2018). https://doi.org/10.1007/978-3-319-77935-5_18
11. Krings, S.: Towards infinite-state symbolic model checking for B and Event-B. Ph.D. thesis, Universitäts-und Landesbibliothek der HHU Düsseldorf (2017)
12. Kwiatkowska, M., Norman, G., Parker, D.: The PRISM benchmark suite. In: Proceedings QEST 2012, pp. 203–204. IEEE CS Press, September 2012
13. Pelánek, R.: BEEM: benchmarks for explicit model checkers. In: Bošnački, D., Edelkamp, S. (eds.) SPIN 2007. LNCS, vol. 4595, pp. 263–267. Springer, Heidelberg (2007). https://doi.org/10.1007/978-3-540-73370-6_17
14. Stump, A., Sutcliffe, G., Tinelli, C.: StarExec: a cross-community infrastructure for logic solving. In: Demri, S., Kapur, D., Weidenbach, C. (eds.) IJCAR 2014. LNCS (LNAI), vol. 8562, pp. 367–373. Springer, Cham (2014). https://doi.org/10.1007/978-3-319-08587-6_28
15. Sutcliffe, G.: The TPTP problem library and associated infrastructure. From CNF to TH0, TPTP v6.4.0. J. Autom. Reason. **59**(4), 483–502 (2017)

Refinement and Verification
of Responsive Control Systems

Karla Morris[1($\boxtimes$)], Colin Snook[2], Thai Son Hoang[2], Geoffrey Hulette[1],
Robert Armstrong[1], and Michael Butler[2]

[1] Sandia National Laboratories, Livermore, CA, USA
knmorri@sandia.gov
[2] ECS, University of Southampton, Southampton, UK

Abstract. Statechart notations with 'run to completion' semantics, are popular with engineers for designing controllers that respond to events in the environment with a sequence of state transitions. However, they lack formal refinement and rigorous verification methods. Event-B, on the other hand, is based on refinement from an initial abstraction and is designed to make formal verification by automatic theorem provers feasible. We introduce a notion of refinement into a 'run to completion' statechart modelling notation, and leverage Event-B's tool support for theorem proving. We describe the difficulties in translating 'run to completion' semantics into Event-B refinements and suggest a solution. We outline how safety and liveness properties could be verified.

Keywords: Run-to-completion · Statecharts · Refinement

1 Introduction

Reactive Statecharts are open systems capable of receiving potentially nondeterministic input. This work, which builds on our previous work [7,8], exposes a srhallow embedding of open Statecharts semantics in Event-B. Statecharts provide a graphical language, generalized from state machines, that is popular with engineers, variants of which appear in Matlab Simulink/Stateflow [6] and the Ansys tools. Particularly attractive is providing accessibility to abstraction/refinement via Rodin/Event-B which has an intuitive metaphor in the Statechart semantics [7,8]. The commercial tools have similar ideas expressed as encapsulation and composition but not entailing any formal guarantees. The hope is that engineers can better understand the origin of proof obligations in refinements and achieve formal guarantees earlier in their designs where it is most tractable.

Related work has developed a number of different semantics all with different purposes and outcomes [2,3,5]. Because our contribution is focused on a mapping to Event-B, safety property preserving refinement is key. Event-B provides not only a definition of refinement but a rubric for finding valid refinements and this

---

Under the terms of Contract DE-NA0003525, there is a non-exclusive license for use of this work by or on behalf of the U.S. Government.

A. Raschke et al. (Eds.): ABZ 2020, LNCS 12071, pp. 272–277, 2020.
https://doi.org/10.1007/978-3-030-48077-6_23

is carried over into the Statecharts work presented here. In our version of Statechart semantics, refinement means a subsetting of traces from an abstraction. This has the beneficial effect of preserving safety properties from abstraction to refinement and permits proofs to be discharged at the highest tractable level of abstraction. It is at the highest level of abstraction that proofs are presumably the easiest to discharge.

2 Background

SCXML is a modelling language based on Harel statecharts [12]. State-Chart XML (SCXML) follows a 'run to completion' semantics, where trigger events may be needed to enable transitions. Trigger events are queued when they are raised, and then one is de-queued and consumed by firing all the transitions that it enables, followed by any (un-triggered) transitions that then become enabled due to the change of state caused by the initial transition firing. This is repeated until no transitions are enabled, and then the next trigger is de-queued and consumed. There are two kinds of triggers: internal triggers are raised by transitions and external triggers are raised by the environment (nondeterministically for the purpose of our analysis). An external trigger may only be consumed when the internal trigger queue has been emptied.

Event-B is a formal method for system design [1,4]. It uses *refinement* to introduce system details gradually into the formal model. An Event-B model contains two parts: *contexts* and *machines*. Contexts contain *carrier sets*, *constants*, and *axioms* constraining the carrier sets and constants. Machines contain *variables* v, *invariants* I(v) constraining the variables, and *events*. An event consists of a guard denoting its enabled-condition and an action defining the value of variables after the event is executed. In general, an event e has the form: any t where G(t, v) then S(t, v) end where t are the event parameters, G(t, v) is the guard of the event, and S(t, v) is the action of the event.

Machines can be refined by adding more details. Refinement can be done by extending the machine to include additional variables (*superposition refinement*) representing new features of the system, or by replacing some (abstract) variables by new (concrete) variables (*data refinement*).

UML-B State-machines provides a diagrammatic modelling notation for Event-B in the form of state-machines and class diagrams [9–11]. The diagrammatic models relate to an Event-B machine and generate or contribute to parts of it.

Each state is encoded as a boolean variable and the current state is indicated by one of the boolean variables being set to TRUE. An invariant ensures that only one state is set to TRUE at a time. Events change the values of state variables to move the TRUE value according to the transitions in the state-machine.

While the UML-B translation deals with the basic data formalisation of state-machines it differs significantly from the semantics discussed in this manuscript. UML-B adopts Event-B's simple guarded action semantics and does not have a concept of triggers and run-to-completion. Here we make use of UML-B's state-machine translation but provide a completely different semantic by generating a behaviour into the underlying Event-B events that are linked to the generated UML-B transitions.

3 Run to Completion

The run to completion semantics is specified via an abstract basis that is extended by the model. The specification of this basis consists of an Event-B *context* and *machine* that are the same for all input models and are refined by the specific output of the translation. This allows us to introduce an abstract behaviour of transitions queueing and using triggers which is gradually refined to introduce the actual triggering and transitions of the specific example being modelled. It would not otherwise be possible for newly introduced transitions to modify the abstract queues. The basis context introduces a set of all possible triggers, which is partitioned into internal and external triggers (e.g. FutureInternalTrigger and FutureExternalTrigger respectively), some of which will be introduced in future refinements. Each refinement partitions these trigger sets further to introduce concrete triggers, leaving a new abstract set to represent the remaining triggers yet to be introduced.

The basis machine declares variables that correspond to the internal and external queues, the dequeued trigger and a flag that signals when a run to completion macro-step has been completed (no un-triggered transitions are enabled). The abstract event futureTriggeredTransitions represents a combination of transitions that are triggered by the trigger presently ready to dequeue, dt. The actions of these transitions may also raise triggers of their own.

In the process of refining a model, a designer takes advantage of the non-determinism in the abstraction to introduce new triggers and statechart behaviour that refines abstract events. By default a run may non-deterministically complete at any stage until no un-triggered transitions are enabled (when completion is the only choice left). This allows for future refinements that may strengthen the guards of transitions (e.g. by introducing new nested states as the source of a transition) Such guard strengthening refinements correspond to earlier (i.e. weaker) completion, hence the need to allow for this behaviour in the abstraction. When a refinement level is reached for which the designer wants to verify a property that relies on a particular control response within the current run, early completion must be disallowed. This is done by specifying (as an annotation in the SCXML model) that the transitions involved in the run are *finalized*. The SCXML translation tool will then automatically strengthens the guards of the completion events to ensure that the run to completion sequence is not interrupted early by non-deterministic behaviour.

The translation of a specific SCXML model extends that described in [7,8] with the following additions:

Trigger Queues in Basis: The encoding of trigger queues in the abstract basis machine has been improved so that triggers are properly dequeued before potential use, which allows triggers to be discarded if the controller cannot respond to them. This more accurately reflects the SCXML semantics.

Finalisation: Transitions can be flagged as finalised which means their guards can not be strengthened in subsequent refinements. This allows them to be 'enforced' when they are enabled (i.e. completion cannot occur until they have fired) which is needed for verification.

Restricted Raising of Internal Triggers: Once a trigger is introduced it must immediately be raised at that refinement level by any transitions that wish to do so. It cannot be raised in later refinements except by newly introduced transitions. This restriction was necessary to make simulation more useful by removing non-deterministic raising of triggers in anticipation of refinements.

Context Instantiation: The axioms of the basis context, that allow future triggers to be added, have been improved so that ProB[1] can automatically create an instantiation.

A tool to automatically translate SCXML models into UML-B has been produced.

4   Statechart Refinement

Our system includes three refinement rules.

1. Guard conditions on a transition can be strengthened; this can done by adding textual guards to the transition, or changing the source of the transition to a nested state.
2. Transitions can have additional actions, provided they do not modify variables appearing in the abstraction; this can be accomplished by adding textual action to the transition or by changing the target to nested state.
3. A statechart can be embedded within a state of another statechart – sometimes called hierarchical composition or hierarchical refinement.

Via the translation explained in Sect. 3, these rules rely on the usual Event-B proof obligations to ensure that they do indeed yield refinements in the Event-B semantics.

5   Verification of Safety Properties

In a state-chart model we naturally wish to verify properties P, about other parallel statechart regions and auxiliary data, that are expected to hold true in a particular state S. Hence, all of the safety properties that we consider are of the form: S=TRUE $\Rightarrow$ P, where the antecedent is implicit from the containment of P within S.

SCXML models represent components that respond to received triggers and are not perfectly synchronised with changes in the monitored properties. Hence, P may be temporarily violated until the system responds by leaving the state

---

[1] ProB is an animator, constraint solver and model checker for the B-Method. https://www3.hhu.de/stups/prob.

S in which the property is expected to hold. To cater for this we express P in a modified form P' that allows time for the response to take place. There are two forms of reaction that can be used to exit S; a) an untriggered transition, or b) a transition that is triggered by an internally raised trigger. For a), the modified property P' becomes P ∨ *untriggered transitions are not complete*, and for b) P' becomes P ∨ *trigger* t *is in the internal queue or dequeued* (where t is the internal trigger raised when the violation of P is detected).

6   Verification of Control Responses

It is sometimes possible to construct a model that satisfies some invariant (e.g. safety) properties, but does not behave in a useful way. Therefore, as well as verifying invariant properties, we would like to verify the system's responsiveness. More specifically in this case, we want to ensure that the controller responds to external triggers to make appropriate modifications to the system variables. These kind of live responses can not be verified by proof of invariants since they are temporal properties. Instead, we can express the property in Linear Temporal Logic (LTL) and use the ProB model checker to verify it.

In general, our liveness properties will have the following form:

$$G([external\_trigger\_event] \Rightarrow F\{predicate\}),$$

where the predicate concerns variables v that the system maintains, and may refer to old values old(v) that existed when the external trigger occurred. To specify a liveness property to be verified, a special LTL element is added to the SCXML model with attributes, property (a string of the above form) and refinement (an integer indicating the refinement level at which the property should be verified). The translator generates a separate 'branch' refinement for each LTL property to be verified. In this special refinement, history variables are added to record the value at the state when the external trigger occurs, of any variables that are referenced as 'old' values. A text file is automatically generated containing the LTL property to be checked.

In this generated version, an assumption of strong fairness is added for all other events in the model. (This assumption is stronger than necessary since some events will not affect the outcome, but is easier to generate and is sufficient for our verification aim).

$$SF[e1] \wedge SF[e2]... \Rightarrow G([external\_trigger \_event] \Rightarrow F[predicate])$$

This property is then verified using the LTL facility of the ProB model checker.

7   Conclusion

Statecharts are useful and widely used by engineers for modelling the design of control systems that respond to sensed changes in the environment. Event-B provides an effective language for formally verifying properties via incremental refinements. However, it is not straightforward to apply the latter to the former. We have developed a technique for introducing refinement of Statecharts that can be translated to Event-B for verification. Invariant properties about the expected coordination of states can be added and are interpreted with additional allowance for the reactions to take place in the control system. Such invariants prove automatically with the existing Rodin theorem provers. We use an LTL model checker as a complementary process for verifying expected reactions to environmental triggers.

In future work we intend to formalise the semantics of our extended SCXML notation in order to define its notion of refinement and correspondence to Event-B.

References

1. Abrial, J.-R.: Modeling in Event-B: System and Software Engineering. Cambridge University Press, Cambridge (2010)
2. Syriani, L.L.E., Sousa, V.: Structure and behavior preserving statecharts refinements. Sci. Comput. Program. **170**(15), 45–79 (2019)
3. Harel, D.: Statecharts: a visual formalism for complex systems. Sci. Comput. Program. **8**(3), 231–274 (1987)
4. Hoang, T.S.: An introduction to the Event-B modelling method. In: Romanovsky, A., Thomas, M. (eds.) Industrial Deployment of System Engineering Methods, pp. 211–236. Springer, Heidelberg (2013)
5. Maraninchi, F.: The Argos language: graphical representation of automata and description of reactive systems. In: IEEE Workshop on Visual Languages (1991)
6. MATLAB. 9.7.0.1190202 (R2019b). The MathWorks Inc., Natick, Massachusetts (2019)
7. Morris, K., Snook, C.: Reconciling SCXML statechart representations and Event-B lower level semantics. In: HCCV - Workshop on High-Consequence Control Verification (2016)
8. Morris, K., Snook, C., Hoang, T.S., Armstrong, R., Butler, M.: Refinement of statecharts with run-to-completion semantics. In: Artho, C., Ölveczky, P.C. (eds.) FTSCS 2018. CCIS, vol. 1008, pp. 121–138. Springer, Cham (2019). https://doi.org/10.1007/978-3-030-12988-0_8
9. Said, M.Y., Butler, M., Snook, C.: A method of refinement in UML-B. Softw. Syst. Model. **14**(4), 1557–1580 (2015)
10. Snook, C.: iUML-B statemachines. In: Proceedings of the Rodin Workshop 2014, Toulouse, France, pp. 29–30 (2014). http://eprints.soton.ac.uk/365301/
11. Snook, C., Butler, M.: UML-B: formal modeling and design aided by UML. ACM Trans. Softw. Eng. Methodol. **15**(1), 92–122 (2006)
12. W3C. State chart XML SCXML: State machine notation for control abstraction, September 2015. http://www.w3.org/TR/scxml/

Articles Contributing to the Case Study

Adaptive Exterior Light and Speed Control System

Frank Houdek[1] and Alexander Raschke[2]([✉])

[1] Research and Development, Mercedes-Benz AG, Sindelfingen, Germany
`frank.houdek@daimler.com`
[2] Institute of Software Engineering, Ulm University, Ulm, Germany
`alexander.raschke@uni-ulm.de`

1 Introduction

This case study continues the successful series of case studies for formal specification and verification of the ABZ conference series, which started with the landing gear system [1] and expanded with the hemodialysis medical device [4] and the European Train Control System (ETCS)[2] in the following years. This document describes two systems from the automotive domain: an adaptive exterior light system (ELS) and a speed control system (SCS). This specification is based on the SPES XT running example [3]. Besides their general architectures, the requirements of the software based controllers are described. Both systems are only loosely coupled, which makes it possible to handle them independently.

Conventions. Throughout this document, we use the following conventions to better distinguish different terms: **Main functions** are set in bold, *sub-functions* are italicized. Predefined `signals` are written in typewriter and for the values of signals we use a font without serifs.

The structure of the document is as follows: First, the general hardware architecture of a modern car is sketched in Sect. 3. Then, the adaptive exterior light system is described in Sect. 4, followed by the requirements of the speed control system (Sect. 5). For each of the systems, the user interface, the needed sensors and the available actuators are described before the different features are explained in detail. In Sect. A, all available signals and their value ranges are summarized in a table.

2 Disclaimer

The example in this document is inspired from real-world systems as they are available in many recent cars. However it is important to note that the given description does not describe a current or past real-world system of any vehicle of the Daimler AG.

A. Raschke et al. (Eds.): ABZ 2020, LNCS 12071, pp. 281–301, 2020.
https://doi.org/10.1007/978-3-030-48077-6_24

3   General Architecture

A modern car offers many different safety and comfort functions. Most of them are nowadays realized in software running on a bunch of electronic control units (ECUs) with connected actuators and sensors. These ECUs are connected via several bus and network techniques like CAN, LIN, or FlexRay (depending on the needed band width and reliability). The avoidance of a single central unit is three-fold: First, the risks with a single-point of failure are reduced, second, the limited space and energy of a car restricts the possible technologies, and third, there is the constant need to balance the weight and space consumption of wiring harness with space and weight consumption of decentral control units that are placed nearby the actuators. Despite the pressure to realize more and more in software, some functions are still implemented in hardware. For example, in this specification it is assumed that the detection of a defective bulb is realized by a corresponding electronic circuit.

Additional to the complexity of a distributed system, each car can be configured individually, either by law restrictions of different countries or by customer's preferences. For example, the rear direction indicator in USA and Canada is realized by a blinking red tail light, whereas in Europe it is an extra yellow light.

Figure 1 presents an exemplary excerpt of a connection diagram for the two systems described in this case study. In this case study, we do not focus on the communication between the different ECUs which is necessary because of the distribution of each functionality over several ECUs. For example, to realize left blinking, the body controller front, the door control unit left, and the body controller rear must be involved to execute the commands given by the steering column switch module. In this case study, we focus on the functionalities and simplify reality by allowing signals to be read and commands to be sent directly.

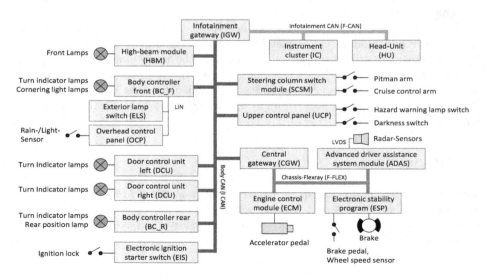

Fig. 1. System overview

In order to save costs, the software of each control unit is parameterized with the different necessary configurations according to the country specification and the individual order. In the context of this case study, the following parameters are defined. They must be taken into account for the formal specification.

- `driverPosition` holds the information, if the car is configured for left-hand or right-hand traffic.
- The Boolean `armoredVehicle` indicates, if the current car is an armored vehicle or not.
- The `marketCode` parameter specifies the market for which the car is to be built. Some example codes are: 001 = USA, 002 = Canada, 003 = EU.

4   Adaptive Exterior Light System

The headlights of a modern car are no longer simply switched on and off by a simple mechanical switch, but the exterior light system integrates various subsystems, like the control of turn signals and comfort functions such as a cornering light. Specifically the following light system functions, among others, are described in detail in this study:

- **Turn Signal**: Control of the driving direction indicators.
- **Low beam headlights**: Control of the low beam headlights. If *daytime running light* is activated, low beam headlights are active all the time and *ambient light* illuminates the vehicle surrounding while leaving the car during darkness. The function low beam headlight also includes *parking light*.
- **Cornering light**: Control of additional headlights that illuminate the cornering area separately when turning left or right.
- **Adaptive high beam**: Control of the high beam headlights.
- **Emergency brake light**: Following drivers are warned by a flashing brake light in case of an emergency brake.

In the following sections, we first introduce the user interface, necessary sensors and the attached actuators of an exterior light system.

4.1   User Interface

The car driver can control the different functions of the lighting system by several buttons and switches, which are described in the following.

The light rotary switch has the following positions: Off, Auto, On (see Fig. 2). The light rotary switch position is transmitted via the signal `lightRotarySwitch`.

The control lever attached to the steering column is called pitman arm and allows for the following movements (see Fig. 3). The pitman arm position is transmitted via the signals `pitmanArmForthBack` and `pitmanArmUpDown`.

- By pushing away from the driver ④ (backward): Permanent activation of the adaptive high beam (with pitman arm engaged).

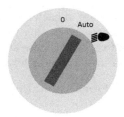

Fig. 2. Light rotary switch

- By pulling towards the driver ① (forward): Temporary activation of the high beam (without engaging, so-called flasher).
- By moving up or down ②/③: Temporary or permanent activation of the direction indicator to the left or right. The temporary activation (so called tip-blinking) happens by a deflection of about 5° (Downward5, Upward5), the permanent activation (engage) by about 7° deflection (Downward7, Upward7). The engagement ends either by manually bringing the pitman back to neutral position or automatically by a mechanical reset mechanism if the steering wheel has been turned more than 10°.
- The neutral position of the pitman arm is signaled by Neutral.

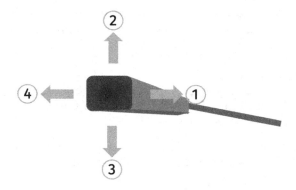

Fig. 3. Pitman arm with four directions of movement

The Hazard Warning Light Switch (see Fig. 4, hazardWarningSwitchOn) is just like the Darkness Switch (only available at armored vehicles, see Fig. 5, darknessModeSwitchOn) a simple toggle switch which turns on the corresponding function when pushed (value True) and turns it off when pushed again (value False).

The user can activate or deactivate the functions *daytime running light* and *ambient light* in the instrument cluster settings menu (which is not described in this specification). The instrument cluster settings are transmitted via daytimeLights and ambientLighting.

Fig. 4. Hazard warning light switch

Fig. 5. Darkness switch (only armored vehicles)

4.2 Sensors

Besides the elements that can be manipulated by the user, several sensors are necessary to provide the desired features.

- Status and position of the key (and thus the information, if the ignition is on). This information is transmitted via keyState and has the values NoKeyInserted, KeyInserted, KeyInIgnitionOnPosition.
- Engine status engineOn (True, False).
- Brightness of the environment brightnessSensor, offering the measured outside brightness in values 0 to 100000.
- Deflection of the brake pedal brakePedal, where 0 means no deflection and 225 means a maximum deflection of 45°.
- Available battery voltage voltageBattery, measured in 0.1 V.
- Angle of the steering wheel steeringAngle.
- Information about the status of the doors (open or closed). For the sake of simplicity there is only the information available if all doors are closed or not (via allDoorsClosed).
- A camera to detect oncoming vehicles, signaled via oncommingTraffic. The state of the camera (Ready, Dirty, NotReady) is signaled via cameraState.
- The current vehicle speed is available via currentSpeed.
- If the reverse gear is engaged, reverseGear becomes True.

4.3 Actuators

Figure 6 schematically shows the possible positions A (front), B (exterior mirror), C (rear), and D (rear center) of exterior lighting elements of a vehicle.

The following lighting actuators[1] are installed at the given positions (each left and right, except D which exists only once):

- Direction indicator (blinker) (A, B, C), controlled via the signals blinkLeft and blinkRight.
- Headlights for low beam headlight (A), controlled via lowBeamLeft and lowBeamRight.
- Headlights for high beam headlight (A), controlled via highBeamOn to activate and deactivate the high beam, highBeamRange to control the high beam luminous, and highBeamMotor to control the high beam illumination distance.

---

[1] Details about the design of the lighting elements are regulated by the directive 93/92/EEC.

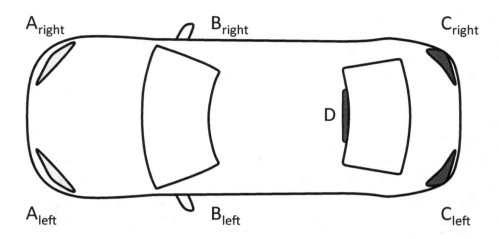

Fig. 6. Schematic position of the exterior lighting elements

- Lamp for cornering light left or right (integrated in front bumper) (A), controlled via `corneringLightLeft` and `corneringLightRight`.
- Brake lamp (C, D), controlled via `brakeLight`
- Tail lamp (C), controlled via `tailLampLeft` and `tailLampRight`.
- Reverse lamp (C), controlled via `reverseLight`.

Cars that are sold in USA or Canada do not have a separate direction indicator at position C. Here, the tail lamps take on the task of the rear indicator lamps.

4.4 Functional Requirements

This section lists the functional requirements for the different functions of the adaptive light system. These functions are not completely independent of each other. Moreover, they interfere at several points, mainly because of the shared use of the given actuators.

Direction Blinking. The function direction blinking defines different ways to indicate the desired direction of the driver at crossings or at lane changes. It is only available, if the ignition is on (KeyInIgnitionOnPosition).

ELS-1 *Direction blinking left*: When moving the pitman arm in position "turn left" ③, the vehicle flashes all left direction indicators (front left, exterior mirror left, rear left) synchronously with pulse ratio bright to dark 1:1 and a frequency of $1.0\,\text{Hz} \pm 0.1\,\text{Hz}$ (i.e. 60 flashes per minute $\pm$ 6 flashes).

ELS-2 *Tip-blinking left*: If the driver moves the pitman arm for less than $0.5\,\text{s}$ in position "Tip-blinking left", all left direction indicators (see Req. ELS-1) should flash for three flashing cycles.

ELS-3 If the driver activates the pitman arm in another direction or activates the hazard warning light switch during the three flashing cycles of the tip-blinking, the tip-blinking cycle must be stopped and the requested flashing cycle must be started (i.e. direction blinking, tip-blinking, or hazard warning light, depending on the interrupting request).

ELS-4 If the driver holds the pitman arm for more than 0.5 s in position "tip-blinking left", flashing cycles are initiated for all direction indicators on the left (see Req. ELS-1) until the pitman arm leaves the position "tip-blinking left".

ELS-5 *Direction blinking right* and *tip-blinking right*: Analogous to the left side (see Req. Req. ELS-1 to Req. ELS-4).

ELS-6 For cars sold in USA and Canada, the daytime running light must be dimmed by 50% during direction blinking on the blinking side.

ELS-7 If the driver activates the pitman arm during the three flashing cycles of tip-blinking for the same direction again, only the current flashing cycle is completed and then the new command is processed (either three flashing cycles due to tip-blinking or constant direction blinking).

Hazard Warning Light. Tightly coupled with the direction blinking is the hazard warning light, which requirements are described in the following.

ELS-8 As long as the hazard warning light switch is pressed (active), all direction indicators flash synchronously. If the ignition key is in the ignition lock, the pulse ratio is bright to dark 1:1. If the ignition key is not in the lock, the pulse ratio is 1:2.

ELS-9 The adaptation of the pulse ratio must occur at the latest after two complete flashing cycles.
Note: The reduction of the pulse is performed due to energy saving reasons, such that, in case of an emergency situation, the hazard warning light is active as long as possible before the car battery is empty.

ELS-10 The duration of a flashing cycle is 1 s.

ELS-11 A flashing cycle (bright to dark) must always be completed, before a new flashing cycle can occur.
Note: By the fact, that a flashing cycle must always be completed, a "switching" behavior of the indicator is avoided. Thus, for example a change of the pitman arm from "tip-blinking" to "direction blinking" or back has no visible effect.

ELS-12 When hazard warning is deactivated again, the pitman arm is in position "direction blinking left" or "direction blinking right" ignition is On, the direction blinking cycle should be started (see Req. ELS-1).

ELS-13 If the warning light is activated, any tip-blinking will be ignored or stopped if it was started before.

Low Beam Headlights and Cornering light. The function low beam headlights includes the functions *daytime running light*, *ambient light*, and *parking light*.

ELS-14 If the ignition is On and the light rotary switch is in the position On, then low beam headlights are activated.

ELS-15 While the ignition is in position KeyInserted: if the light rotary switch is turned to the position On, the low beam headlights are activated with 50% (to save power). With additionally activated ambient light, ambient light control (Req. ELS-19) has priority over Req. ELS-15. With additionally activated daytime running light, Req. ELS-15 has priority over Req. ELS-17.

ELS-16 If the ignition is already off and the driver turns the light rotary switch to position Auto, the low beam headlights remain off or are deactivated (depending on the previous state). In case of conflict, Req. ELS-16 has priority over Req. ELS-17 (i.e. the later manual activitiy overrules running daytime light if ignition is KeyInserted). If ambient light is active (see Req. ELS-19), ambient light delays the deactivation of the low beam headlamps.

ELS-17 With activated *daytime running light*, the low beam headlights are activated after starting the engine. The daytime running light remains active as long as the ignition key is in the ignition lock (i.e. KeyInserted or KeyInIgnitionOnPosition). With additionally activated ambient light, ambient light control (Req. ELS-19) has priority over daytime running light.

ELS-18 If the light rotary switch is in position Auto and the ignition is On, the low beam headlights are activated as soon as the exterior brightness is lower than a threshold of 200 lx. If the exterior brightness exceeds a threshold of 250 lx, the low beam headlights are deactivated. In any case, the low beam headlights remain active at least for 3 s.

ELS-19 Ambient light prolongs (keeps low beam headlamps at 100% if they have been active before) the activation of low beam headlamps (as ambient light) if ambient light has been activated, engine has been stopped (i.e. keyState changes from KeyInIgnitionOnPosition to NoKeyInserted or KeyInserted) and the exterior brightness outside the vehicle is lower than the threshold 200 lx. In this case, the low beam headlamps remain active or are activated. The low beam headlights are deactivated or parking light is activated (see Req. ELS-28) after 30 s. This time interval is reset by
  – Opening or closing a door
  – Insertion or removal of the ignition key

ELS-20 — *Deleted requirement* —

ELS-21 With activated darkness switch (only armored vehicles) the ambient lighting is not activated. As long as the darkness switch is activated, it supresses low beam headlights due to ambient light.

ELS-22 Whenever the low or high beam headlights are activated, the tail lights are activated, too.

ELS-23 In USA or Canada, tail lights realize the direction indicator lamps. In case of direction blinking or hazard blinking, blinking has preference against normal tail lights.

ELS-24 *Cornering light*: If the low beam headlights are activated and direction blinking is requested, the cornering light is activated, when the vehicle drives slower than 10 km/h. 5 s after passing the corner (i.e. the direction blinking is not active any more for 5 s), the cornering light is switched off in a duration of 1 s (gentle fade-out). Activating cornering light means that if driving to the left is indicated, the left cornering light is activated. If driving to the right is indicated, the right cornering light shall be activated.

ELS-25 With activated darkness switch (only armored vehicles) the cornering light is not activated.

ELS-26 The cornering light is also activated, if the direction blinking is not activated, but all other constraints (see Req. ELS-24) are fulfilled and the steering wheel deflection is more than $\pm 10°$.

ELS-27 If reverse gear is activated, both opposite cornering lights are activated.

ELS-28 *Parking light.* The parking light is the low beam and the tail lamp on the left or right side of the vehicle to illuminate the vehicle if it is parked on a dark road at night. The parking light is activated, if the key is not inserted, the light switch is in position On, and the pitman arm is engaged in position left or right (②/③). To save battery charge, the parking light is activated with only 10% brightness of the normal low beam lamp and tail lamp. An active ambient light (see Req. ELS-19) delays parking light.

ELS-29 The normal brightness of low beam lamps, brake lights, direction indicators, tail lamps, cornering lights, and reverse light is 100%.

Manual High Beam Headlights. The low beam light is designed in such a way that it does not dazzle oncoming traffic. On country roads in particular, however, it is useful to illuminate a larger area when there is no oncoming traffic. High beam light fulfills this purpose.

ELS-30 The headlamp flasher is activated by pulling the pitman arm, i.e. as long as the pitman arm is pulled ①, the high beam headlight is activated.

ELS-31 If the light rotary switch is in position On, pushing the pitman arm to ④ causes the activation of the high beam headlight with a fixed illumination area of 220 m and 100% luminous strength (i.e. `highBeamMotor` = 7 and `highBeamRange` = 100).

Adaptive High Beam Headlights. Frequent switching of the high beam is tiring for the driver. With the help of a built-in camera, which detects oncoming vehicles, this task can be automated so that the driver has better illumination of the road as often as possible without endangering oncoming traffic. In addition, the high beam headlight is optimized to always illuminate the appropriate area according to the current speed.

ELS-32 If the light rotary switch is in position Auto, the adaptive high beam is activated by moving the pitman arm to the back ④.

ELS-33 If adaptive high beam headlight is activated and the vehicle drives faster than 30 km/h and no light of an advancing vehicle is recognized by the camera, the street should be illuminated within 2 s according to the characteristic curve in Fig. 7 (for light illumination distance) and Fig. 8 (for luminous strength).

ELS-34 If the camera recognizes the lights of an advancing vehicle, an activated high beam headlight is reduced to low beam headlight within 0.5 s by reducing the area of illumination to 65 m by an adjustment of the headlight position as well as by reduction of the luminous strength to 30%.

ELS-35 If no advancing vehicle is recognized any more, the high beam illumination is restored after 2 s.

ELS-36 The light illumination distance of the high beam headlight is within 100 m and 300 m, depending on the vehicle speed (see characteristic curve in Fig. 7).

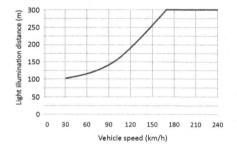

Fig. 7. Characteristic curve of the high beam headlight illumination distance depending on the vehicle speed

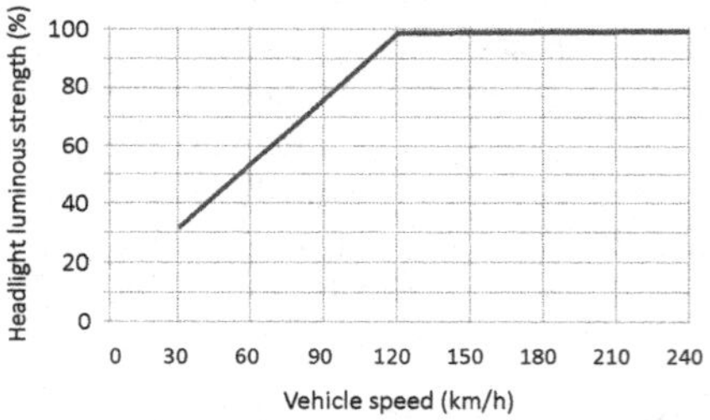

Fig. 8. Characteristic curve of the high beam headlight luminous depending on the vehicle speed

ELS-37 If an adaptive cruise control is part of the vehicle, the light illumination distance is not calculated upon the actual vehicle speed but the target speed provided by the advanced cruise control.

ELS-38 If the pitman arm is moved again in the horizontal neutral position, the adaptive high beam headlight is deactivated. The illumination of the street is reduced immediately (i.e. without gentle fade-out) to low beam headlights.

Emergency Brake Light. For safety reasons, it is important to indicate braking to the drivers behind the vehicle. Studies have shown that a flickering brake light during an emergency stop shortens the reaction time of the following driver.

ELS-39 If the brake pedal is deflected more than 3°, all brake lamps have to be activated until the deflection is lower than 1° again.

ELS-40 If the brake pedal is deflected more than $40.0°$ (i.e. full-brake application), all brake lamps flash with pulse ratio bright to dark 1:1 and a frequency of $6 \pm 1\,\text{Hz}$ (i.e. 360 ± 60 flashes per minute). The flashing stops only when the brake pedal is in its neutral position again (i.e. `brakePedal` $= 0$).

Reverse Light. Indicates that the reverse gear in engaged, i.e. the vehicle will move backwards.

ELS-41 The reverse light is activated whenever the reverse gear is engaged.

Fault Handling. A malfunctioning lighting system is safety critical and must therefore be avoided. E.g. the failure of individual lamps is checked using a hardware circuit and indicated to the driver accordingly. In the following we describe how the software should react to over- or subvoltage in order to guarantee the most important functionality for as long as possible.

ELS-42 A subvoltage is present if the voltage in the vehicle electrical system is less than $8.5\,\text{V}$. With subvoltage, the adaptive high beam headlight is not available.

ELS-43 If the light rotary switch is in position Auto and the pitman arm is pulled, the high beam headlight is activated (see Req. ELS-31) even in case of subvoltage.

ELS-44 With subvoltage the ambient light is not available.

ELS-45 With subvoltage the cornering light is not available.

ELS-46 With subvoltage an activated parking light is switched off.

ELS-47 An overvoltage is present if the voltage in the vehicle electrical system is more than $14.5\,\text{V}$. With overvoltage, activated lights must not exceen the maximum light intensity of $(100 - (\texttt{voltage} - 14.5) \cdot 20)\%$. This reduction serves the protection of the illuminant (protection from "burning out").

ELS-48 With overvoltage, the illumination area requirements do not need to be respected (see Req. ELS-33 and Req. ELS-36). Instead, illumination area is fixed to $220\,\text{m}$.

ELS-49 If the camera is not Ready, adaptive high beam headlights is not available. This means, if `cameraState` is unequal Ready, light rotary switch is in position Auto and the pitman arm is in position ④, manual high beam headlights are activated (see Req. ELS-31), which means that high beam headlights are activated with a fixed illumination area of $220\,\text{m}$ and 100% luminous strength (i.e. `highBeamMotor` $= 7$ and `highBeamRange` $= 100$).

5 Speed Control System

The speed control system is a comfort function that tries to maintain or adjust the speed of the vehicle according to various external influences. In various traffic situations, this relieves the driver, who no longer has to keep the gas pedal in the corresponding position with his right foot. It includes the following user functions:

- **Cruise Control:** The vehicle automatically maintains a set speed independently of the distance to other vehicles. Here, the driver is in charge to maintain safety distance.
- **Adaptive Cruise Control:** The vehicle maintains the distance to the preceding vehicle including braking until a full standstill and starting from a standstill.
- **Distance Warning:** The vehicle warns the driver visually and/or acoustically if the vehicle is closer to the car ahead than allowed by the safety distance.
- **Emergency Brake Assist:** The vehicle decelerates in critical situations to a full standstill.
- **Speed Limit:** The vehicle does not exceed a set speed.
- **Sign Recognition:** The vehicle sets the speed limit automatically according to the recognized signs.
- **Traffic Jam Following:** The vehicle accelerates from a standstill when the preceding vehicle departs.

Similar to the exterior light system, the speed control system provides a specific user interface, uses sensors and controls actuators, which are described in the following sections.

5.1   User Interface

Cruise Control Lever (Fig. 9). The cruise control lever combines the functionality for the cruise control and the speed limiter. It is a little bit smaller than the pitman arm lever and is mounted below it on the steering wheel switch module. The cruise control lever also contains the rotary switch with which the safety distance can be set (see Req. SCS-24). The lever always returns to the neutral position when not touched by the user. The position of the cruise control lever is signaled via SCSLever.
The following movements are possible with the lever:

- By pulling towards the driver ① (Forward): The cruise control is activated with the current speed as the desired speed or the last saved desired speed.
- By moving up or down ②/③: The desired speed is increased/decreased in several steps.
- By pushing the lever away from the driver ④ (Backward): The cruise control is deactivated.
- By turning the head ⑥: The safety distance (safetyDistance) for the adaptive cruise control is modified in three steps (see Req. SCS-24, values 2 s, 2.5 s, 3 s).

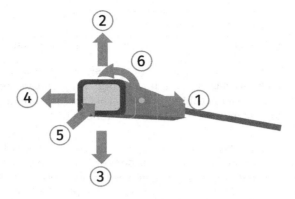

Fig. 9. Speed limiting lever integrated in the cruise control lever

– The cruise control lever can be used as speed limiting lever by pushing the button at the head ⑤ of the cruise control lever. The position of the button is signalled via `speedLimiterSwitchOn`. If the lever controls the speed limit function, an orange LED integrated in the cruise control lever is on (implemented by hardware). The movements have similar functions as for the cruise control (activation, setting of the speed limit, deactivation).

Brake Pedal. The brake pedal is mounted in the footwell area of the driver. Its position is signaled via `brakePedal`.

Gas Pedal. The gas pedal is mounted in the footwell area of the driver. Its position is signaled via `gasPedal`.

Instrument Cluster. The user can activate or deactivate the functions *traffic sign detection* and *adaptive cruise control* in the instrument cluster settings menu (which is not described in this specification). The instrument cluster settings are transmitted via `trafficSignDetectionOn` and `cruiseControlMode`.

5.2 Sensors

The following sensors are connected to the system in order to enable the driver assistance system.

– Status and position of the key (and thus the information, if the ignition is on). This information is transmitted via `keyState` and has the values NoKeyInserted, KeyInserted, KeyInIgnitionOnPosition.
– Engine status `engineOn` (True, False).
– Deflection of the brake pedal `brakePedal`, where 0 means no deflection and 225 means a maximum deflection of 45°.
– A radar system that measures the distance to the nearest obstacle. The state of the radar sensors is reported via `rangeRadarState`, its obstacle detection via `rangeRadarSensor`.

5.3 Actuators

The following actuators are controlled by the speed control system:

- The *engine* is controlled by the two inputs gasPedal and setVehicleSpeed.
  The engine applies the maximum of both inputs if speedLimiterSwitchOn =
  False. If speedLimiterSwitchOn = True, the engines applies setVehicle
  Speed it this value is greater 0, otherwise gasPedal.
  Please note:
  (1) The scales of gasPedal and setVehicleSpeed are different. A maximum
  gasPedal (=45°) is equal to maximum setVehicleSpeed.
  (2) There is no direct relation of gasPedal (or setVehicleSpeed) to the
  vehicle speed. In fact, as long as gasPedal (or setVehicleSpeed) is greater
  0, the vehicle accelerates. This acceleration is reduced by the inertia of the
  vehicle and limited by the car physics that result in a maxmimum speed of
  approx. 250 km/h.
  (3) The maximum acceleration is approx. 3 m/s^2.
- The *brake* is controlled by the system in order to decelerate or even emergency brake if necessary via brakePressure. The maximum brake-implied
  decelaration is approx. 6 m/s^2. For sake of simplicity it may be assumed that
  the deceleration d can be determined via brakePedal as

$$d = \frac{\texttt{brakePedal}}{37.5°\, s^2/m}$$

- An *acoustic warning* and a *visual warning* are given in dangerous situations
  via acousticWarningOn and visualWarningOn.

5.4 Software Functions

Setting and Modifying Desired Speed. This section describes how to set
and modify the desired speed both for adaptive cruise control and (normal)
cruise control. When changing the desired speed, the instrument cluster displays
the current value. This is not covered in this specification.

SCS-1 After engie start, there is no previous desired speed. The valid values
for desired speed are from 1 km/h to 200 km/h.

SCS-2 When pulling the cruise control lever to ①, the desired speed is either
the current vehicle speed (if there is no previous desired speed) or the
previous desired speed (if already set).

SCS-3 If the current vehicle speed is below 20 km/h and there is no previous
desired speed, then pulling the cruise control lever to ① does not
activate the (adaptive) cruise control.

SCS-4 If the driver pushes the cruise control lever to ② up to the first resistance level (5°) and the (adaptive) cruise control is activated, the
desired speed is increased by 1 km/h.

SCS-5 If the driver pushes the cruise control lever to ② above the first resistance level ($7°$, beyond the pressure point) and the (adaptive) cruise control is activated, the desired speed is increased to the next ten's place.

Example: Current desired speed is $57\,\text{km/h}$ $\longrightarrow$ new desired speed is $60\,\text{km/h}$.

SCS-6 Pushing the cruise control lever to ③ reduces the desired speed accordingly to Req. SCS-4 and Req. SCS-5. The lowest desired speed that can be set by pushing the cruise control lever beyond the pressure point is $10\,\text{km/h}$.

SCS-7 If the driver pushes the cruise control lever to ② with activated cruise control within the first resistance level ($5°$, not beyond the pressure point) and holds it there for $2\,\text{s}$, the desired speed of the cruise control is increased every second by $1\,\text{km/h}$ until the lever is in neutral position again.

Example: Current desired speed is $57\,\text{km/h}$ $\longrightarrow$ new desired speed is $58\,\text{km/h}$ (due to Req. SCS-4), after holding $2\,\text{s}$, desired speed is set to $59\,\text{km/h}$, after holding another second, desired speed is set to $60\,\text{km/h}$, after holding another second, desired speed is set to $61\,\text{km/h}$, etc.

SCS-8 If the driver pushes the cruise control lever to ② with activated cruise control through the first resistance level ($7°$, beyond the pressure point) and holds it there for $2\,\text{s}$, the speed set point of the cruise control is increased every $2\,\text{s}$ to the next ten's place until the lever is in neutral position again.

Example: Current desired speed is $57\,\text{km/h}$ $\longrightarrow$ new desired speed is $60\,\text{km/h}$ (due to Req. SCS-5), after holding $2\,\text{s}$, desired speed is set to $70\,\text{km/h}$, after another $2\,\text{s}$, desired speed is set to $80\,\text{km/h}$, after holding another $2\,\text{s}$, desired speed is set to $90\,\text{km/h}$, etc.

SCS-9 If the driver pushes the cruise control lever to ③ with activated cruise control within the first resistance level ($5°$, not beyond the pressure point) and holds it there for $2\,\text{s}$, the desired speed of the cruise control is reduced every second by $1\,\text{km/h}$ until the lever is in neutral position again.

Example: Current desired speed is $57\,\text{km/h}$ $\longrightarrow$ new desired speed is $56\,\text{km/h}$ (due to Req. SCS-6) after holding $2\,\text{s}$, desired speed is set to $55\,\text{km/h}$, after another second, desired speed is set to $54\,\text{km/h}$, after holding another second, desired speed is set to $53\,\text{km/h}$, etc.

SCS-10 If the driver pushes the cruise control lever to ③ with activated cruise control through the first resistance level ($7°$, beyond the pressure point) and holds it there for $2\,\text{s}$, the speed set point of the cruise control is increased every $2\,\text{s}$ to the next ten's place until the lever is in neutral position again.

SCS-11 If the (adaptive) cruise control is deactivated and the cruise control lever is moved up or down (either to the first or above the first resistance level), the current vehicle speed is used as desired speed.

SCS-12 Pressing the cruise control lever to ④ deactivates the (adaptive) cruise control. `setVehicleSpeed = 0` indicates to the car that there is no speed to maintain.

Cruise Control. The following requirements describe the simple cruise control system without adaption to the traffic situation which is the basis for the adaptive cruise control system. The distinction between cruise control and adaptive cruise control is made via `cruiseControlMode`.

SCS-13 The cruise control is activated using the cruise control lever according to Reqs. SCS-1 to SCS-12.

SCS-14 As long as the cruise control is activated, the vehicle maintains the current vehicle speed at the desired speed without the driver having to press the gas pedal or the brake pedal.

SCS-15 If the driver pushes the gas pedal and by the position of the gas pedal more acceleration is demanded than by the cruise control, the acceleration setting as demanded by the driver is adopted.
Note: This handling is done by the engine autonomously.

SCS-16 By pushing the brake, the cruise control is deactivated until it is activated again.

SCS-17 By pushing the control lever backwards, the cruise control is deactivated until it is activated again.

Adaptive Cruise Control. In the adaptive cruise control mode, maintenance of the speed does not only depend on the desired speed but also vehicles ahead. For this purpose, the desired speed of the driver must be distinguished from the target speed of the control system. The Reqs. SCS-13 to SCS-17 still hold except SCS-14. The distinction between cruise control and adaptive cruise control is made via `cruiseControlMode`.

SCS-18 When the driver enables the cruise control (by pulling the cruise control lever or by pressing the cruise control lever up or down), the vehicle maintains the set speed if possible.

SCS-19 The adaptive cruise control desired speed is controlled using the cruise control lever according to Reqs. SCS-1 to SCS-12.

SCS-20 If the distance to the vehicle ahead falls below the specified speed-dependent safety distance (see Req. SCS-24), the vehicle brakes automatically. The maximum deceleration is $3\,\text{m/s}^2$.

SCS-21 If the maximum deceleration of $3\,\text{m/s}^2$ is insufficient to prevent a collision with the vehicle ahead, the vehicle warns the driver by two acoustical signals ($0.1\,\text{s}$ long with $0.2\,\text{s}$ pause between) and by this demands to intervene.

SCS-22 If the distance to the preceding vehicle increases again above the speed-dependent safety distance, the vehicle accelerates with a maximum of $1\,\text{m/s}^2$ until the set speed is reached.

Example: Fig. 10 shows an exemplary situation with a desired speed of 120 km/h. At the beginning, the car drives at this speed until another car appears with 80 km/h. The adaptive cruise control decelerates to 80 km/h with a maximum deceleration of 3 m/s^2. If this is not sufficient, two acoustical signals warn the driver. As soon as the vehicle in front accelerates to 100 km/h, the adaptive cruise control also accelerates with a maximum of 1 m/s^2. When the vehicle in front finally accelerates to a speed of more than 120 km/h the adaptive cruise control increases the speed back to 120 km/h.

SCS-23 If the speed of the preceding vehicle is 20 km/h or below, the distance is set to 2.5 s·`currentSpeed`, down to a standstill. When both vehicles are standing the absolute distance is regulated to 2 m. When the preceding vehicle is accelerating again, the distance is set to 3 s · `currentSpeed`. This distance is valid until the vehicle speed exceeds 20 km/h, independent of the user's input via the distance level (turning the cruise control lever head).

SCS-24 By turning the cruise control lever head, the distance to be maintained to the vehicle ahead can be selected. Three levels are available: 2 s, 2.5 s and 3 s. The desired level only applies within the velocity window > 20 km/h. Below this level, the system autonomously sets the distance according to Req. SCS-23.

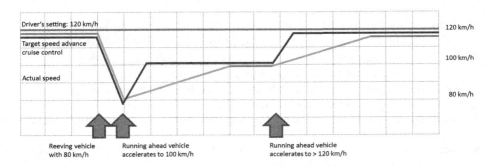

Fig. 10. Illustration of the difference between "actual speed", "desired speed", and "target speed" of the adaptive cruise control

Distance Warning. The adaptive cruise control system has to calculate the distance (time) to the vehicle ahead and has to issue the following warnings depending on the calculated value:

SCS-25 A visual warning is activated if the actual distance is less than $(current\ speed/3.6) \cdot 1.5$.

SCS-26 An acoustic alarm is activated if the actual distance is less than $(current\ speed/3.6) \cdot 0.8$.

Emergency Brake Assistant. The emergency brake assistant initiates braking in critical situations.

SCS-27 The emergency brake assistant must be available in the following speed windows: 0–60 km/h, for emergency braking to stationary obstacles, 0–120 km/h on moving obstacles.

SCS-28 The time necessary to perform braking to standstill is determined by the value for the maximum deceleration. If an object is ahead of the vehicle and the time until an impact is less or equal to the time until a standstill plus 3 s, three acoustic signals are given (0.1 s long with 0.05 s pause between) is issued and the brakes are activated by 20% (i.e. 1.2 m/s^2). If the time until an impact is less or equal to the time until a standstill plus 1.5 s, the brake is activated by 60% (i.e. 3.6 m/s^2. If the time until an impact is less or equal to the time until standstill then the brake is activated at 100% (i.e. 6 m/s^2). In case that both adaptive cruise control (see Req. SCS-20) and the emergency brake assistand request braking, the higher deceleration value shall be applied.

Speed Limit. The speed limit function prevents the driver from accidentally driving faster than a preset desired speed. In case of emergency, the driver can overrule the speed limit.

SCS-29 The speed limiter mode is activated by pressing the button at the head of the control lever.

SCS-30 An active speed limit function of the cruise lever is indicated by an orange LED integrated in the control lever (realized in hardware).

SCS-31 Activating speed limit desired speed and modifying the desired speed is done according to Reqs. SCS-1 to SCS-12.

SCS-32 As long as the speed limit function is activated, the current speed must not exceed the set speed limit.

SCS-33 By pressing the gas pedal beyond 90% the speed limit is temporarily deactivated.

SCS-34 When the pressure on the gas pedal decreases below 90%, the speed limit is automatically activated again.

SCS-35 An active speed limit can be deactivated by either pushing the cruise control lever backwards ④ or by pushing the head of the cruise control lever ⑤.

Traffic Sign Detection. If a road sign is indicating a speed limit with active traffic sign detection (controlled by `trafficSignDetectionOn`), the desired speed is modified by the recognized traffic sign value.

SCS-36 Traffic sign detection is active, while adaptive cruise control is active and the driver has activated traffic sign detection in the instrument cluster.

SCS-37 With active traffic sign detection and gas pedal in position 0, a recognized traffic sign sets the desired speed to the detected value.

SCS-38 A later manual modification of the desired speed via the cruise control lever (see Reqs. SCS-1 to SCS-12) modifies the desired speed again.

Hint: The desired speed is determined by the latest modification: A user setting via cruise control lever is overruled by a later traffic sign detection and this is again overruled by a later modification via cruise control lever.

SCS-39 If traffic sign detection recognizes Unlimited, the new desired speed is set to

- 120 km/h, if the previous desired speed has been lower than 120 km/h
- the desired speed d_{man}, where d_{man} is the last manually set desired speed that has been higher than 120 km/h

Note: For the sake of simplicity, country dependence and road type dependence has been omitted.

Fault Handling and General Properties. A malfunctioning speed control system might be safety critical and must therefore be avoided. E.g. a wrong detection of the distance to the car in front could lead to dangerous situations. These situations should be avoided with the following requirements.

SCS-40 The radar system carries out a self-test at each start and also continuously checks the plausibility of the values of the various sensors. If one of the values is found to be extremely close, the status is set to "Dirty". During the self-test and with other errors (strong fluctuations, very different values of the individual sensors) the status is set to "NotReady".

SCS-41 If the radar sensor self-test device reports a fault (Dirty or NotReady), all systems depending on the distance to the vehicle must be suspended and the driver must be warned by an appropriate light in the instrument cluster (not part of this specification). In this case, the self-test of the radar system is restarted every 10 min.

SCS-42 The gas or brake pedal depressed by the driver must always be able to override a target speed specified by the system.

SCS-43 If the system performs a brake action, the brake lights must be activated as if the brake pedal has been pressed by the driver (see light system specification).

A   Interface

The following table defines all signals that either reflect the determined input of the various user interfaces and sensors or are used to control the actuators. For the sake of simplicity, all signals are available all the time. There are no timeouts or delays.

Signal identifier	Description	Value range
keyState	Status of ignition key	NoKeyInserted, KeyInserted, KeyInIgnitionOnPosition
engineOn	Status of engine	True, False
allDoorsClosed	Status of vehicle doors	True, False
gasPedal	Deflection of the gas pedal from the neutral position	Resolution: 0.2° Value range: 0–225 (0.0–45.0 °)
brakePedal	Deflection of the brake pedal from the neutral position	Resolution: 0.2° Value range: 0–225 (0.0–45.0 °)
reverseGear	Status of the reverse gear	True, False
voltageBattery	Available battery voltage	Resolution: 0.1 V Value range: 0–500 (0.0–50.0 V)
currentSpeed	Current vehicle speed in km/h	Resolution: 0.1 km/h Value range: 0–5000 (0.0–500.0 km/h)
steeringAngle	Steering angle (deflection of the steering wheel)	0 = sensor is calibrating 1–410 = steering wheel rotation to the left (Resolution: 1° starting from 10° deflection) 411–510 = steering wheel rotation to the left (Resolution: 0.1° for 0°–10° deflection) 511–513 = steering wheel in neutral position 514–613 = steering wheel rotation to the right (Resolution: 0.1° for 0°–10° deflection) 614–1022 = steering wheel rotation to the right (Resolution: 1° starting from 10° deflection)
daytimeLights	True, if option is selected in instrument cluster	True, False
ambientLighting	True, if option is selected in instrument cluster	True, False
lightRotarySwitch	Status of light rotary switch	Off, Auto, On
pitmanArmForthBack	Status of pitman arm regarding high beam (horizontal position)	Neutral, Backward, Forward
pitmanArmUpDown	Status of pitman arm regarding blinker (vertical position)	Neutral, Downward5, Downward7, Upward5, Upward7
hazardWarningSwitchOn	Status hazard warning switch	True, False
darknessModeSwitchOn	Status darkness switch (only armored vehicles)	True, False
brightnessSensor	Measurement of rain/light sensor regarding brightness	Resolution: 1 lx Value range: 0–100000
cameraState	Status of camera	Ready, Dirty, NotReady
oncomingTraffic	Advancing vehicle detected	True, False
brakeLight	Brake light command	0–100%
blinkLeft	Perform left blinking	0–100%
blinkRight	Perform right blinking	0–100%
lowBeamLeft	Low beam command left	0–100%
lowBeamRight	Low beam command right	0–100%
taillampleft	Tail lamp command left	0–100%
taillampright	Tail lamp command right	0–100%
highBeamOn	High beam command	True, False

Signal identifier	Description	Value range
highBeamRange	High beam light range (brightness)	0–300 desired light range
highBeamMotor	Desired position for high beam motor	0–14 desired position: 0 = 65 m 1 = 100 m 2–14 = 120–360 m (20 m step size)
corneringLightLeft	Cornering light left	0–100%
corneringLightRight	Cornering light right	0–100%
reverseLight	Reverse light command	0–100%
SCSLever	Position of cruise control lever	Neutral, Downward5, Downward7, Upward5, Upward7, Forward, Backward
safetyDistance	Safety distance level (turning knob at SCSLever)	2 s, 2.5 s, 3 s
speedLimiterSwitchOn	Status speed limiter switch	True, False
rangeRadarState	status of long-range radar sensors	Ready, Dirty, NotReady
rangeRadarSensor	Evaluation of long-range radar sensor	0 = no dectected obstacle in the travel corridor 1–200 = distance in meters of obstacle detected in the travel corridor 255 = radar state is Dirty or NotReady
cruiseControlMode	Operation mode of cruise control	1 = (normal) cruise control, 2 = adaptive cruise control
trafficSignDetectionOn	Operation mode of traffic sign detection	True, False
detectedTrafficSign	Speed limit of observed traffic sign	None, 20–130, Unlimited
setVehicleSpeed	Used to control the engine via cruise control	0–100
brakePressure	The pressure of the brake shoes	0–100%
acousticWarningOn	Acoustic warning command	True, False
visualWarningOn	Visual warning command	True, False
driverPosition	Vehicle configuration of driver position	LeftHandDrive, RightHandDrive
armoredVehicle	True, if vehicle is armored	True, False
marketCode	The market region for which the car is built for	001 = USA, 002 = Canada, 003 = EU,

References

1. Boniol, F., Wiels, V.: The landing gear system case study. In: Boniol, F., Wiels, V., Ait Ameur, Y., Schewe, K.-D. (eds.) ABZ 2014. CCIS, vol. 433, pp. 1–18. Springer, Cham (2014). https://doi.org/10.1007/978-3-319-07512-9_1
2. Hoang, T.S., Butler, M., Reichl, K.: The hybrid ERTMS/ETCS level 3 case study. In: Butler, M., Raschke, A., Hoang, T.S., Reichl, K. (eds.) ABZ 2018. LNCS, vol. 10817, pp. 251–261. Springer, Cham (2018). https://doi.org/10.1007/978-3-319-91271-4_17
3. Houdek, F.: Automotive example: exterior lighting and speed control. In: Pohl, K., Broy, M., Daembkes, H., Hönninger, H. (eds.) Advanced Model-Based Engineering of Embedded Systems, pp. 13–19. Springer, Cham (2016). https://doi.org/10.1007/978-3-319-48003-9_1
4. Mashkoor, A.: The hemodialysis machine case study. In: Butler, M., Schewe, K.-D., Mashkoor, A., Biro, M. (eds.) ABZ 2016. LNCS, vol. 9675, pp. 329–343. Springer, Cham (2016). https://doi.org/10.1007/978-3-319-33600-8_29

Modelling an Automotive Software-Intensive System with Adaptive Features Using ASMETA

Paolo Arcaini[1] , Silvia Bonfanti[2(✉)] , Angelo Gargantini[2] , Elvinia Riccobene[3] , and Patrizia Scandurra[2]

[1] National Institute of Informatics, Tokyo, Japan
arcaini@nii.ac.jp
[2] Department of Economics and Technology Management, Information Technology and Production, Università degli Studi di Bergamo, Bergamo, Italy
{silvia.bonfanti,angelo.gargantini,patrizia.scandurra}@unibg.it
[3] Dipartimento di Informatica, Università degli Studi di Milano, Milan, Italy
elvinia.riccobene@unimi.it

Abstract. In the context of automotive domain, modern control systems are software-intensive and have adaptive features to provide safety and comfort. These software-based features demand software engineering approaches and formal methods that are able to guarantee correct operation, since malfunctions may cause harm/damage. Adaptive Exterior Light and the Speed Control Systems are examples of software-intensive systems that equip modern cars. We have used the Abstract State Machines to model the behaviour of both control systems. Each model has been developed through model refinement, following the incremental way in which functional requirements are given. We used the ASMETA tool-set to support the simulation of the abstract models, their validation against the informal requirements, and the verification of behavioural properties. In this paper, we discuss our modelling, validation and verification strategies, and the results (in terms of features addressed and not) of our activities. In particular, we provide insights on how we addressed the adaptive features (the adaptive high beam headlights and the adaptive cruise control) by explicitly modelling their software control loops according to the MAPE-K (Monitor-Analyse-Plan-Execute over a shared Knowledge) reference control model for self-adaptive systems.

1 Introduction

Modern control systems, like those in the automotive domain, are software-intensive, have adaptive features, and must be reliable. Formal methods can be applied in order to improve their development and guarantee their correct operational behaviour and safety assurance. In this paper, we report our experience in applying the Abstract State Machine (ASM) formal method to the Adaptive Exterior Light (ELS) and the Speed Control Systems (SCS), which are examples of software-intensive systems that equip modern cars. We used the ASMETA framework, which provides a wide tool support to

P. Arcaini is supported by ERATO HASUO Metamathematics for Systems Design Project (No. JPMJER1603), JST. Funding Reference number: 10.13039/501100009024 ERATO.

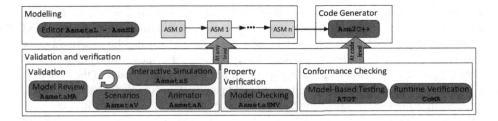

Fig. 1. ASM-based development process

ASMs. ASMETA tries to combine the formality of ASMs with a tools set that supports the editing, simulation and animation of the abstract models, their validation against the informal requirements, and the verification of behavioural properties. Moreover, ASMETA adopts a refinement-based process, and employs several constructs in order to allow modularity.

To address the adaptive features of the case study, i.e., the adaptive high beam headlights and the adaptive cruise control, we exploited the concept of self-adaptive ASMs [5], which allows modelling adaptation logics in terms of MAPE-K (Monitor-Analyse-Plan-Execute over a shared Knowledge) feedback control loops.

The paper is structured as suggested by the *call for paper* of the case study. The following subsection briefly presents the ASM formal method and its supporting tool-set ASMETA. Section 2 explains our modelling strategy. Details about our models and how they capture the requirements are provided in Sect. 3. We have applied several *validation* and *verification* (V&V) activities that are presented in Sect. 4. Section 5 discusses some observations that we draw from this experience, together with some limits of our approach. Section 6 concludes the paper.

1.1 The ASM Method and the ASMETA Tool-Set

Basic Definition. ASMs [10,11] are an extension of Finite State Machines (FSMs) where unstructured control states are replaced by *states* comprising arbitrary complex data (i.e., domains of objects with functions defined on them), and *transitions* are expressed by transition rules describing how the data (state function values saved into *locations*) change from one state to the next. ASM models can be read as "pseudocode over abstract data" which comes with a *well defined semantics*: at each computation step, all transition rules are executed in parallel by leading to simultaneous (consistent) updates of a number of locations.

Modelling Process and Tools. ASMs allow an iterative design process, shown in Fig. 1, based on model refinement. Tools supporting the process are part of the ASMETA (ASM mETAmodeling) framework[1] [4]. Requirements modelling starts by developing a high-level model called *ground model* (ASM 0 in Fig. 1). It is specified by reasoning on the informal requirements (generally given as a text in natural language) and using terms of the application domain, possibly with the involvement of all stakeholders. The ground model should *correctly* reflect the intended requirements and should

---

[1] http://asmeta.sourceforge.net/.

be *consistent*, i.e., without possible ambiguities of initial requirements. It does not need to be *complete*, i.e., it may not specify some given requirements. The ground model and the other ASM models can be edited in AsmEE by using the concrete syntax AsmetaL [14]. Starting from the ground model, through a sequence of *refined* models, further functional requirements can be specified until a complete model of the system is obtained. The refinement process allows to tackle the system complexity, and to bridge, in a seamless manner, specification to code. At each refinement level, already at the level of the ground model, different V&V activities can be applied. In Sect. 4, we explain in detail how model validation and property verification are performed in ASMETA tools.

Model to code transformation are supported for C++ code [9], and *conformance checking* is possible to check if the implementation, if externally provided, conforms to its specification. The tool ATGT [13] can be used to automatically generate tests from ASM models and, therefore, to check the conformance offline; CoMA [3], instead, can be used to perform runtime verification, i.e., to check the conformance online.

1.2   Distinctive Features of the Modelling Approach

Machine and Modules. As better explained in Sect. 2, in our modelling activity we strongly make use of *modularization*. To concretely support it, we exploited the concept of *ASM module* as introduced in [11] and provided by the language of the ASMETA tool set. Specifically, an *ASM module* contains the declaration and definitions of domains, functions, invariants, and rules; an ASM *machine* is an ASM module that additionally contains a (unique) *main rule* representing the starting point of the machine execution, and an *initial state*. The keyword asm introduces the main ASM while the keyword module indicates a module.

Self-adaptive ASMs. To model the adaptive features of the two automotive subsystems, we exploit the concept of MAPE-K (a sequence of four computations Monitor, Analyze, Plan, and Execute over a shared Knowledge) feedback control loop [16] commonly used to structure the adaptation logic of self-adaptive software systems. To this end, we adopt *self-adaptive Abstract State Machines* (self-adaptive ASMs) as defined in [5] to formalize the sequential execution of the four MAPE computations of a MAPE-K loop in terms of ASM transitions rules (see Sect. 3.1, *CarSystem003*).

2   Modelling Strategy

We here describe the general strategy adopted while modelling the ELS and the SCS using ASMs. We explain how our model is structured, how the structure relates to the requirements, the model purpose (in terms of properties addressed and not addressed by our solution), and our formalization approach.

Model Structure. The ELS and SCS sub-systems are loosely coupled (i.e., they work in parallel and share some external signals), so we handle their requirements independently. More specifically, for each subsystem, we developed an ASM specification through a sequence of refined models, following the incremental way in which functional requirements of the software based controllers are described in the requirements document [15]. The models are numbered from 1 to 9: models from CarSystem001 to CarSystem004 refer to the ELS, while the SCS is modelled from CarSystem005 to CarSystem007. CarSystem008 merges the two systems and CarSystem009 introduces the faults handling and general properties.

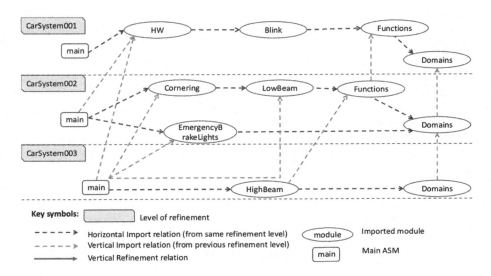

Fig. 2. ASM structure of the first 3 levels

Since, at a given refinement level, only some parts of the model were refined, to keep track of maintained and refined parts and to relate abstract and refined sub-models, we structure an ASM model in *modules*. Therefore, at each level, an ASM results in the horizontal and vertical composition of ASM modules (by exploiting the import module feature). For example, Fig. 2 illustrates the ASM models structure in terms of horizontal and vertical imports for the first three refined levels of the *CarSystem* related to the ELS subsystem. Level 1 consists of an ASM *CarSystem001main* that imports module *CarSystem001HW* that imports module *CarSystem001Blink*, till the final module *CarSystem001Domains* is imported[2]. At this level, module relations are all horizontal imports. Similarly, ASM *CarSystem002main* imports (horizontally) module *CarSystem002Cornering* from the same refinement level, and it imports (vertically) module *CarSystem001HW* from the previous refinement level. Note that module *CarSystem002Domains* imports module *CarSystem001Domains*, since the former enlarges (as expected during refinement) the latter. Vertical relations can be also module refinement relations. For example, Fig. 3 focuses on level 4 and illustrates the ASM *CarSystem004main* that imports, among the others, the module *CarSystem004HW* refining module *CarSystem001HW* from level 1, and module *CarSystem004Cornering* refining module *CarSystem002Cornering* from level 2. Other depicted import/refine-

---

[2] Note that common functions and common domains are declared and defined into specific modules (*CarSystem00XFunctions* and *CarSystem00XDomains*) which are imported by others.

ment relations are self-explanatory. More details on the refinement levels and corresponding models are given in Sect. 3.

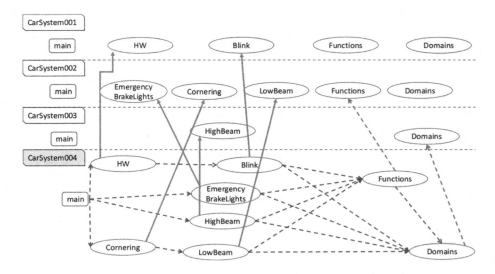

Fig. 3. ASM structure of level 4

To model the overall behaviour resulting from the union of the behaviours of all distributed software we simply exploit the notion of *parallel ASMs* [7]. Therefore, we model software running on ECUs as parallel algorithms working in sequential global time. Whenever necessary (as stated in the requirements document) and to avoid interference, the machine follows only one or a restricted subset of all possible parallel execution paths by using the *mutual exclusive guards pattern* at the level of rules, i.e., the simultaneous activation of certain rules is avoided by modelling them as conditional rules with mutual exclusive guards.

Model Purpose. The proposed ASM model is primarily tailored to the formalization and analysis of functional aspects of the ELS and SCS subsystems in order to provide guarantees of their operational correctness. Modelling and validation activities revealed, however, some statements where the requirements were wrong or unambiguous. We had the possibility to check our doubts with the chairs (working as domain experts) and we got corrected version of the requirements. Further details on these missing/ambiguous aspects are given in Sect. 5.

Except for some features not addressed by our solution (see below), our model(s) capture(s) all requirements described in the document [15]. Each refined model was analyzed using different techniques (see Sect. 4), considering also the description of the operational scenarios that are given as annex part of the requirements. The most important properties addressed by our solution are:

– On the ELS subsystem: *a)* The priority of hazard warning over blinking is guaranteed; *b)* Low beam headlights are turned on and off as required; *c)* Priority of ELS-19 over other requirements has been addressed when the ambient light is activated; *d)* High beam headlights are automatically turned on/off when the light rotary switch is set to Auto; *e)* When subvoltage/overvoltage occurs the system reacts as required.

- On the SCS subsystem: *a)* (Adaptive) cruise control desired speed is set as required: the adaptive cruise control sets automatically the speed to reach the target based on different factors like the current speed and the speed of the vehicle ahead. *b)* Emergency brake intervenes to avoid collisions; *c)* Speed limit and traffic sign detection set the threshold speed when they are activated by the user.

ELS-18 If the light rotary switch is in position Auto and the ignition is On, the low beam headlights are activated as soon as the exterior brightness is lower than a threshold of 200 lx. If the exterior brightness exceeds a threshold of 250 lx, the low beam headlights are deactivated. In any case, the low beam headlights remain active at least for 3 seconds.

```
macro rule r_LowBeamHeadlights =
  ...
  if (lightRotarySwitch = AUTO and engineOn(keyState) and
                        brightnessSensor<200) then
    if ((not lowBeamLightingOn)) then r_LowBeamTailLampOnOff[100]
    endif
  endif
  if (lightRotarySwitch = AUTO and engineOn(keyState) and
                        brightnessSensor>250) then
    if (lowBeamLightingOn and passed3Sec) then
      r_LowBeamTailLampOnOff[0]
    endif
  endif
```

Code 1. Translating text into rules

Not Addressed Features. The main feature not addressed by our solution is the time management. We are not able to deal with continuous time, although a notion of reactive timed ASMs has been proposed [17], there is no tool support for it. To overcome this limitation, we assume that a monitored function notifies the system whether an interval time is passed. For example, requirement ELS-18 states that low beam headlights remain active at least for 3 s. As shown in Code 1, we have introduced a monitored function passed3Sec that notifies if 3 s had elapsed since the low beam headlights received the command to be turned off. A further not addressed feature is the frequency of blinking. Since the model does not support continuous time it is not possible to set the direction lamps state (ON or OFF) every second (the duration of a flashing cycle is 1 s). Due to this limitation, we have introduced an enumerative value that indicates the current pulse ratio and we have supposed that the Head-Unit sets the state of direction lamp given the pulse ratio value.

Requirements Formalization Approach. To understand and specify the behaviour, we started from the textual description of the systems, and we tried to express the text in terms of transition rules, by supplying the necessary definitions of domains and functions. In naming functions, domains and transition rules, we have used an domain-specific terminology that can be understood by the stakeholders. Furthermore, for the functions defined in the tables at the end of the document of specification we have used the names proposed. An example of requirement formalization is shown in Code 1 that reports the requirement ELS–18 and the corresponding ASM rule. Sometimes the requirements are not independent of each other, for this reason more than one requirement is modelled by one rule.

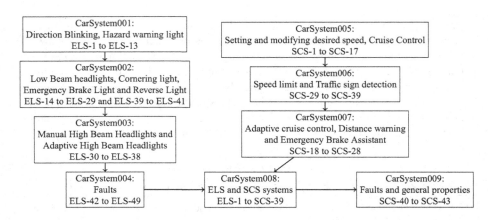

Fig. 4. Chain of refined models and captured requirements

3   Model Details

In this section, we present the result of our modelling of the Adaptive Exterior Light and Speed Control systems[3].

We have proceeded through refinement of the two systems independently and finally we have merged them together to obtain the complete system. Figure 4 shows the model refinement chain and lists the requirements introduced in each model.

Table 1. Models dimension

	Functions				Rules	
	Monitored	Controlled	Derived	Static	n rules declarations	n rules
CarSystem001	4	12	6	1	14	103
CarSystem002	14	26	14	4	26	218
CarSystem003	18	31	24	5	33	244
CarSystem004	19	31	31	5	33	252
CarSystem005	8	4	1		10	78
CarSystem006	12	8	2		14	125
CarSystem007	21	17	6		24	183
CarSystem008	36	46	36	5	56	433
CarSystem009	36	46	37	5	56	433

Table 1 shows the model dimensions in terms of number of functions and rules and the traceability table between requirements and rules/functions (see Table 2). In the sequel, we will show some parts of the models for each subsystem and then we will explain the merging of two systems.

---

[3] Artifacts are at https://github.com/fmselab/ABZ2020CaseStudyInAsmeta.

Table 2. Traceability table between requirements and rules/functions

	CS001	CS002	CS003	CS004	CS005	CS006	CS007	CS008	CS009
Exterior Light System									
Direction Blinking									
ELS-1 to ELS-7	■								
Hazard Warning									
ELS-8 to ELS-13	■								
Low Beam Headlights Cornering Light									
ELS-14 to ELS-29		■							
Manual Hight Beam Headlights									
ELSL-30 to ELS-31		■							
Adaptive High Beam Headlights									
ELS-32 to ELS-38		■							
Emergency Brake Light									
ELS-39 to ELS-40		■							
Reverse Light									
ELS-41		■							
Fault Handling									
ELS-42 to ELS-49				■					
Speed Control System									
Setting and modifying desired speed									
SCS-1 to SCS-12									
Cruise Control									
SCS-13 to SCS-17					■				
Adaptive Cruise Control									
SCS-18 to SCS-24							■		
Distance Warning									
SCS-25 to SCS-26							■		
Emergency Brake Assistant									
SCS-27 to SCS-28							■		
Speed Limit									
SCS-29 to SCS-35							■		
Traffic Sign Detection									
SCS-36 to SCS-39							■		
Fault Handling and General Properties									
SCS-40 to SCS-43									■

3.1 Adaptive Exterior Light System

ELS has been modelled into 4 refinement steps which are explained below.

CarSystem001. This model describes the functions of direction blinking and hazard warning. The critical features found in this modelling phase are the following: *a)* the hazard warning has the priority over direction blinking (ELS-3); *b)* the tail lamp is used as an indicator for cars sold in USA and Canada (ELS-6). Code 2 shows that if hazard warning request is activated and any kind of blinking is running, blinking must be stopped and hazard warning is started. In case of direction blinking, the value of pitman arm is stored to restart the request as soon as hazard warning is deactivated; if the pitman arm is moved back to neutral position, the request is cancelled. The tail lamp status is updated to FIX or BLINK. It is BLINK only if blinking is active and car is sold in USA or Canada, otherwise it is FIX. Moreover, the value of light is dimmed by 50% during blinking (see Code 3). To address the requests from the pitman arm, we have defined three functions, *pitmanArmUpDown* for the incoming request, *pitmanArmUpDown_RunnReq* for the running request and *pitmanArmUpDown_Buff* to save the incoming request if it cannot be satisfied in the current state. When the running request has been processed, the request in the buffer is executed unless a new one arrives.

```
//Hazard Warning
macro rule r_HazardWarningLight =
  If (hazardWarningSwitchOn_Runn) then
      ...
  else // If HW is not running
      ...

  if (hazardWarningSwitchOn) then
    par
      if (pitmanArmUpDown_RunnReq !=
                          NEUTRAL_UD) then
        r_InterruptBlinking[]
      endif
      hazardWarningSwitchOn_Start := true
      //If request from pitman arm UpDown arrives
      // save it into the buffer
      //If UPWARD7 or DOWNWARD 7
      r_SavePitmanArmUpDownReq[]
    endpar
  else
    r_DirectionBlinking[]
  endif
      ...

  endif
```

Code 2. Hazard warning has priority over direction blinking

```
//True if the car is sold in USA or CANADA
function tailLampAsIndicator = (marketCode = USA or
                          marketCode = CANADA)

//Set tail lamp status
macro rule r_setTailLampLeft ($value in
LightPercentage , $status in TailLampStatus) =
  par
    tailLampLeftStatus := $status
    tailLampLeftBlinkValue := $value
  endpar

macro rule r_BlinkLeft ($value in LightPercentage,
                          $pulse in PulseRatio) =
  par
    blinkLeft := $value
    blinkLeftPulseRatio := $pulse
    //ELS-6
    if (tailLampAsIndicator) then
      if ($value = 0) then
        r_setTailLampLeft[0,FIX]
      else
        r_setTailLampLeft[50,BLINK]
      endif
    endif
  endpar
```

Code 3. Tail lamp as indicator during direction blinking and hazard warning

CarSystem002. This introduces the low beam headlights and cornering light, emergency brake light, and reverse light functions. Each of these functions is modelled in an ASM module. Common functions and domains are extended starting from those defined in the *CarSystem001*, while hazard warning and direction blinking are unchanged. Requirements ELS-15, ELS-16, ELS-17, and ELS-19 are interconnected because ELS-19 has the priority over the others if ambient light activated. We have defined a guard called *ambientLightingAvailable* which is true if ambient lighting is activated, the vehicle is not armoured and darkness mode is switched off. Some requirements state that the system performs an action if function X changes its value from state s to state $s + 1$. To detect the value change, we store the value of function X in the previous state ($X\_Previous$) and we compare that value with the current value. When the model detects a value change, the system acts as defined by rules. An example is the requirement ELS-19: the low beam headlamps are activated if engine has been stopped. We have captured the key state mutation by checking the value of *keyState_Previous* compared to *keyState*.

CarSystem003. This step introduces the control features for the manual and adaptive high beam headlights (ELS-30 to ELS-38). We first modelled the manual control of the high beam headlights and then, in the same refinement level, the adaptive one. The control variables are the high beam luminous strength (a percentage) and the illumination distance (expressed in meters).

In manual mode (ELS30-ELS31), the user can set a fixed illumination area of 220 m and 100% of luminous strength, or activate the high beam headlights temporary (so-called *flasher*). We had to make the following assumptions due to missing requirements: (*i*) a maximum illumination area of 360 m and 100% of luminous strength in the flasher mode; (*ii*) the key is inserted or the engine is on to activate high beam in a fixed way.

```
macro rule r_MAPE_HBH =
par
    r_Monitor_Analyze_HBH[]
    if adaptiveHighBeamDeactivated then
        highBeamOn := false
    endif
endpar

macro rule r_Monitor_Analyze_HBH =
if adaptiveHighBeamActivated then
par
    if drivesFasterThan(currentSpeed,300)
        and not oncomingTraffic then
        r_IncreasingPlan_HBH[]
    endif
    if oncomingTraffic then
        r_DecreasingPlan_HBH[]
    endif
endpar
endif
```

```
macro rule r_Execute_HBH ($setHighBeam in Boolean,
$setHighBeamMotor in HighBeamMotor,
$setHighBeamRange in HighBeamRange) =
r_set_high_beam_headlights[$setHighBeam,
            $setHighBeamMotor,$setHighBeamRange]

macro rule r_IncreasingPlan_HBH =
    let ($d = lightIlluminationDistance(calculateSpeed),
    $l = luminousStrength(calculateSpeed)) in
            r_Execute_HBH[true,$d,$l] endlet

macro rule r_DecreasingPlan_HBH = r_Execute_HBH[true,30,0]

macro rule r_set_high_beam_headlights($v in Boolean,
$d in HighBeamMotor, $l in HighBeamRange) =
par
    highBeamOn := $v
    highBeamMotor := $d
    highBeamRange := $l
endpar
```

Code 4. MAPE loop may start and stop the adaptive high beam headlight

In adaptive mode (ELS32-ELS38), the illumination of the road is automated depending on the incoming vehicles (as detected by a built-in camera) and optimized to illuminate the appropriate area according to the vehicle speed, the last being the current speed of the vehicle or the target speed provided by the advanced cruise control in case it is activated. The illumination distance and the luminous strength are adjusted according to characteristic curves provided in the requirements document. In order to calculate such values, we had to reverse engineered the formulas as suggested in the additional information provided with the specification document. We modelled this adaptive behaviour in terms of a MAPE-K feedback control loop that starts with the rule *r_MAPE_HBH* (see Code 4). A control loop in self-adaptive systems is a sequence of four computations: Monitor-Analyse-Plan-Execute (MAPE) over a knowledge base. In self-adaptive ASMs [5], it is modelled by means of four rules, one per MAPE computation, while the knowledge is modelled by means of functions, since in ASMs system memory is represented in terms of functions. In our case, the MAPE loop consists of the following rules invoked in a waterfall manner within one single ASM-step machine (see Code 4): *r_Monitor_Analyze_HBH*, where monitor and analyze computations are modelled as a unique activity; *r_IncreasingPlan_HBH* and *r_DecreasingPlan_HBH* to plan the adaptation if necessary: light illumination distance and luminous strength are increased or decreased according to the vehicle speed; *r_Execute_HBH* to set the values as planned: *highBeamOn* to activate/deactivate the high beam, *highBeamRange* and *highBeamMotor* for the high beam luminous strength and illumination distance.

function subVoltage = (currentVoltage < 85)	function overVoltage = (currentVoltage > 145)

macro rule r_CorneringLights =
//ELS—46 cornering lights on if no subvoltage
  if (not subVoltage) **then**
  //Turn on cornering light
  ...
  else
  //ELS—46 cornering lights off if subvoltage
  if (corneringLightRight != 0 or
    corneringLightLeft != 0) **then**
    r_CorneringLightsOff[]
  endif
endif
	function = (100−(currentVoltage−145)∗20)
	
	function setOverVoltageValueLight($value **in** Integer) =
	if (overVoltage and $value > overVoltageMaxValueLight) **then**
	  overVoltageMaxValueLight **else** $value **endif**
	
	macro rule r_setTailLampLeft ($value **in**
	LightPercentage , $status **in** TailLampStatus) =
	par
	  tailLampLeftStatus := $status
	  tailLampLeftBlinkValue := setOverVoltageValueLight($value)
	endpar

Code 5. Subvoltage handling   **Code 6.** Overvoltage handling

CarSystem004. This modelling phase introduces fault handling, in particular how the software reacts to overvoltage or subvoltage. When subvoltage is present, some functionalities like cornering light and parking light are not available. This has been addressed by adding a guard that, in case of subvoltage (the voltage value is less than 8.5 V), disables them (see Code 5). In case the voltage is more than 14.5 V, the system is in overvoltage. The maximum value of lights is computed by the *setOvervoltage-ValueLight* function: it returns the minimum between current light value and the value calculated by *overVoltageMaxValueLight* function. In this step of refinement, we have refined the modules whose behaviour is affected by the voltage value.

3.2 Speed Control System

The final model of Speed Control System has been addressed through three steps of refinement explained in the following.

CarSystem005. In the first model of SCS, we have implemented the functionalities of cruise control and setting and modifying desired speed (see *CarSystem005Desired-SpeedCruiseC* module). Desired speed and target speed are modified based on SCS lever position when (adaptive) cruise control is activated.

CarSystem006. This step of refinement introduces the speed limit and traffic sign detection functionalities. If traffic sign detection is on, the target speed is modified by the recognized traffic sign value. The speed limit modifies the desired speed which must not be exceeded by the current speed.

CarSystem007. This step of refinement introduces the adaptive cruise control and distance warning from the vehicle ahead (from SCS-18 to SCS-26), and the brake assistant (from SCS-27 to SCS-28) to initiate braking in critical situations. Similarly to the adaptive high beam headlights (see refinement CarSystem003), we modelled the adaptive behaviour of the cruise control in terms of a MAPE-K feedback control loop that monitors the distance from the vehicle ahead, plans and executes acceleration/deceleration automatically, including braking until a full standstill and starting from a standstill.

3.3 Merging ELS and SCS Models

Once we have developed the ELS and SCS separately, we have merged them to obtain a model that includes both systems.

CarSystem008. This step of refinement has been obtained easier than what we expected due to the modularity followed in the previous steps of refinement. Once the main module (*CarSystem008main*) has been defined, we have simply imported the modules previously developed. All rules are executed in parallel and no inconsistent update has been found because the systems are independent of each other, they have only common inputs. A schema that shows how we have imported modules is available on-line.

CarSystem009. We have introduced the requirements from SCS-40 to SCS-43. The dangerous situations in SCS-40, SCS-41 and SCS-42 are already managed by the model; in this step of refinement we have integrated SCS-43 by refining the guards of CarSystem004EmergencyBrakeLights module. The brake lights are activated either by the brake pedal pressed by the user or the system activates the emergency brake.

4   Validation and Verification

Validation and verification are supported by a set of ASMETA tools. In this section, we report results and tools used for each activity, and explain the changes to the models that resulted from the validation and the verification.

Validation. Model validation helps to ensure that the specification really reflects the intended requirements, and to detect faults and inconsistencies as early as possible with limited effort. While writing models, we have started the validation activity by using the animator AsmetaA [8] which uses tables to convey information about states and their evolution. We have performed *interactive animation* that consists in providing inputs (i.e., values of monitored functions) to the machine and observing the computed state. The animator, at each step, performs *consistent updates checking* to check that all the updates are consistent (in an ASM, two updates are inconsistent if they update the same location to two different values at the same time), and invariant checking.

With the increasing complexity of the ELS and SCS system models, we have intensely used the scenario-based validation AsmetaV [12] that allows to build and execute *scenarios* of expected system behaviours. In scenario-based validation, the designer provides a set of scenarios specifying the expected behaviour of the models (using the textual notation Avalla [12]). These scenarios are used for validation by instrumenting the simulator AsmetaS [14]. During simulation, AsmetaV captures any check violation and, if none occurs, it finishes with a *PASS* verdict. Avalla provides constructs to express execution scenarios in an algorithmic way, as interaction sequences consisting of actions committed by the user to set the environment (i.e., the values of monitored/shared functions), to check the machine state, to ask for the execution of certain transition rules, and to enforce the machine itself to make one step as reaction of the user actions. Code 7 shows an example of scenario: it specifies the behaviour of the second validation sequence for the exterior light provided as part of the documentation[4]. More scenarios are available on our online repository, including all those provided with the case study document.

---

[4] See the *ValidationSequences_v1.8.xlsx* document in the web site of the case study.

```
//Light switch to AUTO −> lights remain off
set lightRotarySwitch := AUTO;
set keyState := NOKEYINSERTED;
set hazardWarningSwitchOn := false;
set darknessModeSwitchOn := false;
set brightnessSensor := 100;
set pitmanArmUpDown := NEUTRAL_UD;
set reverseGear := false;
set brakePedal := 0;
step
//Ignition to ON position −> no effect on light
check tailLampLeftBlinkValue = 0;
check tailLampLeftFixValue = 0;
check tailLampRightBlinkValue = 0;
check tailLampRightFixValue = 0;
//Cornering lights off
check corneringLightRight = 0;
check corneringLightLeft = 0;
//Low beam headlight
check lowBeamLeft = 0;
check lowBeamRight = 0;
check reverseLight = 0;
check brakeLampLeft = 0;
check brakeLampRight = 0;
check brakeLampCenter = 0;
```

```
//Turn the light rotary switch ON
set lightRotarySwitch := ON;
step
//Ignition to KeyInserted −> lights reimain off
set keyState := KEYINSERTED;
step
//Engine start −> light on with 100%
set keyState := KEYINIGNITIONONPOSITION;
step
check tailLampLeftBlinkValue = 0;
check tailLampLeftFixValue = 100;
check tailLampRightBlinkValue = 0;
check tailLampRightFixValue = 100;
//Cornering lights off
check corneringLightRight = 0;
check corneringLightLeft = 0;
//Low beam headlight
check lowBeamLeft = 100;
check lowBeamRight = 100;
check reverseLight = 0;
check brakeLampLeft = 0;
check brakeLampRight = 0;
check brakeLampCenter = 0;
.
```

Code 7. Scenario for normal light, no daytime light, ambient light, night

Although interactive and scenario-based simulations are very useful to get a fast understanding of the developed models and quickly detect possible modelling errors, they do not allow to perform an exhaustive check. Therefore, we performed *model review* using the AsmetaMA tool [2], as a complementary validation technique: it is a form of static analysis to determine if a model has sufficient *quality* attributes (as minimality, completeness, consistency). This automatic activity can find problems that could pass undetected during interactive simulation and scenario validation, which cannot be exhaustive and perform only some system executions. For example, CarSystem003 has a rule that decides when to switch on the parking lights (parkingLightON := true). By introducing requirement ELS-46, however, we added a rule to switch the lights off in case of subvoltage (parkingLightON := false). In the first version of the CarSystem004, the two rules sometimes conflicted, so leading to an inconsistent update. Model review allowed us to spot this problem, that we solved by introducing a guard to avoid the conflict.

Verification. Formal verification of ASMs is possible by means of the tool AsmetaSMV [1], and both *Computation Tree Logic* (CTL) and *Linear Temporal Logic* (LTL) formulas are supported. To perform model checking with NuSMV (the model checker AsmetaSMV is built on) that requires a finite state space, we reduce infinite domains in the original models to finite domains.

For the case study, some properties can be naturally derived from the requirements. The typical form of a requirement is "when/if . . . then . . ." describing that when something happens (a given external input received, a given state condition, etc.), some actions must be taken. Such kind of properties are naturally translated in temporal properties as $\Box(\phi \rightarrow \bigcirc(\psi))$ or $\Box(\phi \rightarrow \Diamond(\psi))$. However, these kinds of properties (in particular the first one) are also reflected in the structure of ASM rules derived from the requirements, that take the form of **if** ... **then** ... **else** ... **endif** rules. In this case, temporal properties assume the form of *redundant specifications* that can be used to enforce the model and make it robust against possible wrong future modifications of the model. An example of such kind of property that we developed related to requirement ELS-18 is the CTL property: **ag**((lightRotarySwitch = AUTO and engineOn and brightnessSensor < 200)implies **ax**(lowBeamLightingOn)).

In addition to these straightforward properties, we specified more general properties that are not directly related to single requirements. For example, we specified the following three properties:

- both direction indicators blink iff the car is in hazard warning:
  ag((blinkLeft != 0 and blinkLeftPulseRatio != NOPULSE and blinkRight != 0 and blinkRightPulseRatio != NOPULSE)= hazardWarningSwitchOn_Runn)
- if tail lamps are blinking, the car is not European:
  ag((tailLampLeftStatus=BLINK or tailLampRightStatus=BLINK)implies marketCode!=EU)
- the market code of a car cannot be changed:
  forall $c **in** MarketCode **with** (marketCode = $c implies **ag**(marketCode = $c))

5 Discussion

We here provide a more detailed discussion about our experience in modelling and analysing the case that took totally less than one month (around one week for understanding the requirements, and the remaining time for the model development and V&V). We report flaws we discovered in the requirements documents, features that we would have expected in the documents, and also missing functionalities of our framework that would have been helpful (besides the temporal aspects discussed in Sect. 2).

Scenarios. The scenarios provided with the case study requirements turned out to be very useful as, in some cases, allowed us to clarify some misunderstanding we had when we developed the models starting from the requirements document. For example, observing the scenarios, we realized that desiredSpeed and targetSpeed are two separate entities that can assume different values with two updating policies, while, only reading the requirements document, we thought that the user could modify both at the same time. However, in other cases, the description of the scenarios and the description of the requirements were inconsistent. For some of such inconsistencies, it was clear that the document to trust was the scenario description; for example, the requirements document uses in an improper way the terms *desired*, *target*, and *set* speed, sometimes using them interchangeably; the scenarios document, instead, clearly distinguishes them and allows to observe their different roles. Other inconsistencies, instead, were less easy to disambiguate. This was the case of *traffic sign detection* for which the requirements document states that only the target speed is modified when the sign is recognized; however, scenario 6 of *Validation Sequences Speed* shows a case in which the car, when detects the sign, modifies both desired and target speed.

Requirements Coverage. Since the scenarios turned out to be so useful, we would have liked to have a more exhaustive validation sequences in the informal documentation in order to build a set of scenarios *covering* all the requirements; using the coverage feature of our validator AsmetaV, we realized that this is not the case. For example, requirements ELS-42 to ELS-47 are not covered by any validation sequence and therefore by any of our scenario. Moreover, when doing this coverage checking, we realized a limit of our coverage tool that can only provide coverage information at the level of macro rule (similarly to call coverage). This coarse grained level of coverage may be not informative enough in situations in which requirements are mapped at the level of, e.g., branches of conditional rules. As future work, we plan to extend our coverage evaluator to provide information as decision and condition coverage.

Scenario Derivation and Animation. As said previously, we have extensively used scenario-based validation during the modelling activity. We realized that a better integration with model animation [8] would permit to save animation sessions in terms of scenarios, and also to animate existing scenarios. For this work, we have developed the former technique that allows us to export into Avalla scripts the animations we perform, and re-execute them later when changing the model (in a kind of regression testing using "record and replay").

Parametric ASMs. The systems described in the case study actually represent a family of systems that can be configured on the base of the market and type of car. Different configurations lead to different behaviours of the systems. In order to model all these features, we had to introduce *flags* to be set in the initial state: this reduces the readability and maintainability of the models. It would be useful to have a *parametric version* of the ASM model, similarly to what done for software programs using Software Product Lines. A recent approach has been proposed in this context for ASM [6], and we plan to consider it in future usages.

Implementation. In modelling the case study, we did not consider any implementation. However, the ASMETA framework provides support in this sense. First of all, a translator to C++ [9] is available; it could be applied to generate a first prototype of the implementation, which could be then further extended by developers. Instead, if a system implementation is available, *conformance checking* approaches can be applied, in terms of *model-based testing* and *runtime verification* both supported by ASMETA.

6   Conclusions

We have presented the specification, validation, and property verification of an automotive system by using ASMs. We have discussed our experience in modelling an adaptive exterior light and the speed control systems that equip modern cars, also addressing the adaptive features of the two systems in terms of MAPE-K feedback control loops. We have also shown how to write and run scenarios and verify properties addressed by our models. We have found some misunderstanding in the document of requirements, because the described behaviour was different from the behaviour expected in validation sequences. We have found some limitations in our tools, e.g., the coverage evaluator provides only coverage in terms of macro rules, that we plan to overcome with future improvements. On the other hand, this case study has provided us the opportunity to test our framework in terms of robustness and user experience in modelling complex systems. We have noticed that the framework is particularly suitable to handle the increasing complexity of the models: the support for modularization (at the level of modelling and scenario construction) allows producing refined model and refined scenarios with limited effort.

References

1. Arcaini, P., Gargantini, A., Riccobene, E.: AsmetaSMV: a way to link high-level ASM models to low-level NuSMV specifications. In: Frappier, M., Glässer, U., Khurshid, S., Laleau, R., Reeves, S. (eds.) ABZ 2010. LNCS, vol. 5977, pp. 61–74. Springer, Heidelberg (2010). https://doi.org/10.1007/978-3-642-11811-1_6
2. Arcaini, P., Gargantini, A., Riccobene, E.: Automatic review of abstract state machines by meta property verification. In: Proceedings of the Second NASA Formal Methods Symposium (NFM 2010), pp. 4–13. NASA (2010)
3. Arcaini, P., Gargantini, A., Riccobene, E.: CoMA: conformance monitoring of Java programs by abstract state machines. In: Khurshid, S., Sen, K. (eds.) RV 2011. LNCS, vol. 7186, pp. 223–238. Springer, Heidelberg (2012). https://doi.org/10.1007/978-3-642-29860-8_17
4. Arcaini, P., Gargantini, A., Riccobene, E., Scandurra, P.: A model-driven process for engineering a toolset for a formal method. Softw. Pract. Exp. **41**, 155–166 (2011)
5. Arcaini, P., Riccobene, E., Scandurra, P.: Formal design and verification of self-adaptive systems with decentralized control. ACM Trans. Auton. Adapt. Syst. **11**(4), 25:1–25:35 (2017)
6. Benduhn, F., Thüm, T., Schaefer, I., Saake, G.: Modularization of refinement steps for agile formal methods. In: Duan, Z., Ong, L. (eds.) ICFEM 2017. LNCS, vol. 10610, pp. 19–35. Springer, Cham (2017). https://doi.org/10.1007/978-3-319-68690-5_2
7. Blass, A., Gurevich, Y.: Abstract state machines capture parallel algorithms. ACM Trans. Comput. Log. **4**, 578–651 (2003)
8. Bonfanti, S., Gargantini, A., Mashkoor, A.: AsmetaA: animator for abstract state machines. In: Butler, M., Raschke, A., Hoang, T.S., Reichl, K. (eds.) ABZ 2018. LNCS, vol. 10817, pp. 369–373. Springer, Cham (2018). https://doi.org/10.1007/978-3-319-91271-4_25
9. Bonfanti, S., Gargantini, A., Mashkoor, A.: Design and validation of a C++ code generator from abstract state machines specifications. J. Softw. Evol. Proc. **32**(2), e2205 (2020)
10. Börger, E., Raschke, A.: Modeling Companion for Software Practitioners. Springer, Heidelberg (2018). https://doi.org/10.1007/978-3-662-56641-1
11. Börger, E., Stärk, R.: Abstract State Machines: A Method for High-Level System Design and Analysis. Springer, Heidelberg (2003). https://doi.org/10.1007/978-3-642-18216-7
12. Carioni, A., Gargantini, A., Riccobene, E., Scandurra, P.: A scenario-based validation language for ASMs. In: Börger, E., Butler, M., Bowen, J.P., Boca, P. (eds.) ABZ 2008. LNCS, vol. 5238, pp. 71–84. Springer, Heidelberg (2008). https://doi.org/10.1007/978-3-540-87603-8_7
13. Gargantini, A., Riccobene, E., Rinzivillo, S.: Using spin to generate tests from ASM specifications. In: Börger, E., Gargantini, A., Riccobene, E. (eds.) ASM 2003. LNCS, vol. 2589, pp. 263–277. Springer, Heidelberg (2003). https://doi.org/10.1007/3-540-36498-6_15
14. Gargantini, A., Riccobene, E., Scandurra, P.: A metamodel-based language and a simulation engine for abstract state machines. J. UCS **14**(12), 1949–1983 (2008)
15. Houdek, F., Raschke, A.: Adaptive exterior light and speed control system (2019)
16. Kephart, J.O., Chess, D.M.: The vision of autonomic computing. Computer **36**(1), 41–50 (2003)
17. Slissenko, A., Vasilyev, P.: Simulation of timed abstract state machines with predicate logic model-checking. J. Univers. Comput. Sci. **14**(12), 1984–2006 (2008)

Validating Multiple Variants of an Automotive Light System with Electrum

Alcino Cunha, Nuno Macedo$^{(\boxtimes)}$, and Chong Liu

INESC TEC and Universidade do Minho, Braga, Portugal
nfmmacedo@di.uminho.pt

Abstract. This paper reports on the development and validation of a formal model for an automotive adaptive exterior lights system (ELS) with multiple variants in Electrum, a lightweight formal specification language that extends Alloy with mutable relations and temporal logic. We explore different strategies to address variability, one in pure Electrum and another through an annotative language extension. We then show how Electrum and its Analyzer can be used to validate systems of this nature, namely by checking that the reference scenarios are admissible, and to automatically verify whether the established requirements hold. A prototype was developed to translate the provided validation sequences into Electrum and back to further automate the validation process. The resulting ELS model was validated against the provided validation sequences and verified for most of requirements for all variants.

1 Introduction

Electrum [10] is a state-based modelling language that extends the structural definitions and first-order relational logic of Alloy [8] with mutable relations and (past and future) linear temporal logic (LTL) operators. Its companion Analyzer [2], itself an extension of the Alloy Analyzer, provides support for validation – through scenario animation – and verification – through two automatic model checking backends, one bounded and another complete. Both animation instances and verification counter-examples are presented back to the user in a unified graphical interface. The combination of first-order and temporal logic makes Electrum well-suited to address systems rich in both structural and dynamic properties, such as automotive software product lines with architectural and behavioural variability. To further ease the feature-oriented design of software families, language extensions to Alloy have also been proposed [1,9].

This paper reports the modelling and subsequent validation and verification of an *adaptive exterior lights* system (ELS) with multiple variants in Electrum[1], carried out as an answer to the ABZ'20 call for case study submissions, following the successful submission to ABZ'18 [4]. The employed approach – which we

---

[1] All resources relevant for the ELS case study are available at https://github.com/haslab/Electrum2/wiki/ELS.

© Springer Nature Switzerland AG 2020
A. Raschke et al. (Eds.): ABZ 2020, LNCS 12071, pp. 318–334, 2020.
https://doi.org/10.1007/978-3-030-48077-6_26

hope can be applied to similar signal-based systems – is presented in Sect. 2. As described in Sect. 3, we have been able to model most ELS requirements by finding an abstraction sweet-spot – in particular for real-time issues. The Electrum language is presented throughout this section as needed. Section 4 describes two explored approaches to modelling multiple variants, one in pure Electrum and another using language extension for feature-oriented design [9]. The ELS model was validated against all the provided validation sequences [7], and verified for most of the ELS requirements, as described in Sect. 5. To ease validation, a prototype was developed to translate tabular validation sequences into Electrum and back for inspection by domain experts. Lastly, Sect. 6 discusses issues identified in the requirements and limitations of the followed approach.

2 ˙ Modelling Strategy

The main goal of this work was to validate the ELS requirements by checking their feasibility and consistency for all valid variants. We started by modelling a single variant of the ELS as a (rough) state machine against which the validation sequences were tested and the requirements subsequently verified. An Electrum model contains both the system specification and the analysis commands, thus our model, described in detail in Sect. 3, is structured as follows:

Environment the available input and output signals, their acceptable values, and possible restrictions to their evolution [6, §4.1–4.3].

ELS state machine a predicate calculating the state of output signals (mostly) from the current state of the input ones, allowing alternative behaviours; inferred from the requirements [6, §4.4].

Animation scenarios simple state sequences, and associated run commands, that exercise the ELS for preliminary validation and regression testing.

Reference scenarios the encoding of the provided validation sequences [7], and associated run commands, with imposed inputs and expected outputs, for validating the modelled ELS state machine; a prototype was developed to translate them from the provided tabular format [7].

Visual elements elements ignored by the analyses but aiding the visualization of scenarios (accompanied by a theme, stored in a separate file).

Requirement assertions the formalization of the requirements [6, §4.4] in temporal logic, and associated check commands to automatically verify them; these assess the overall consistency of the ELS requirements.

The ELS has both structural – that introduce additional signals – and behavioural – that change certain signal outcomes – variability points [6, §3], which Electrum is well-suited to address. Once a single variant was modelled and validated, two strategies were explored to address the remaining variants, as described in Sect. 4: one based on an Electrum idiom where features and variability points are modelled in plain Electrum, and another adapting a language extension developed by us for Alloy [9] where variability points are annotated with features. This introduces another component in the ELS model for imposing the set of valid variants – the *feature model*.

As expected, the development of these components was not sequential but rather iterative as new ELS functions were added to the model. This process was applied to all 9 main ELS functions divided in 48 requirements as of version 1.17 [6], for all 12 valid variants (although only 4 effectively have distinct behaviour) and to all 9 validation sequences of version 1.7 [7]. This work focused on the ELS, but we believe a similar approach could be followed for the *speed control system* (SCS) [6], although the SCS is richer in continuous aspects, which would require additional abstractions. It should also be noted that the authors had no particular domain knowledge, and that the process was solely based on the provided reference material [6,7] and discussions with the case study chair.

The most challenging features of the ELS were those dealing with real-time aspects and the integer nature of the signals. For both we developed proper abstractions – respectively arbitrary duration events and value discretization, described in the next section – that still allowed us to address most requirements. Only requirements requiring arithmetic operations were not addressed at all.

3   The ELS Model

This section describes the main features of the ELS model developed for the simplest variant, that is, when the vehicle is not armoured and aimed at the EU market (the driver position does not affect the ELS).

System Environment. The ELS follows a typical architecture that communicates with the external world through input – from the user interface and sensors – and output signals – to actuators. Our model mimics this architecture so that the translation can be streamlined. In Electrum, likewise Alloy, structure is introduced through the declaration of *signatures* – sets of uninterpreted atoms – and *fields* declared within them – relations of arbitrary arity between signatures. A hierarchy on signatures can be imposed through simple *inclusion* or through *extension*, in which case children must be disjoint; signatures can also be declared as *abstract*, meaning all atoms must belong to its children. Signatures and fields can be restricted by simple *multiplicity* constraints. Lastly, both may be *static* (by default) or declared as *variable*, in which case their state may change over time.

The ELS environment specification declares signals, the values that can be assigned to such signals, and how these assignments are represented in time. Signals form a static hierarchy starting from an (abstract) Signal signature. Although the ELS signals are "flat", related signals are often better handled together. Thus, for instance, all light signals are aggregated in an abstract signature Light, and left and right low beam signals in the abstract signature LowBeam. At the bottom of the hierarchy are the concrete signals themselves as singleton signatures (multiplicity **one**), such as LowBeamRight and LowBeamLeft, whose names match those specified in the reference documents. The hierarchy relevant for the low beams function is encoded in Electrum as

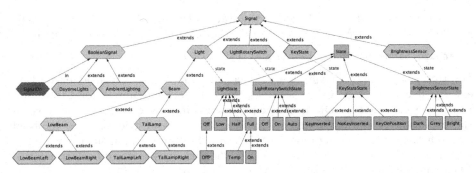

Fig. 1. Meta-model of the system environment model for low beam headlights.

```
abstract sig Light extends Signal { var state : one LightState }
abstract sig Beam, ... extends Light {}
abstract sig LowBeam, TailLamp extends Beam {}
one sig LowBeamLeft, LowBeamRight extends LowBeam {}
one sig TailLampLeft, TailLampRight extends TailLamp {}
```

where field `state` will be explained shortly. To simplify the modelling process, we distinguish Boolean signals (`BooleanSignal`, a sub-signature of `Signal`) from the others. Those relevant for the low beams function are declared as:

```
abstract sig BooleanSignal extends Signal {}
one sig AmbientLighting, DaytimeLights extends BooleanSignal {}
```

Although signals are integer, most requirements simply test whether they are within certain ranges. Thus, to keep the model manageable and avoid state explosion, we discretize the values of each signal into those ranges relevant for the requirements. For instance, it is only relevant to detect whether the ambient brightness levels are below 200, over 300 or between the two, while low beam headlights are only set to 20%, 50% or 100% intensity [6, 4.4]. Thus only these distinct classes of values are encoded in our model. Values form a hierarchy matching that of the signals, topped by `State`, whose direct children group the states of related signals, such as `LightState` for `Light` signals. The next layer provides the discretized values, such as `Off`, `Low`, `Half` or `Full` for beam intensity. Lastly, since our model abstracts real-time aspects, occasionally we require additional temporal context regarding the state of the signals. For instance, when low beams are activated due to ambient darkness, they must remain active for 3 s even if ambient brightness is detected (ELS-18); thus, within `Full` beam intensity we distinguish between this temporary state (`Temp`) and permanent activation (`On`). Part of this hierarchy relevant for the low beam function is encoded as:

```
abstract sig LightState extends State {}
abstract sig Full, Off extends LightState {}
one sig Half, Low extends LightState {}
one sig On, Temp, ... extends Full {}
one sig OffP, ... extends Off {}
```

Lastly, we model the evolution of the state of the signals. For Boolean signals a variable sub-signature `SignalOn` will contain at each state all active signals:

```
var sig SignalOn in BooleanSignal {}
```

For the other signals, a variable field called `state` will contain at each state exactly **one** respective value, such as the one declared above for `Light` and respective `LightState`. Often the requirements impose certain restrictions on the evolution of the environment. In `Electrum` these restrictions are imposed through *facts*, representing model axioms, which can contain arbitrary temporal constraints. In the ELS, e.g., a fact forces the pitman arm to go back to neutral when the steering wheel returns to the vertical position [6, §4.1].

We have encoded the over 30 signals of the ELS in this manner, including those of the user interface [6, §4.1], the sensors [6, §4.2] and the actuators [6, §4.3]. An excerpt of the resulting environment signature hierarchy for the low beam headlights function is depicted in Fig. 1 as generated by the `Analyzer`. Dashed elements are variable and singleton signatures are in thicker lines (some state names are abbreviated). Throughout the rest of the paper we will rely on this function to demonstrate the features of the developed model.

State Machine. Next we derived a state machine from the ELS requirements. `Electrum` formulas are written in relational linear temporal logic with transitive closure. Relational expressions combine signatures and fields (and constants, namely the universe of atoms **univ**, the unary empty relation **none** and the identity relation **iden**) with typical set theory operators such as union (+), intersection (&), difference (--), Cartesian product (→), binary relation overriding (++), and relational join (.). In `Electrum` everything is seen as a relational expression, so `s.state` can be used to retrieve the current state of a concrete signal `s` or all the states of a set of signals `s`. Primed expressions can be used to refer to their value in the succeeding state, e.g., `s.state'` for the next value of `s.state`. Atomic formulas either test relational expressions for inclusion **in** or equality = or are simple multiplicity tests. So, `s` **in** `SignalOn` tests whether a Boolean signal `s` is currently active, and `s.state` **in** `v` whether signal `s` currently has value `v` (if `s` singleton). Complex formulas are composed by Boolean operators (e.g. **not**, **and**, **or**, **iff**, **implies** or **implies**-then- **else**), first-order operators (e.g. **all** or **some**), and future (unary **after**, **always** or **eventually**, or binary **until** or **releases**) and past (unary **before**, **historically** or **once**, or binary **since** or **triggered**) linear temporal logic operators. Predicates and functions can be defined for auxiliary formulas and expressions and let-expressions for local definitions.

A predicate is defined for each function encoding the expected behaviour, which are subsequently called in a fact that enforces the full state machine. For the low beam headlights function, this predicate mostly restricts the succeeding state of the low beam headlights given the current state of the other signals. For instance, if the light rotary switch (`LightRotarySwitch.state`) is set to `LSOn` while the key (`KeyState.state`) is in the ignition on position, the succeeding state of the low beams is set to `On` (ELS-14):

```
KeyState.state in KeyInIgnitionOnPosition and
  LightRotarySwitch.state in LSOn implies LowBeam.state' in On
```

Expression `LowBeam.state` aggregates the state of both the left and right low beams; since every light must have a `state` assigned, `LowBeam.state in On` sets both to full intensity. As a more complex example, consider ELS-17 that specifies daytime running lights, which activate the low beams when the engine is started until the key is removed from ignition, unless ambient light control is also active:

```
DaytimeLights in SignalOn and
  KeyState.state not in KeyInIgnitionOnPosition or
  (LowBeam.state in On and KeyState.state in KeyInserted and
    AmbientLighting not in SignalOn) implies LowBeam.state' in On
```

In our model, real-time is abstracted away and no particular duration is imposed to states, meaning that within a certain interval of time an arbitrary number of events may occur. This affects the modelling of events with a bounded duration, since we must identify when the trace is within that bound and allow multiple steps within. For this purpose, such events are explicitly identified in our state but not forced to last any particular number of states. For instance, the mandatory 3 s for automatic low beams (ELS-18) is identified by the state `Temp`; when brightness is detected, the low beams may be turned `Off` or the `Temp` state propagated. This could be encoded in the following relational formula:

```
let low = LowBeam.state |
  LightRotarySwitch.state in LRSAuto and
    KeyState.state in KeyInIgnitionOnPosition implies
      one low' and
      BrightnessSensor.state in Dark implies
        low' in low.(univ→Temp+Temp→On++On→On) else
      BrightnessSensor.state in Bright implies
        low' in low.(univ→Off+Temp→Temp) else
      BrightnessSensor.state in Grey and low not in Temp implies
        low' in low.(iden+Temp→On)
```

Here `low` abbreviates the state of both left and right low beams and we rely on relational operators to specify alternative updates. For instance, expression `univ→Off+Temp→Temp` relates every state with `Off` and additionally `Temp` with itself; thus, `low.(univ→Off+Temp→Temp)` returns `Temp` and `Off` when the current state is `Temp` and solely `Off` otherwise. This allows the exploration of transitions with different durations: either low beams activation remains within the 3 s, or the 3 s are exceeded and they are deactivated. Formula `one low'` guarantees that left and right beams are updated consistently (i.e., with the same value). Liveness properties then guarantee that the system eventually evolves. In Electrum arbitrary temporal constraints can be imposed, this one taking the shape:

```
low in Temp implies eventually low not in Temp
```

This strategy was employed to model all the ELS main functions – direction blinking, hazard warning light, low beams, cornering lights, manual and adaptive high beams, emergency brake and reverse lights, and fault handling.

4 Handling Variability

The ELS assumes the existence of variability points, namely the market region, whether it is an armoured vehicle and the driver position (although this last does not affect the behaviour of the ELS) [6, §3]. The model described in the previous section represented a single ELS variant, and multiple independent models could be developed in such a way for each of the valid variants. However, such a strategy has poor maintainability and will not scale as the number of features increase. Electrum is sufficiently flexible to support systems with structural and behavioural variability points and effectively model families of software products. However, such idioms may be cumbersome, error-prone, and reduce comprehension, so to explore alternative approaches we implemented in Electrum an annotative language extension to natively support feature-oriented design. This extension was previously developed for Alloy but its adaptation to Electrum was straightforward. This section describes the design of the ELS family of products in both approaches, which allow simultaneously specifying and analysing all the 12 ELS variants. For both approaches, we assume the variant presented in the previous section to be the base variant, which is extended into a multi-variant model.

A Pure Electrum Idiom. The first step in both approaches is to encode the feature model – the possible features and the constraints over them denoting the valid variants. When relying on a variability idiom, this is done by making features first-class elements of the model. A possibility is to create a signature (here, Feature) with an atom for each available feature (through singleton sub-signatures, such as EU or ArmoredVehicle for the ELS). A sub-signature then contains a particular selection of these features, representing the variant under analysis (here, Variant). Lastly, a fact restricts which variants are considered valid, in the case of ELS forcing a single market to be selected through a multiplicity test:

```
fact FeatureModel { one (EU+USA+Canada) & Variant }
```

To model architectural variability, conditional signatures and fields can be assigned a loose multiplicity that is restricted depending on the variant under analysis. In the ELS the darkness mode switch only exists on armoured vehicles, so its multiplicity is set to **lone** (at most one such signal exists), and then a fact forces its existence exactly when the respective feature is selected:

```
fact darknessModeSwitchOn {
some DarknessModeSwitchOn iff ArmoredVehicle in Variant }
```

Behavioural variability can be modelled by testing which features are selected in Variant and adapting the relevant transitions of the state machine predicates. In the case of low beams, for instance, ambient lights should be ignored with active darkness mode in armoured vehicles (ELS-21), so the pre-condition for activating them when the engine is started (ELS-19) is adapted to:

```
not (ArmoredVehicle in Variant and DarknessModeSwitchOn in SignalOn)
and AmbientLighting in SignalOn and BrightnessSensor.state in Dark
and before KeyState.state in KeyInIgnitionOnPosition and
KeyState.state not in KeyInIgnitionOnPosition implies
    LowBeam.state' in Temp
```

Notice that since features are regular signatures, it may become difficult to identify which parts of the predicate are variability points. It may also led to unpredictable issues if the architectural variability is not handled with care: the distracted developer could simply write `DarknessModeSwitchOn in SignalOn` to test whether darkness mode is active without testing the feature presence, which is always true in variants without feature `ArmoredVehicle` since `DarknessModeSwitchOn` is empty, thus permanently disabling ambient lighting.

For an example regarding the USA and Canada market variants, during direction blinking, for instance, the intensity of daytime running lights (ELS-17) must be reduced to half in the respective side (ELS-6), so the transition shown in the previous section would be adapted to:

```
DaytimeLights in SignalOn and ... implies
    LowBeamLeft.state' in
        (some (USA+Canada) & Variant and BlinkLeft.state' not in OffP)
            implies Half else On and
    LowBeamRight.state' in
        (some (USA+Canada) & Variant and BlinkRight.state' not in OffP)
            implies Half else On
```

where the state of the blinking lights `BlinkLeft` and `BlinkRight` is tested in case the USA or Canada markets are selected.

A Colourful Electrum Extension. Approaches to explicitly introduce variability in a system usually fall in two categories: *compositional* approaches where features are implemented as distinct code units which are then composed when creating a variant, and *annotative* approaches where the code is annotated to dictate which fragments will appear in each variant. Both compositional [1] and annotative [9] approaches have been proposed to enable feature-oriented design in Alloy, the latter by us relying on colourful annotations that have been shown to improve understandability [5]. Annotative approaches are better suited for small granularity variability points, which in our experience is often the case in Alloy/Electrum, such as the examples above where one needs to change part of a formula or expression rather than replace the predicate altogether.

In our lightweight annotative approach model elements can be marked with features, identified by a digit, to control their presence/absence without obfuscating the code. Positive and negative annotations are introduced, respectively, by delimiters ⑦ and ❶ for $1 \leq i \leq 9$, and colour highlighted by the Analyzer. These can be nested, representing the conjunction of presence conditions, and be applied to most declarations or branches of certain operators (namely conjunction, disjunction, intersection and union). Semantically, when the presence conditions are not met the element is interpreted as the neutral element of the respective operator. For instance, in ①p① **and** ❷q❷, p is only tested in variants with feature 1, and q in those without feature 2, being replaced by *true* otherwise.

The multi-variant ELS model under this extension uses five feature annotations, one for each variability point. To model the feature model one can rely on annotated facts to forbid certain variants. For the ELS this could be achieved by the following fact, which mimics the colour highlighting of the Analyzer:

```
fact FeatureModel {
  // ① USA, ② Canada, ③ EU, ④ Armored, ⑤ DriverPosition
  ①②some none②① and ② some none ②
  ① some none ③① and ①②③some none③②① }
```

where, for instance, ①②**some none**②① forbids the coexistence of USA and Canada market codes, and ❶❷❸**some none**❸❷❶ forces the selection of at least one market code[2]. At the level of abstraction of Electrum, feature models are usually small and simple to encode with facts like the one above, but we are studying whether dedicated support for encoding feature models is necessary.

Architectural variability is trivially modelled, as one may mark the signature (or field) declaration with the relevant annotations, as in the case of the darkness mode switch signal, that only exists for armoured vehicles:

```
④one sig DarknessModeSwitchOn extends BooleanSignal ④
```

One type rule imposed by colourful Electrum is that element calls must respect the annotations in which they were declared, thus guaranteeing that they are never called in variants where the element is absent. Thus, the interaction between ELS-19 and ELS-21 would now be encoded as:

```
④not DarknessModeSwitchOn in SignalOn④ and
AmbientLighting in SignalOn and BrightnessSensor.state in Dark and
... implies LowBeam.state' in Temp
```

In variants without ④ this test will be disregarded (i.e., interpreted as *true*). The same mechanism can be applied to relational expressions. For instance, the interaction of ELS-17 and ELS-6 for USA and Canada markets is encoded as:

```
DaytimeLights in SignalOn and ... impliesLowBeamLeft.state' in
③On③+❸BlinkLeft.state' not in OffP implies Half else On③ and
LowBeamRight.state' in
③On③+❸BlinkRight.state' not in OffP implies Half else On③
```

where the beams are always set to On in the EU market, but in other markets (through the negative ❸) the state of blinking lights is tested. A union branch is interpreted as the empty relation when the presence conditions do not hold.

---

[2] Electrum, like Alloy, does not natively support Boolean constants, so **some none** is commonly used to denote a trivially unsatisfiable formula.

5   Validation and Verification

The Analyzer is able to execute animation and verification commands. Both instances and counter-examples are graphically depicted in a visualizer that can be customized for improved interpretation. This section describes how these functionalities were used to validate and verify the ELS model.

5.1   Animation and Validation

Validation Scenarios. Animation commands are defined through **run** instructions, which can be provided arbitrary constraints that must hold for the generated instances. This allows the quick definition of scenarios for early validation, which are also useful as regression tests as the model evolves. For the ELS we have defined over 60 such scenarios exercising simple behaviours of the system. We follow an idiom where one predicate defines the evolution of the environment (state of input signals) and another the expected behaviour of the system (state of output signals). For instance, to test basic low beam headlights sub-functions such as having the light rotary switch set to on with key inserted, a predicate is defined to encode the behaviour of the relevant input signals:

```
pred LowBeam2Env {
  always AmbientLighting not in SignalOn
  always KeyState.state in KeyInserted
  let lrs = LightRotarySwitch.state |
  lrs in LSOff;always lrs in LSOn }
```

where **always** *p* forces *p* to hold in all states of the trace and *p*;*q* abbreviates *p* **and after** *q*, an operator introduced precisely to ease scenario specification [4]. A predicate then encodes the expected outcome of the ELS for these inputs:

```
pred LowBeam2Exp {
  LowBeam.state in OffP;always LowBeam.state in Half }
```

This predicate states that the beams should be activated with intensity reduced to half. Lastly, a command to generate this scenario by enforcing the environment and the expected behaviour (in the succeeding state, since output signals are calculated from the previous state) is defined:

```
run LowBeam2 { LowBeam2Env and after LowBeam2Exp } for 5 Time
```

Commands must have scopes assigned to signatures, but in our ELS model all signatures are exactly bound, since all signals and possible states are known a priori. For bounded model checking – more efficient and thus better suited for validation – the maximum number of states that form a trace must also be provided (the scope of **Time**). Since this is a simple scenario that bound is set to 5. Once instances are generated, the user is able to iterate over alternative scenarios for which the constraints hold. Scenario exploration operations (see the toolbar of Fig. 2) include changing the configuration (here, the selected variant), the initial state, or the current transition [3].

In the multi-variant ELS models one is able to restrict which subset of variants should be analysed. As an example, let us consider the animation of the effect of darkness mode when ambient lighting is activated. In the Electrum variability idiom the part of this environment predicate could be specified as:

```
ArmoredVehicle in Variant
let key = KeyState.state |
   key in KeyInIgnitionOnPosition;always key in KeyInserted
always AmbientLighting in SignalOn
always DarknessModeSwitchOn in SignalOn
```

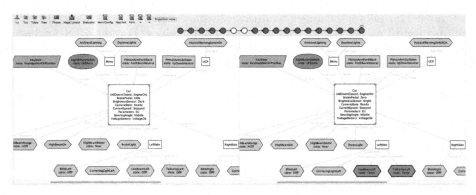

Fig. 2. A step of sequence 1 in the Analyzer under the developed theme. (Color figure online)

which includes the selection of the feature ArmoredVehicle and the behaviour of the DarknessModeSwitchOn. The same scenario in the colourful extension would instead be specified as:

```
let key = KeyState.state |
   key in KeyInIgnitionOnPosition;always key in KeyInserted
always AmbientLighting in SignalOn
④always DarknessModeSwitchOn in SignalOn④
```

where the behaviour of the darkness mode switch is annotated with the corresponding feature. The execution of this scenario must then also be restricted to only variants where feature 4 is selected. In colourful Electrum this is defined through the command scope as:

```
run LowBeam19 {
   LowBeam19Env and after LowBeam19Exp } with ④ for 5 Time
```

Theme Customizations. In our experience, the proper graphical representation of instances is key to promote the interpretation of the model among interested parties. Inheriting from Alloy, the Analyzer depicts instances as graphs, applying a graph representation algorithm and distributing nodes among layers, obliviously of the underlying semantics of the nodes and edges. Themes may be defined to ease interpretation. From our experience the most useful customizations are simply changing the colour, shape or label of elements, hiding elements, showing relations as edges or attributes, and inverting edges (the easiest way to change the shape of the graph). Visualization can also be projected over a signature, focusing the visualization on the elements related to the selected atom. These customizations are hierarchical, meaning that subsets of elements may inherit the parameters of their parents or change them. Although simple, these features can become extremely powerful given another key functionality of the visualizer – after analysis, and during the creation of the graph, auxiliary functions defined in the model are introduced into the instance. These can be of arbitrary arity, and thus can represent subsets of atoms or new relations between them.

In our ELS model we have used such features to produce a visualization such as that of Fig. 2. Since the signals are mostly flat, we introduce elements to somehow layout signals according to their role in the system. Singleton signatures – which do not affect the solving process since they are exactly bound and not referred elsewhere – simulate the vehicle architecture, such as the Car itself or the driver's Menu

```
one sig Car, LeftSide, RightSide, Menu, UCP {}
```

Auxiliary relations (defined as functions with zero arguments) then connect such elements to signals, such as assigning the sensors to the car (which are set to be shown as attributes of Car rather than edges) or the lights to the respective side of the car, and can be defined as follows:

```
fun _lightsensor : Car → BrightnessSensorState {
  Car → BrightnessSensor.state }
fun _actuators : univ → univ {
  LeftSide → (BlinkLeft+LowBeamLeft+TailLampLeft+...) +
  RightSide → (BlinkRight+LowBeamRight+TailLampRight+...) }
```

Auxiliary sets grouping together signals under certain states were also defined to ease the theme customization. For instance, all active signals are grouped so that they can easily be painted with a distinguishing colour (yellow in Fig. 2):

```
fun _on : set univ { state.Full+state.(LSOn+LSAuto)+SignalOn+... }
```

The theme file is available alongside the model specification.

5.2   Reference Validation Sequences

To effectively validate the developed model we checked its behaviour against that of the reference validation sequences [7]. These are complex – each step specifying the value of all the over 30 input and output signals, with some containing over 20 steps – rendering their manual codification infeasible. Thus, we implemented a prototype to automatically translate tabular data that represents signal values over time into Electrum and back. This validator is able to $i)$ given a sequence of input and output signals, report whether it is a valid execution in our model; and $ii)$ given a sequence of only input signals, generate possible executions of the output signals to be subsequently validated by domain experts.

We implemented the prototype so that the process could be reproducible for other signal-based systems. Thus, besides the sensor data, two additional pieces of information must be provided to the validator for each specific application: i) how the signal values should be discretized; and ii) the presence conditions for signals. For our prototype, this information is passed in the header of the tabular data, as depicted in Table 1 for validation sequence 1 of the ELS (note that this is only an excerpt of the codification of the more than 30 signals over 17 steps). Single-value ranges are assumed to have the same lower- and upper-bound. It also assumes, as described in Sect. 3, that all signals are leaves of the hierarchy on Signal with the exact same name as that of the sequence header, and that elements representing the discretized values are at the second layer of the State hierarchy, again with the same name as the discretization in the header.

Table 1. Snippet of tabular data provided to our validator for sequence 1.

...	Time	ambient Lighting	darknessMode SwitchOn	lightRotary Switch	brightnessSensor	marketCode	armored Vehicle	...	lowBeam Left	...
...		0=False; 1=True	0=False; 1=True	0=Off; 1=Auto; 2=On	0-199=Dark; 200-250=Grey; 251-100000=Bright	1=USA; 2=Canada; 3=EU	1=True; 0=False	...	0=Off; 10=Low; 50=Half; 100=Full	...
...			armored Vehicle=True					...		...
...	...	...	...	...	...	...	...	...	...	...
...	0:03	0	0	1	500	3	0	...	0	...
...	0:04	0	0	1	200	3	0	...	0	...
...	0:05	0	0	1	199	3	0	...	100	...
...	...	...	...	...	...	...	...	...	...	...

```
1  let s1 = not AmbientLighting in SignalOn |
2    always s1
3  let s1 = not DarknessModeSwitchOn in SignalOn |
4    always s1
5  let s1 = LightRotarySwitch.state in LSAuto, s0 = LightRotarySwitch.state in LSOn,
6    s2 = LightRotarySwitch.state in LSOff |
7    s2;s2;s2;s1;s1;s1;s1;s1;s1;s1;s0;s0;s1;s0;s0;s0;always s1
8  let s0 = BrightnessSensor.state in Dark, s1 = BrightnessSensor.state in Grey,
9    s2 = BrightnessSensor.state in Bright |
10   s2;s2;s2;s2;s1;s0;s2;s0;always s2
11 EU in Variant
12 ArmoredVehicle not in Variant
13 ...
14 after {
15   let s2 = LowBeamLeft.state in LightLow, s3 = LowBeamLeft.state in LightOff,
16     s1 = LowBeamLeft.state in LightHalf, s0 = LowBeamLeft.state in LightFull |
17   s3;s3;s3;s3;s3;s0;s0;s0;s3;s3;s0;s3;s3;s1;s3;s2;always s3
18   ... }
```

Fig. 3. Electrum encoding of the sequence from Table 1.

The translation can then be streamlined as follows. The presence/absence of a Boolean signal s can simply be stated as s in SignalOn and s not in SignalOn, respectively, while the state of the others is encoded as s.state in v for a discretized value v. Sequences of signal states are encoded using the operator ;, and let-expressions are used to simplify this codification. The particular variant of the sequence must also be encoded. The validator currently implements only the pure variability idiom, forcing the exact value of signature Variant.

The resulting predicates resemble the one in Fig. 3 for the sequence from Table 1 (including steps that have been omitted for simplicity). The expected variant (ll.11–12) and both the sequence of input (ll. 1–10) and output (ll. 15–17) signals are encoded, relying on let-expressions for improved readability (recall that unlike the validation sequences, our output signals are only updated in the succeeding state, hence the **after**). At the last state an **always** operator is applied, since outputs are expected to stabilize when inputs do. Although the reference sequences provide timestamps for the events (the first column), these are ignored since real-time is abstracted in our model.

Figure 2 depicts the outcome of running this predicate (with **Time** scope determined from the length of the sequence), particularly the transition where the brightness is below the threshold and the low beam headlights are activated. We were able to model all 9 validation sequences of version 1.8 and show that they hold for our ELS model, except for concrete values for the high beam illumination distance and strength in sequence 9 (ELS-33) due to arithmetic operations.

5.3   Requirement Verification

The last step of the process was to effectively verify whether the requirements hold for the modelled ELS. In Electrum assertions (**assert**) can be specified in full relational temporal logic, which the Analyzer is instructed to verify (within given scopes) with **check** commands.

As an example, consider requirement ELS-14, stating that whenever the engine is on and the light switch set to on, low beams will be active. This can be specified in the following temporal assertion:

```
assert ELS14 {
  always (KeyState.state in KeyInIgnitionOnPosition and
    LightRotarySwitch.state in LSOn implies LowBeam.state' in Full) }
```

For a more complex example, consider ELS-17, stating that with daytime running light but without ambient light, the low beams are activated until the engine is turned off. This can be encoded as:

```
assert ELS17 {
  let keyPos = KeyState.state in KeyInIgnitionOnPosition,
      amb = AmbientLighting in SignalOn,
      day = DaytimeLights in SignalOn |
  always (day and not amb) implies always (
    (LowBeam.state' in Full+Half until not keyPos) or always keyPos) }
```

stating that in traces where daytime running light is active but not ambient lighting, the engine is turned off and the low beams are deactivated (temporal operator **until**) or the engine remains on forever.

We were able to check most ELS requirements except for the limitations discussed in the following section. The described checks (that verify the property for all variants at once) take around 6 s and 10 s, respectively, using the bounded engine of Electrum under the Glucose SAT solver and for 15 Time in a commodity 2,3 GHz Intel Core i5 with 16 GB RAM. More complex requirements – like those including periodic events such as ELS-2 and ELS-4 – take around 1 min.

6 Results Discussion

The Reference Document. Throughout the development of the ELS model we encountered 14 issues with the reference documents, mostly during modelling and preliminary validation, and when running the reference sequences. We reported them to the case study chair who promptly replied. Of the first 4 reported issues, 3 resulted in fixes to the reference document (version 1.11); unfortunately, at the time of submission no new version has been released after the other interactions (unofficially, at least 3 resulted in validation sequence fixes). Roughly, the issues encountered were either with the

Environment model inconsistencies or missing features related to the signals detected in the early modelling process (e.g., the lack of a signal for the middle brake light, making it impossible to flash (ELS-40); or inconsistent representations of the pitman arm signals when it was split into two distinct signals for vertical and horizontal movement);

Behavioural model ambiguities detected in the requirements while modelling and animating the state machine (e.g., conflicting requirements where the precedence is not explicitly stated, such as whether ELS-18 or ELS-19 has priority on low beam behaviour; ambiguous nomenclature, such as what activating high beams means for the 3 relevant signals; or under-specified behaviour, such as the beam intensity of tail lamps);

Validation sequences inadmissible sequences, meaning that the expected output signals could not be achieved from the input signals in our model (e.g., tail lamps not being activated or not blinking in sequence 7).

It must also be noted that, since the modelling and validation process was iterative, some requirement ambiguities were clarified by observing the reference sequences. For instance, it is not clear from ELS-22 that when tail lamps are activated, they are so with the same intensity as that of the low beams, but the sequences showed that to be the case (e.g., in ELS-15).

In our experience, there were two main sources of confusion in the requirements. One has to do with the blinking lights and the nature of the dark cycles: it was not clear under which situations, if any, such cycles should be interrupted, and under which situations do they impact the tail lamps. The second has to do with high beam headlights, which are controlled by 3 distinct signals: it is often not clear what it means to activate the high beams and how the 3 signals should be updated and again how they relate to the intensity of the tail lamps.

The Followed Approach. As already stated, we only failed to address requirements requiring arithmetic operations (ELS-33 for calculating the illumination distance and luminous strength of high beams, and ELS-47 for calculating the maximum light intensity under over-voltage) since concrete integer values are not represented. The abstracted time also renders reasoning about real-time requirements infeasible, such as ELS-10 enforcing the duration of blinking cycles to 1 s, or the part of ELS-18 forcing the activation of the automatic low beams for 3 s. Some features were simplified to avoid additional internal states, namely the gentle fade-out of cornering lights (ELS-24) or the flashing of emergency brake lights (ELS-40). ELS-37, dealing with the interaction with the SCS, has been disregarded. Requirements related to periodic events – such as the bright and dark cycles of blinking lights – proved to be the most cumbersome to specify.

The multiple variants of the ELS requirements motivated the implementation of the feature annotations for Electrum and its Analyzer. Since the ELS is not particularly rich in variability, we did not find multi-variant modelling in a pure Electrum idiom to be unmanageable, but it did affect the comprehension of the model. In general, the colourful Electrum model is easier to understand. The exception is the axiomatization of the feature model, and we are already studying sensible ways to improve it, that we also expect to be useful in more advanced feature-oriented analysis procedures. The complexity of the case study also helped us identify additional operators whose annotation would be useful in colourful Electrum – namely, if-then-else expressions common in the definition of state machines, when certain branches are only relevant in certain variants.

Acknowledgements. The authors would like to thank Frank Houdek for helping clarifying the requirements. This work is financed by National Funds through the Portuguese funding agency, FCT – Fundação para a Ciência e a Tecnologia, within project UIDB/50014/2020. The third author was financed by the ERDF – European Regional Development Fund through the Operational Programme for Competitiveness and Internationalisation – COMPETE 2020 Programme and by National Funds through the FCT, within project POCI-01-0145-FEDER-016826.

References

1. Apel, S., Scholz, W., Lengauer, C., Kästner, C.: Detecting dependences and interactions in feature-oriented design. In: ISSRE, pp. 161–170. IEEE (2010)
2. Brunel, J., Chemouil, D., Cunha, A., Macedo, N.: The electrum analyzer: model checking relational first-order temporal specifications. In: ASE, pp. 884–887. ACM (2018)
3. Brunel, J., Chemouil, D., Cunha, A., Macedo, N.: Simulation under arbitrary temporal logic constraints. In: F-IDE@FM. EPTCS, vol. 310, pp. 63–69 (2019)
4. Cunha, A., Macedo, N.: Validating the hybrid ERTMS/ETCS level 3 concept with Electrum. Int. J. Softw. Tools Technol. Transfer **22**, 281–296 (2020). https://doi.org/10.1007/s10009-019-00540-4
5. Feigenspan, J., et al.: Do background colors improve program comprehension in the #ifdef hell? Empir. Softw. Eng. **18**(4), 699–745 (2013). https://doi.org/10.1007/s10664-012-9208-x

6. Houdek, F., Raschke, A.: Adaptive exterior light and speed control system, v1.17 (2019)
7. Houdek, F., Raschke, A.: Validation sequences for ABZ case study "adaptive exterior light and speed control system", v1.8 (2019)
8. Jackson, D.: Software Abstractions: Logic, Language, and Analysis, revised edn. MIT Press, Cambridge (2012)
9. Liu, C., Macedo, N., Cunha, A.: Simplifying the analysis of software design variants with a colorful alloy. In: Guan, N., Katoen, J.-P., Sun, J. (eds.) SETTA 2019. LNCS, vol. 11951, pp. 38–55. Springer, Cham (2019). https://doi.org/10.1007/978-3-030-35540-1_3
10. Macedo, N., Brunel, J., Chemouil, D., Cunha, A., Kuperberg, D.: Lightweight specification and analysis of dynamic systems with rich configurations. In: SIGSOFT FSE, pp. 373–383. ACM (2016)

Modelling and Validating an Automotive System in Classical B and Event-B

Michael Leuschel$^{(\boxtimes)}$, Mareike Mutz, and Michelle Werth

Institut für Informatik, Universität Düsseldorf, Universitätsstr. 1,
40225 Düsseldorf, Germany
michael.leuschel@hhu.de

Abstract. We have modelled parts of the ABZ automotive case study
using the B-method. For the early phases of modelling we have used the
classical B for software, while for proof we have used Event-B and Rodin.
It is maybe surprising that classical B's machine inclusion mechanism
along with operation calls can be used for modular system modelling.
Moreover, for one particular style of modelling, the result can then be
translated to superposition refinement with event extension in Event-B.
Before conducting the proof, we have validated our models using model
checking and animation with visualizations. The graphical visualizations
were constructed using a new plugin (VisB) which helped uncover errors
and transforms our model into an executable, interactive reference specification which can be examined by users without formal background.

1 Introduction and Background

In this work we have used both classical B and Event-B as modelling languages
with PROB and RODIN for tool support. The classical B-method [1] is a formal
method rooted in predicate logic, set theory, arithmetic. B arose out of Z, with
a focus on tool support and successive refinement to derive provably correct
software out of high-level specifications. This initial version of B, supported by
the tool ATELIER-B, is now called classical B or also "B for software".

Event-B was developed to enable systems modelling. It also tried to correct a
few issues in classical B, simplifying the language (e.g., making it easier to parse
and removing the complex inclusion mechanism) and trying to make refinement
proofs easier and more scalable. The main addition though is a more flexible
refinement concept targeted at systems modelling rather than software development. The foundations of Event-B are laid out in the book [2]. Event-B is
supported by the RODIN platform [3], which we have used for proving. In our
development of the case study we made heavy use of the animator and model
checker PROB [10] which supports both classical B and Event-B. PROB is also
available as a plugin for RODIN.

The goals we set ourselves were as follows:

- Validate the usability of the new visualization plugin VISB for PROB, in
  particular whether images from the case study description [7] can be reused
  with little effort.

© Springer Nature Switzerland AG 2020
A. Raschke et al. (Eds.): ABZ 2020, LNCS 12071, pp. 335–350, 2020.
https://doi.org/10.1007/978-3-030-48077-6_27

– Examine whether systems modelling can be conducted in classical B.
– Examine whether a componend-based model in classical B can be converted to Event-B for use by the RODIN for proof.

Indeed, in earlier work we have often realised that classical B can also be used for systems modelling and may have some advantages. However, for proving systems, Event-B and RODIN provide a nice platform and simpler[1] proof obligations than classical B. We have modelled a subset of the lighting system, as well as a subset of the adaptive cruise control system. For the latter some insights found during modelling have led to improvements in the case study specification (see Sect. 6). In this article we only show details of the lighting system models, which we have developed using a development methodology described in Sect. 2, which consisted of using classical B during the exploration phase and Event-B for the proving phase.

The contributions of this article thus consist in showing the usefulness of visualisation using a new lightweight SVG-based tool, comparing classical B and Event-B for systems modelling, and illustrating a development methodology. Together with the visualisation, our model provides an executable reference specification for a subset of the specification, which can be used by domain experts without formal methods training. We plan on extending this subset in future.

2   Modelling Methodology and Strategy

The methodology we used was inspired by earlier experience on modeling and validating railway systems (e.g., [4,5]). We divided the modelling into three phases (see Fig. 1):

1. an **exploratory** modelling phase where one tries to understand the specification and domain and identifies important subcomponents.
2. a **synthesis** phase, where the subcomponents are integrated to form ever larger subsystems and the integration is at least partially verified. We could have named this also the *combination* or *structuring* phase.
3. an exhaustive **verification** phase, where the completed system is now formally proven to be safe and functionally correct.

The exploratory phase 1 is dominated by editing and animation. Here it is important that one can quickly change the model, and one wants to use a maximally powerful language. Hence, in our case, we have used classical B with its rich substitution language for this phase. Also, the fact that we have a textual representation (which was versioned in Git) allows for easy editing and collaboration. Animation and graphical visualization are vital to check the presence of desired functionality, but also to check for unexpected behaviour and to gain

---

[1] E.g., due to the absence of an if-then-else substitution and the use of witnesses the refinement proof obligations are much simpler.

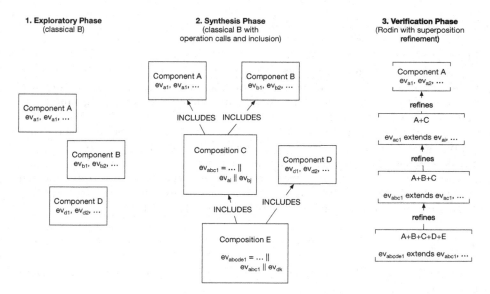

Fig. 1. Machines and events in the three modelling phases

a better understanding of the problem domain. To cite Bryan Cantrill:[2] *"The visual cortex is unparalleled at detecting patterns."* and *"The value of visualization is not merely providing answers but especially provoking new questions."* The outcome of phase 1 is a first decomposition of the system into functional subcomponent models which can be animated.

In phase 2 the model starts to stabilise, but we still experimented with various ways to integrate the components. In a refinement-based approach (e.g., Event-B) one has to decide upon a particular refinement order in which components are integrated, which can be sometimes difficult to find and tedious to change. Hence we again stayed in the classical B paradigm where the powerful machine inclusion features allowed us to experiment with various ways of assembling the entire system (see also Fig. 1). To ensure the proper functioning we start to write safety invariants in this phase. The verification in this phase was dominated by (safety) model checking and trace replay in the animator. The outcome of phase 2 is a functional composition into an overall system which can be animated and where various safety invariants can be checked using model checking.

The last phase is dominated by proof. Here, the RODIN platform's proof system is a key technology and the simplicity of Event-B's proof obligations can pay off. In this phase the decomposition of the system has stabilised and it was hence possible to transform the classical B model into a linear refinement hierarchy. This requires translating machine inclusion to machine refinement and operation calls to event extension (see Fig. 1 and Sect. 4). For functional aspects of the system we complemented proof with LTL model checking.

---

[2] https://www.slideshare.net/bcantrill/visualizing-systems-with-statemaps.

3   Classical B Modelling Details

Initially we had modelled a subset of the cruise control system, which led to a few interesting insights and issues (see also Sect. 6). However, to illustrate our modelling strategy (Sect. 2) and new visualization tool (Sect. 2) we then modelled a subset of the light system. Our models are available at: https://github.com/hhu-stups/abz2020-models.

Due to time restrictions we were unable to model the complete case study. We concentrated on the management of the blinkers, i.e., the blinkLeft and blinkRight actuators as influenced by the vertical position of the Pitman arm (PitmanArmUpDown sensor), the hazard warning switch (hazardWarningSwitchOn sensor), as well as obviously by timing constraints. The engineOn and keyState sensor values were also used.

3.1   Components

At the end of the exploratory phase (see Sect. 2) we had separated out the system into three components (see Fig. 2):

1. A small component Sensors to manage the state of all sensors of the system and their updates.
2. A component BlinkLamps for managing the blinking cycle and the actuators blinkLeft and blinkRight, based on a logical state variable stipulating which blinkers are active and how often blinking should be repeated.
3. The PitmanController composed component which reacts to changes in the sensors and sets the logical state variable of the BlinkLamps components.
4. A GenericTimer component which manages current time evolution and manages hard deadlines. There are actually two versions of this component, one for animation or simulation and one for finite state model checking.
5. The main PitmanController_TIME system, which adds treatment of real time issues to the PitmanController model.

3.2   Blink Lamps Controller

The purpose of the BlinkLamps component is to separate the management of the blink lamps from a logical state of the system. This logical state is modelled by a variable active_blinkers which is a subset of {left_blink,right_blink}, i.e., it has four possible logical states (from all blinkers off to all blinkers on). This model does not worry about real time and deadlines yet; it manages various lamp states using an internal state variable onCycle but, e.g., does not stipulate how long lamps should remain on or off and does not have a variable representing time at all. The model also manages a remaining_blinks variable, which stipulates how often the blinker still has to run independently (used later to implement tip blinking). A special value denotes unlimited continuous blinking. A small part of the model is shown in Listing 1.1. Safety invariants ensure the consistency

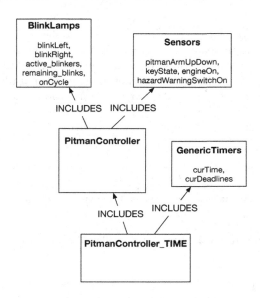

Fig. 2. Hierarchy of final classical B model with variables

between the logical variables `active_blinkers` and `remaining_blinks` with the blink lamp actuators. Early versions of our model had several errors in that respect, e.g., when events such that turning an engine on or off changed the system's state without consistently updating the actuators.

```
1   MACHINE BlinkLamps ...
2   DEFINITIONS
3     blinkersOff == (blinkLeft=lamp_off & blinkRight=lamp_off)
4   INVARIANT
5     active_blinkers <: BLINK_DIRECTION &
6     remaining_blinks : BLINK_CYCLE_COUNTER &
7     blinkLeft: LAMP_STATUS & blinkRight : LAMP_STATUS & onCycle: BOOL &
8     /*@label "SAF1" */
9     ((remaining_blinks=0 & blinkersOff) <=> active_blinkers={}) & ...
10    /*@label "SAF6" */
11    (onCycle=TRUE & active_blinkers/={} => not(blinkersOff))
12  OPERATIONS
13    SET_BlinkersOn(direction,rem) = PRE direction: BLINK_DIRECTION & rem:
          BLINK_CYCLE_COUNTER & rem /= 0
14    THEN
15      active_blinkers := {direction} || remaining_blinks := rem ||
16      IF direction=right_blink THEN
17          blinkLeft := lamp_off ||
18          blinkRight := cycleMaxLampStatus(onCycle)
19      ELSE
20          blinkLeft := cycleMaxLampStatus(onCycle) ||
21          blinkRight := lamp_off
22      END
23    END; ...
```

Listing 1.1. Blink Management Machine

3.3 Integrating with Pitman Controller

The `PitmanController` reacts to changes in the sensors and sets the logical blinking state using the operations provided by the `BlinkLamps` machine. For example, Listing 1.2 shows how this is achieved using two parallel operation calls, one to to `SET_EngineOn` from the `Sensors` machine and one to `SET_BlinkersOn` from `BlinkLamps` shown above in Listing 1.1. It is also interesting to note that the second operation call is wrapped into a conditional.

```
1   ENV_Turn_EngineOn =
2   BEGIN
3     SET_EngineOn ||
4     IF pitmanArmUpDown :PITMAN_DIRECTION_BLINKING &
5        hazardWarningSwitchOn = switch_off THEN
6           SET_BlinkersOn(pitman_direction(pitmanArmUpDown),continuousBlink)
7     END
8   END;
```

Listing 1.2. Pitman Controller Machine Event

3.4 Modelling Time

We have modelled time as a discrete integer variable representing elapsed time in milliseconds. The `GenericTimers` machine partially shown in Listing 1.3 manages a set of deadlines. The parameter `TIMERS` specifies a set of timers for which one can associate individual deadlines using the `AddDeadline` operation. Time can be advanced using the `IncreaseTime` operation, but it cannot proceed beyond a deadline, thus forcing events associated with the deadlines to be executed first. This is a typical scheme to model time in B, but variations and different approaches (e.g., using a controller event executed at fixed intervals).

For our modelling we needed two timers: one for the blinking phases and one for the management of "tip blinking". Future extensions of the model will probably require additional deadlines which can be easily added.

```
1    MACHINE GenericTimers(TIMERS)
2    ...
3    INVARIANT
4      curTime : NATURAL &
5      curDeadlines : TIMERS +-> NATURAL
6    OPERATIONS
7      AddDeadline(timer,deadline) = PRE timer:TIMERS & deadline:NATURAL THEN
8         curDeadlines(timer) := curTime+deadline
9      END;
10     IncreaseTime(delta) = SELECT delta:NATURAL &
11        (curDeadlines/={} => curTime+delta <= min(ran(curDeadlines))) THEN
12        curTime := curTime + delta
13     END; ...
```

Listing 1.3. Generic Timer Machine for Simulation

The complete system model `PitmanController_TIME` then combines the untimed pitman controller with the `GenericTimers`. The Listing 1.4 shows how a deadline is set for the tip blinking event. When the deadline expires without the `pitmanArmUpDown` sensor changing the tip blinking is converted to a regular direction blinking.

```
1  MACHINE PitmanController_TIME
2  SETS      PTIMERS = {blink_deadline , tip_deadline}
3  INCLUDES Blinking ,
4           GenericTimers (PTIMERS)
5  ...
6  OPERATIONS
7    ENV_Pitman_Tip_blinking_start (newPos) =
8    SELECT newPos : PITMAN_TIP_BLINKING &
9           newPos /= pitmanArmUpDown THEN
10      ENV_Pitman_Tip_blinking_short (newPos ,pitman_direction (newPos)) ||
11      AddDeadline (tip_deadline ,500)
12    END ;
13   ...
```

Listing 1.4. Pitman controller with time

The above version of `GenericTimers` in Listing 1.3 is good for simulation or replaying traces with explicit timing information such as the ones accompanying [7] (see Appendix A). As `curTime` is unbounded, however, it leads to an infinite state space for model checking. Hence we also developed a second "drop-in-replacement" of this machine, which has no `curTime` variable: the deadlines are always rescaled as if the current time was 0 (see Listing 1.5). In our case this was sufficient to produce a finite state model. Model checking the full system with PROB takes about 2.3 s, generating 2095 states and 16472 transitions. (See [12] or [9] for related ways of model checking timed systems in state-based formal methods.)

```
1  MACHINE GenericTimersMC (TIMERS)
2  ...
3  INVARIANT
4    curDeadlines : TIMERS +-> NATURAL
5  OPERATIONS
6    AddDeadline (timer ,deadline) = PRE timer :TIMERS & deadline :NATURAL THEN
7      curDeadlines (timer) := deadline
8    END ;
9    IncreaseTime (delta) = SELECT delta :NATURAL &
10     (curDeadlines /={} => delta <= min(ran(curDeadlines))) THEN
11     curDeadlines := %x.(x:dom(curDeadlines)|curDeadlines (x)-delta)
12   END ;
13   ...
```

Listing 1.5. Generic timer machine for model checking

4 Systems Modelling with Classical B and Translation to EventB

When translating a classical B composition such as the one in Fig. 2, one has to linearise the inclusion into a refinement chain. As you can see in Fig. 3, we have chosen the `BlinkLamps` as the top-level Event-B machine. In the translation, we had to split certain B operations into multiple events (as RODIN does not provide an IF-THEN-ELSE). E.g., the operation `SET_BlinkersOn` from Listing 1.1 is translated into two events: `SET_LeftBlinkersOn` and `SET_RightBlinkersOn`.

Let us now look at the `PitmanController` classical B machine, including `Sensors` and `BlinkLamps`. To translate this inclusion into refinement, we construct a *superposition* refinement of `BlinkLamps`, i.e., we add new variables and

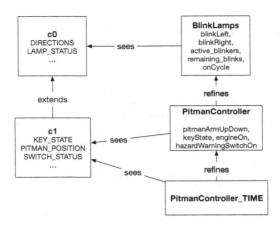

Fig. 3. Hierarchy of final Rodin model, translated from classical B model in Fig. 2

do not remove existing variables.[3] Moreover, as Event-B does not allow operation calls, we need to encode them using event refinement, more precisely refinement using the RODIN **extends** keyword which only adds actions and guards to an existing abstract event.

Let us look again at the operation ENV_Turn_EngineOn in Listing 1.2 from Sect. 3.3 above. It calls SET_EngineOn from Sensors and conditionally calls SET_BlinkersOn from BlinkLamps. This operation is translated into three events *refining* either **skip** or extending SET_LeftBlinkersOn or SET_Right BlinkersOn:

Event $ENV\_Turn\_EngineOn\_Noblink$
    where
      grd1: $engineOn = FALSE \wedge keyState = KeyInsertedOnPosition$
      grd2: $pitmanArmUpDown \notin PITMAN\_DIRECTION\_BLINKING \vee$
          $hazardWarningSwitchOn = switch\_on$
    then
      act1: $engineOn := TRUE$
    end
Event $ENV\_Turn\_EngineOn\_BlinkLeft$ **extends** $SET\_LeftBlinkersOn$
    where
      grd11: $engineOn = FALSE \wedge keyState = KeyInsertedOnPosition$
      grd12: $pitmanArmUpDown \in PITMAN\_DIRECTION\_BLINKING$
      grd13: $pitman\_direction(pitmanArmUpDown) = left\_blink$
      grd14: $hazardWarningSwitchOn = switch\_off$
      grd15: $rem = continuousBlink$
    then
      act11: $engineOn := TRUE$
    end
Event $ENV\_Turn\_EngineOn\_BlinkRight$ **extends** $SET\_RightBlinkersOn$
    where
      grd11: $engineOn = FALSE \wedge keyState = KeyInsertedOnPosition$
      grd12: $pitmanArmUpDown \in PITMAN\_DIRECTION\_BLINKING$
      grd13: $pitman\_direction(pitmanArmUpDown) = right\_blink$
      grd14: $hazardWarningSwitchOn = switch\_off$
      grd15: $rem = continuousBlink$
    then
      act11: $engineOn := TRUE$
    end

One can see that the translation has resulted in code duplication, making it more tedious to adapt the model. A similar issue occurs at the next refinement

---

[3] Machine inclusion in classical B can only add variables, not remove them.

level, where we introduce timing. As again we cannot refine multiple components, the code for the timer logic gets interspersed and duplicated in multiple events. Here we show just two instances in the **PitmanController_TIME_MC** model, which show how the guards and actions are replicated in multiple events.

Event *TIME_BlinkerOn* **extends** *TIME_BlinkerOn*
  any *delta*
  where
      grd21: $delta \in \mathbb{N}$
      grd22: $blink\_deadline \in dom(curDeadlines)$
      grd23: $delta = curDeadlines(blink\_deadline)$
      grd24: $delta = min(ran(curDeadlines))$
  then
      actTm: $curDeadlines := (\lambda x \cdot x \in dom(curDeadlines) \setminus \{blink\_deadline\} \mid$
                $curDeadlines(x) - delta) \cup \{blink\_deadline \mapsto 500\}$
  end
Event *TIME_BlinkerOff* **extends** *TIME_BlinkerOff*
  any *delta*
  where
      grd21: $delta \in \mathbb{N}$
      grd22: $blink\_deadline \in dom(curDeadlines)$
      grd23: $delta = curDeadlines(blink\_deadline)$
      grd24: $delta = min(ran(curDeadlines))$
  then
      actTm: $curDeadlines := (\lambda x \cdot x \in dom(curDeadlines) \setminus \{blink\_deadline\} \mid$
                $curDeadlines(x) - delta) \cup \{blink\_deadline \mapsto 500\}$
  end

If we want to modify the time management we need to edit multiple events in a consistent fashion, which is tedious and error-prone. The classical B model called the same operation in both cases; changing details about how time is handled just means changing or substituting the **GenericTimers** machine. Hence the conversion to Event-B in our methodology (Sect. 2) is delayed until the model has sufficiently stabilised.

Proof Statistics and Insights. We checked the correctness of our translation by checking the state spaces generated by PROB is identical for those three models that can be compared. The results are summarised in Table 1.

Table 1. Size of classical B and Event-B model state spaces and model checking times

Model	States	Transitions	Model checking time
BlinkLamps (B)	31	417	0.10 s
BlinkLamps (Event-B)	31	417	0.07 s
PitmanController (B)	74	514	0.13 s
PitmanController (Event-B)	74	514	0.10 s
PitmanController_TIME_MC (B)	2096	16742	2.39 s
PitmanController_TIME_MC (Event-B)	2096	16742	2.00 s

In summary, the translation resulted in a model with many more events and more duplication, but was worth it in the end. The proving process was relatively painless, the RODIN prover was easy to use. The proof statistics can be found in

Table 2, and the proving process itself helped to uncover a few interesting invariant properties and some issues in our sub-components (which were not reachable by the model checking, but would have appeared if the subcomponents were used differently). Also, the generation and discharging of the well-definedness proof obligations provides another safety guarantee, which is not exhaustively covered by model checking with PROB.

Table 2. Rodin proof statistics

Element Name	Total	Auto	Manual	Rev.	Und.
ABZ2020_v4	**228**	**204**	**24**	**0**	**0**
c0	2	1	1	0	0
c1	1	0	1	0	0
ctimers	0	0	0	0	0
BlinkLamps	72	64	8	0	0
PitmanController	107	100	7	0	0
PitmanController2_TIME	19	17	2	0	0
PitmanController2_TIME_MC	27	22	5	0	0

LTL Model Checking. As the RODIN model contains a series of events for one particular action, like turning the engine on or off, we wanted to check that it is always possible to turn the engine on or off and that there is exactly one event which describes the system's evolution. For this we introduced the relative deadlock and controller state LTL properties in [8], which we used here. One particular property is the following one, which was violated by earlier versions of our translation:

```
1    G not(deadlock(ENV_Turn_EngineOn_BlinkLeft,
2              ENV_Turn_EngineOn_BlinkRight,ENV_Turn_EngineOn_Noblink,
3              ENV_TurnEngineOff_Blink, ENV_TurnEngineOff_Noblink))
```

Listing 1.6. LTL Formula

5   VisB Visualization of the Light System

The core idea of our new PROB plugin called VISB is to have a lightweight visualization engine which can be easily maintained and which make use of graphics generated with off-the-shelf editors. More concretely, VISB makes use of a SVG (scalable vector graphics) file and a JSON markup file. The markup files contains formal model expressions which specify attributes of objects which should be updated according to the current state of a formal model.

The present case study description [7] contains several appealing images, which we kindly obtained from the coordinators of the case study. Luckily, some were already in the SVG format and as such relatively little effort was needed to obtain a first visualisation (after about one hour the first visualisation was

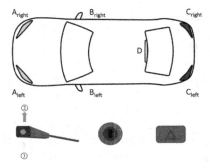

Fig. 4. SVG graphics used for VisB without modifications

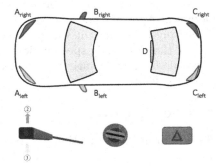

Fig. 5. Erroneous state uncovered via the visualisation

working). The main work was to combine the images into a single one, iron out a few quirks in the SVG files, and change the relevant object identifiers to mnemonic names and then write a JSON glue file referring to these identifiers. The original, unmodified SVG file is shown in Fig. 4 and the start of the JSON glue file in Listing 1.7.

As one can see, it specifies the original SVG file and provides instructions on how to update the attributes of objects (such as **stroke-opacity**) of particular objects (such as **A-right** corresponding to the front left light of the car). The value is given by a B expression which can reference the state of the formal model, in this case the **active_blinkers** variable. The glue file also specifies events which should be executed when certain objects are clicked upon.

```
1  {
2    "svg":"LichtUebersicht_v4.svg",
3    "items":[
4      {
5        "id":"A-right",
6        "attr":"stroke-opacity",
7        "value":"IF right_blink:active_blinkers THEN \"0.5\" ELSE \"1\" END"
8      },
9    ...
```

Listing 1.7. Start of VisB JSON glue file

The effect of applying the VISB glue file to the original SVG can be seen in Fig. 5. This visualization shows the state of the actuators `blinkLeft` and `blinkRight`, as well as the state of the sensors `keyState`, `pitmanArmUpDown`, `hazardWarningSwitchOn` and `engineOn`. It also shows the internal state of the controller, e.g., the `active_blinkers` variable by changing the stroke-opacity and using a light orange fill. This state is an actual error that was spotted thanks to the visualisation: the hazard warning switch is on, but only the right blinkers are on. We can also see that the pitman arm is pushed up, but the controller correctly considers both left and right blinkers to be active, but somehow the blink management component has *not* set `blinkLeft` to the expected value (100).

VISB works for all of PROB's supported state-based formalisms (B, Event-B, Z, TLA+, Alloy), and as such we reused the same visualization for our classical B and Event-B models. A visualisation of the important states of the validation trace 7 from [7] can be found in Appendix A.

In conclusion, with relatively little effort it was possible to generate a visualisation from the graphics of the case study specification. In addition, the visualisation glue file was re-usable for a wide variety of models (several version of the classical B and Event-B models).

6   Description of Issues Uncovered

Some of the issues we found during implementing and especially during model checking the models are:

- Per specification, the system is not commutative. In ELS-13 it is stated that any tip-blinking is ignored or will be stopped by the hazard warning lights when hazard warning is active, respectively activated. So, we do have possible traces where at the same time hazard warning light is switched off and tip-blinking to either side is activated. Depending on which signal gets processed first, we either get three cycles of blinking or none at all. Same holds for switching on the engine and tip-blinking.
- In cruise control: while model checking we got shown errors in the activation of the cruise control as well as in the increasing and decreasing the desired speed, these errors seemed intended or at least directly caused by the specification. Depending on the version of the specification, there were different errors:
  - The desired speed in an activated cruise control could be lowered arbitrarily in previous versions (up to 1.14). This led to invariant violations while model checking, as by pushing down the lever from a desired speed of 0, it was further decreased.
  - The same invariant violation was caused by increasing the desired speed further than said maximum.
  - In the current version (1.17), those are fixed, as some upper and lower thresholds are specified, as well as a minimum velocity when activating it. But as of now, when no previous desired speed is set, the scs is required to be activated at current speed (scs-2) as long as the current speed is above or at 20 km/h. Here an activation at a speed of above 200 km/h should be possible according to the specification, thus directly violating the invariant of being in the range of 1..2000.

- • As we have a reverse gear specified (but our speed being in the range of 0..5000), obviously the speed is measured in absolutes. That means that activation of the speed control system is also possible when in reverse gear (as long as going over 20 km/h in reverse). With possible high previous desired speed, this should either wreck our motor or gearbox fast if it keeps going and accelerating in reverse.
- – There was a problem with the distance in adaptive cruise control, as the distance could be set to f.e. 3 s – with speeds of 240 km/h we would have had to measure distances that are outside our possible range. The latest specification solves this issue by setting a maximum target speed of 200 km/h.
- – Ambient lights: els-19 states the low beam headlight should stay active for 30 more seconds with the interval being reset for example any time a door is opened or shut. We don't have sensors to measure that, as we only have a sensor to state if any door is open. So basically, this requirement has to be changed to either the 30 s cycle being reset continuously until the status of all doors is closed or to the cycle being reset only for the first door opened and respectively the last door shut.

7   Related Work and Conclusion

The importance of animation and visualization seems now accepted; see e.g. the applications [13] and earlier ABZ case studies [5,6,8]. The idea of exploratory phases has also been promoted elsewhere, e.g., in [11].

The modelling approach described in Sect. 2 has been successful. Along the way we gained insights about the relationship between classical B machine inclusion and Event-B refinement. We were successful in re-using existing graphics to generate a custom graphical visualization, which helped us validate the model and allowed us to spot one error at least (cf. Fig. 5).

A weak point of our approach is the manual translation from classical B to Event-B in phase 3. An automated translation of a subset of classical B to Event-B would reduce human effort and avoid errors in the translation. Another alternative would be to stay with ATELIER-B and use its Event-B syntax, which does allow a more powerful substitution language. However, we were much more comfortable with the proving environment of RODIN, hence the effort to translate the models was worth it for us here (and for an industrial system modeling project undertaken by the first author). Finally, why did we not simply do all of the modelling in Rodin to start with? The reasons are the limited tool support for structuring, editing and sharing of models. Future improvements to RODIN could obviate the need to conduct the phases 1 and 2 of Sect. 2 in classical B. These improvements would require improved editing[4] with git integration, improved model composition and decomposition features.

Acknowledgements. We thank Frank Houdek, Alexander Raschke and anonymous reviewers for their useful feedback.

---

[4] CamilleX developed by the Univ. of Southampton is step in the right direction.

A   Trace 7 from Case Study Specification with VisB

For convenience we include a condensed form of the visualisation of trace 7 from https://github.com/hhu-stups/abz2020-models. This shows trace 7 of the case study specification graphically as replayed with PROB and VISB.

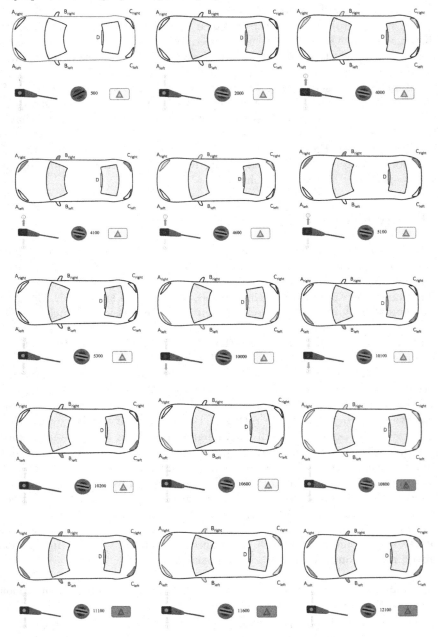

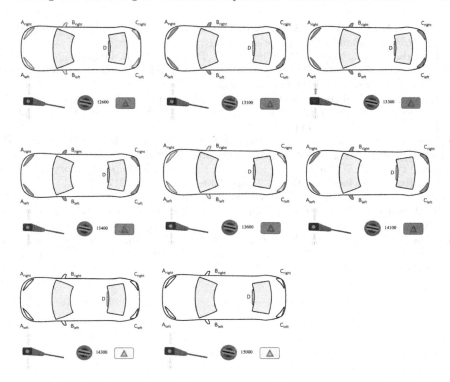

References

1. Abrial, J.-R.: The B-Book. Cambridge University Press, Cambridge (1996)
2. Abrial, J.-R.: Modeling in Event-B: System and Software Engineering. Cambridge University Press, Cambridge (2010)
3. Abrial, J.-R., Butler, M., Hallerstede, S., Voisin, L.: An open extensible tool environment for Event-B. In: Liu, Z., He, J. (eds.) ICFEM 2006. LNCS, vol. 4260, pp. 588–605. Springer, Heidelberg (2006). https://doi.org/10.1007/11901433_32
4. Comptier, M., Leuschel, M., Mejia, L.-F., Perez, J.M., Mutz, M.: Property-based modelling and validation of a CBTC zone controller in Event-B. In: Collart-Dutilleul, S., Lecomte, T., Romanovsky, A. (eds.) RSSRail 2019. LNCS, vol. 11495, pp. 202–212. Springer, Cham (2019). https://doi.org/10.1007/978-3-030-18744-6_13
5. Hansen, D., et al.: Using a formal B model at runtime in a demonstration of the ETCS hybrid level 3 concept with Real Trains. In: Butler, M., Raschke, A., Hoang, T.S., Reichl, K. (eds.) ABZ 2018. LNCS, vol. 10817, pp. 292–306. Springer, Cham (2018). https://doi.org/10.1007/978-3-319-91271-4_20
6. Hoang, T.S., Snook, C., Ladenberger, L., Butler, M.: Validating the requirements and design of a hemodialysis machine using iUML-B, BMotion studio, and co-simulation. In: Butler, M., Schewe, K.-D., Mashkoor, A., Biro, M. (eds.) ABZ 2016. LNCS, vol. 9675, pp. 360–375. Springer, Cham (2016). https://doi.org/10.1007/978-3-319-33600-8_31

7. Houdek, F., Raschke, A.: Adaptive exterior light and speed control system (2019). https://abz2020.uni-ulm.de/case-study
8. Ladenberger, L., Hansen, D., Wiegard, H., Bendisposto, J., Leuschel, M.: Validation of the ABZ landing gear system using proB. STTT **19**(2), 187–203 (2017). https://doi.org/10.1007/s10009-015-0395-9
9. Lamport, L.: Real-time model checking is really simple. In: Borrione, D., Paul, W. (eds.) CHARME 2005. LNCS, vol. 3725, pp. 162–175. Springer, Heidelberg (2005). https://doi.org/10.1007/11560548_14
10. Leuschel, M., Butler, M.J.: ProB: an automated analysis toolset for the B method. STTT **10**(2), 185–203 (2008). https://doi.org/10.1007/s10009-007-0063-9
11. Oda, T., Araki, K., Larsen, P.G.: A formal modeling tool for exploratory modeling in software development. IEICE Trans. **100–D**(6), 1210–1217 (2017)
12. Rehm, J., Cansell, D.: Proved development of the real-time properties of the IEEE 1394 root contention protocol with the event b method. In: ISoLA, pp. 179–190 (2007)
13. Yang, F., Jacquot, J.-P., Souquières, J.: The case for using simulation to validate Event-B specifications. In: Leung, K.R.P.H., Muenchaisri, P. (eds.) APSEC, pp. 85–90. IEEE (2012)

An Event-B Model of an Automotive Adaptive Exterior Light System

Amel Mammar[1]([✉]) [iD], Marc Frappier[2] [iD], and Régine Laleau[3]

[1] SAMOVAR, Institut Polytechnique de Paris, Télécom SudParis, Évry, France
amel.mammar@telecom-sudparis.eu
[2] Laboratoire GRIF, Département d'informatique, Faculté des sciences,
Université de Sherbrooke, Québec, Canada
marc.frappier@usherbrooke.ca
[3] LACL, Université Paris-Est Créteil, Créteil, France
laleau@u-pec.fr

Abstract. This paper introduces an EVENT-B formal model of the adaptive exterior light system for cars, a case study proposed in the context of the ABZ2020 conference. The system describes the different provided lights and the conditions under which they are switched on/off in order to improve the visibility of the driver without dazzling the oncoming ones. The system can be viewed as a lights controller that reads different information form the available sensors (key state, exterior luminosity, etc.) and takes the adequate actions by acting on the actuators of the lights in order to ensure a good visibility for the driver according to the information read. Our model is built using stepwise refinement with the EVENT-B method. We consider all the features of the case study, all proof obligations have been discharged using the RODIN provers. Our model has been validated using PROB by applying the different provided scenarios. This validation has permitted us to point out and correct some mistakes, ambiguities and oversights in the first versions of the case study.

Keywords: Adaptive exterior light system · EVENT-B method · Refinement · Verification

1 Introduction

This paper presents a formal system model of an adaptive exterior light system (ELS) for a car. This system has been proposed as a case study for the ABZ2020 conference. We use EVENT-B to construct and represent this formal model.

The exterior light system subject of this case study has objective to adapt the brightness of the different lights with respect to the status of the car but also the oncoming ones. For that purpose, the cars are equipped with different lights that can be switched on/off under specific conditions. In this paper, we

This work was supported in part by NSERC (Natural Sciences and Engineering Research Council of Canada).

A. Raschke et al. (Eds.): ABZ 2020, LNCS 12071, pp. 351–366, 2020.
https://doi.org/10.1007/978-3-030-48077-6_28

stress more on the modeling of low beams, tail lamps and direction indicators. Roughly speaking, the low beams illuminate the road when the vehicle is running or vehicle surrounding while leaving the car during darkness; tail lamps permit to illuminate the vehicle if it is parked on a dark road at night, whereas the direction indicators allow to inform the following vehicle that the car will turn on the right/left. To control these exterior lights, the driver acts on the different physical elements like the key, the hazard switch etc. The position of the key (NoKeyInserted, KeyInserted, KeyInIgnitionOnPosition) is transmitted to the controller of the lights via the sensor keyState. Similarly, the hazard warning switch, with two positions (On/Off), permits to make both director indicators flashing at the same time.

The rest of this paper is structured as follows. After a brief overview of the EVENT-B method provided in Sect. 2, Sect. 3 presents our modelling strategy. Section 4 describes our model in more details. The validation and verification of our model are discussed in Sect. 5. Section 6 identifies the weaknesses of the requirements document provided for the case study, and the adequacy of the EVENT-B method for constructing a model of this case study. We conclude in Sect. 7.

2   EVENT-B Method

EVENT-B is a formal system modeling notation [1] proposed by Abrial. It allows for the stepwise construction of models using refinement. It is inspired from actions systems originally proposed by Back [2] and extended by several others. An EVENT-B model is made of components of two types: machine and context. A machine consists of events that modify state variables. An event has a set of guards and actions. When the guards hold, the event can be triggered; its actions can then modify the system state. A machine has state invariants that can be proved by discharging proof obligations.

A machine can refine another machine by replacing or adding state variables and by adding new events. System elements can be gradually taken into account through refinement. Existing variables can also be replaced by new variables; to show behavior preservation, a gluing invariant must relate the old variables to the new variables. An event refines an existing event by reducing its nondeterminacy, by strengthening its guards and/or choosing a value v, called *witness*, for a parameter p of the event. In that case, we should replace each occurrence of p by the value v in the guard and the substitution of the event. New events implicitly refine a skip event, so they cannot modify existing variables; they can only modify the added (new) variables. System constants are specified in contexts. A context can extend another context. Invariants are preserved by refinement, so invariants are introduced at the most appropriate step that simplifies their proof.

EVENT-B is supported by the RODIN platform [10], an Eclipse-based tool that provides editors, provers and several other plugins for various tasks (*e.g.*, animation and model checking with PROB [5], integration with UML class diagrams and state machines with UML-B [13]). In this paper, we have used RODIN with the PROB and AtelierB provers plugins. We did not use any other plugin.

In this paper, we report on the use of this formal method for the modeling and the verification of the automotive adaptive exterior light system whose behavior is briefly described in the introduction. The chose of this formal method can be justified by the refinement technique it provides to gradually introduce the details of the system and also its different available support tools for modeling, animating and proving a specification.

3   Modelling Strategy

We reuse the terminology introduced in [11]. A control system interacts with its environment using sensors and actuators. A sensor measures the value of some environment characteristic m, called a *monitored* variable (*e.g.*, the state of the ignition key), and provides this measure (*e.g.*, whether the key is inserted or not) to the software controller as an *input* variable i. In a perfect world, we have $m = i$, but a sensor may fail. The software controller can influence the environment by sending commands, called *output* variable o to actuators. An actuator influences the value of some characteristics of the environment, call a *controlled* variable c. Variables m and c are called *environment variables*. Variables i and o are called *controller variables*. Finally, a controller has its own internal state variables to perform computations. We use EVENT-B state variables to represent environment (*i.e.*, *monitored* and *controlled*) variables, and controller variables. We do not model sensor/actuator failures.

3.1   Control Abstraction

A typical implementation of a control system such as the ELS is either a control loop that reads all input variables at once and then computes all output variables in the same iteration, or it can be driven by interruption triggered when a sensor provides a new value. The body of a control loop represents a single event and state transition. This allows for the definition of priorities between input variable changes. In our model, we use a more abstract approach, as it is common in the EVENT-B style of system modeling. We define one event for each input variable change, which allows for a more modular specification that is easier to prove. This is closer to an interrupt-driven control system. Our EVENT-B abstraction is also a reasonable abstraction for a control loop, considering that in most cases, a single input variable changes between two control loop iterations. The control loop can be derived from our specification by merging all events and defining priorities between events.

3.2   Modeling Structuration

The specification is structured into five refinements steps (five contexts and six machines). At the most abstract level we introduce various kinds of lights controlled by the system. They are declared as constants in Context C0. The considered lights are: the direction indicators (left or right), the low beam headlights (left and right), the tail lamp (left and right), the reverse light (that indicates that the vehicle will move backwards), the brake lights and the cornering lights (that illuminate the cornering area separately when turning left or right). The high beam headlights are considered in Context C4 and Machine M5 since their behavior is different from the other lights, as it can be adaptive. Constant *LigntnessLevel* indicates the high beam light range, as specified in the requirement document [3].

Machine M0 in Fig. 1 contains a unique variable *headingState* that associates a level of brightness to each light declared in Context C0, and a unique event headLightSet that assigns an arbitrary level of brightness to these lights.

MACHINE M0
SEES C0
VARIABLES
    headingState
INVARIANTS
    inv1: $headingState \in HeadLights \rightarrow LigntnessLevel$
EVENTS
Initialisation
    begin
        act1: $headingState := HeadLights \times \{0\}$
    end
Event headLightSet $\widehat{=}$
    any
        hl
    where
        grd1: $hl \in HeadLights \nrightarrow LigntnessLevel$
    then
        act1: $headingState := headingState \ntriangleleft hl$
    end
END

Fig. 1. Machine M0

The first refinement, Machine M1 and Context C1, introduces the elements that the car driver can control and that can have an impact on the state of the lights declared in Context C0, namely the ignition key, the pitman arm, the light rotary switch, the brake pedal and the hazard warning light switch. For each of these elements, there is one event that refines headLightSet and that arbitrary modifies the lights impacted by this element.

Each of the subsequent refinements describes the behavior of particular lights. The choice of the lights taken into account in the refinements is arbitrary. Machine M2 and Context C2 consider the direction indicators, the hazard warning light and the emergency brake light. Machine M3 and Context C3 consider the low beam lights. Machine M4 considers the cornering lights and Machine M5 and Context C4 consider the high beam headlights.

3.3   Formalization of the Requirements

Table 1 relates the components of our model with the requirements listed in [3]. As one can remark, some requirements are specified as invariant whereas others are only considered in the related events. Requirement ELS-10 for instance stating the duration of a flashing cycle does not correspond to an invariant but it is considered in the event flashingDark that makes the current time progress by a unit of time. Specifying such requirements as an invariant would require the introduction of two extra variables to store the starting and the ending moment of the cycle to set that the difference should be equal to a unit of time. Roughly speaking, a timed requirement, an action duration more precisely, is modelled as an event if there is no other requirement that refers to such a duration otherwise an invariant is associated with it. Moreover, let us note that $M3$ is the refinement with the most invariants number because it models several interrelated lights, that is the low beams, the tail lamps, the parking lights etc.

3.4   Modeling of Temporal Requirements

Some properties of the requirements depend on two consecutive states. For example, requirement ELS-16 applies only when the rotary switch is turned to Auto while the ignition is already Off. This requirement can be expressed using an LTL formula as follows:

$$\mathsf{G} \;\; ((keyState \neq KeyInIgnitionOnPosition \wedge lightSwitch \neq Auto)$$
$$\Rightarrow$$
$$\mathsf{X} \; (lightSwitch = Auto \Rightarrow headingState[LowBeams] = 0))$$

Unfortunately EVENT-B does not support the expression of LTL formula as part of the specification even if the PROB model-checker can check LTL formulas on an EVENT-B specification with a finite state space, but it does not terminate for our model on such properties, because of the size of the state space. On the other hand, a proof-based approach for temporal formulas is proposed in [7], but it generates a large number of proof obligations for a model of this size. Thus, we have chosen to express these properties as invariants by adding an extra variable to store the previous value of a state variable that is needed in a two-consecutive-state property. For example, to express ELS-16 as an invariant, we have to say that: (1) the current and previous states of the ignition are not equal to On, (2) the previous state of the switch is different from Auto, and (3) the current state of the switch is equal to Auto, which is represented by the following invariant (Machine M3, Invariant inv18)

$$ELS16 = TRUE \wedge ELS16P = FALSE$$
$$\Rightarrow$$
$$keyState \neq KeyInIgnitionOnPosition \wedge$$
$$keyStateP \neq KeyInIgnitionOnPosition \wedge$$
$$lightSwitch = Auto \wedge lightSwitchP \neq Auto$$

Table 1. Cross-reference between the components of our model and the requirements of [3]

Requirements [3]	Component	Invariant/event
ELS-1, ELS-2, ELS-4, ELS-23	M2	inv5, inv7
ELS-3		movePitmanUD
ELS-5, ELS-23	M2	inv8
ELS-6	M3	inv10
ELS-7	M2	movePitmanUD
ELS-8	M2	inv6, inv8
ELS-10	M2	flashingDark
ELS-11 to ELS-13	M2	movePitmanUD
ELS-14	M3	inv2
ELS-15	M3	inv3
ELS-16	M3	inv4
ELS-17	M3	inv5
ELS-18	M3	inv6, 7, 8, 9
ELS-19	M3	inv10
ELS-21	M3	inv3–5, inv10, inv14
ELS-22	M3	inv11, 12, 13
ELS-24, 25, 26, 27	M4	inv2–inv13
ELS-28	M3	inv14
ELS-29		All invariants defining the brightness level
ELS-30, ELS-31	M5	inv3, 5
ELS-32..38	M5	inv6–11
ELS-39	M2	inv12, 13
ELS-40	M2	inv14
ELS-41	M1	inv12, 13
ELS-42	M5	inv4
ELS-43...49	M5	inv6–11

Variable $ELS16$ represent the satisfaction of the conditions of ELS-16 and it is maintained by event moveSwitchAuto representing the state change of the rotary switch to position Auto. Variable $ELS16P$ represents its previous value. It conditions the invariant to the state change of the rotary switch.

These extra variables storing previous values must obviously be maintained in the events that change the value of the corresponding variable, but also in events that rely on the previous value for making a decision, even if they do not modify the corresponding variable.

4   Model Details

In this section, we briefly describe some specific ways of modelling that characterize our specification. The complete archive of the EVENT-B project is available
in [6].

4.1   Modeling Complex User Interface Elements

There are elements manipulated by the car driver that have several positions
and that control several lights depending on their positions. This is the case of
the key and the light rotary switch. For each of these elements, the position it
can take depends on the current position and thus can be described by a state-
transition diagram. In the more abstract levels, we have chosen to gather all the
possible transitions into a single event because at these levels the invariants do
not depend on a specific position.

Let us take the case of the key. In Context C1, set *keyStates* describes all the
states of the key:

$$partition(keyStates,$$
$$\{NoKeyInserted\}, \{KeyInserted\}, \{KeyInIgnitionOnPosition\})$$

In Machine M1, Variable *keyState* represents the current state of the key,
Variable *keyStateP* contains the previous state of the key and the authorized
transitions are specified in Invariants inv2, inv3:

$$keyState = NoKeyInserted$$
$$\Rightarrow keyStateP = NoKeyInserted \lor keyStateP = KeyInserted$$

$$keyState = KeyInIgnitionOnPosition$$
$$\Rightarrow keyStateP = KeyInIgnitionOnPosition \lor keyStateP = KeyInserted$$

Event moveKey specifies the new state of the key according to its previous
state and restricts the value of the event parameter *hl* to the lights controlled
by the key.

Event moveKey $\widehat{=}$
refines headLightSet
   any
      hl,valkey
  where
      grd1: $hl \in \quad LowBeams \cup tailLamps \cup directionIndicators$
                $\cup \{corneringLightLeft, corneringLightRight\}$
          $\twoheadrightarrow LigntnessLevel$
      grd2: $(keyState = NoKeyInserted \Rightarrow valkey = KeyInserted)$
         $\land (keyState = KeyInserted \Rightarrow valkey \in$
                $\{NoKeyInserted, KeyInIgnitionOnPosition\})$
         $\land (keyState = KeyInIgnitionOnPosition \Rightarrow valkey = KeyInserted)$

then
  act1: $headingState := headingState \triangleleft hl$
  act2: $keyState := valkey$
  act3: $keyStateP := keyState$
  act4: $pitmanArmUDP := pitmanArmUD$
end

In Machine M2, Event moveKey is refined to specify the behavior of the direction indicator and the tail lamps according to the position of the key and the position of the hazard warning switch.

In Machine M3, we have split Event moveKey into four events (*i.e.*, insertKey, insertKeyputIgnitionOn, insertKeyputIgnitionOff, removeKey) to be more precise on the state of the lights according to the position of the key.

Let us take the two events insertKey and insertKeyputIgnitionOn. In Event insertKey, Action act4 specifies that if the hazard warning switch is not activated then the direction indicator is off, otherwise it is on and the two flashing lights are on.

Event insertKey $\widehat{=}$
refines moveKey
 any
  hl
 where
  grd1: $hl \in \quad LowBeams \cup tailLamps \cup directionIndicators$
        $\rightarrow LigntnessLevel$
  grd2: $keyState = NoKeyInserted$
  grd3: ...
  grd4: $hazardWarningSwitchOn = FALSE$
    $\Rightarrow (directionIndicators) \times \{0\} \subseteq hl$
    ...
 with
  valkey: valkey= keyInserted
 then
  act1: $headingState := headingState \triangleleft hl$
  act2: $keyState := KeyInserted$
  act3: $keyStateP := keyState$
  act4: $direcIndFlash :=$
    $\{TRUE \mapsto \{blinkRight \mapsto FALSE, blinkLeft \mapsto FALSE\},$
    $FALSE \mapsto directionIndicators \times \{TRUE\}$
    $\}(bool(hazardWarningSwitchOn = FALSE))$
    ...
end

In Event putIgnitionOn, Action act4 specifies that if the hazard warning switch is not activated then the direction indicator is activated to the left or right according to the position of the pitman arm, otherwise it is on and the two flashing lights are on.

Event putIgnitionOn $\hat{=}$
refines moveKey
  any
     hl
  where
     grd1: $hl \in \quad LowBeams \cup tailLamps \cup directionIndicators$
                $\rightarrow LigntnessLevel$

       ...
  with
     valkey: valkey= KeyInIgnitionOnPosition
  then
     act1: $headingState := headingState \mathbin{\lhd\mkern-9mu-} hl$
     act2: $keyState := KeyInIgnitionOnPosition$
     act3: $keyStateP := keyState$
     act4: $direcIndFlash :=$
         $\{TRUE \mapsto \{blinkRight \mapsto bool(pitmanArmUD \in Upward),$
                   $blinkLeft \mapsto bool(pitmanArmUD \in Downward)\},$
         $FALSE \mapsto directionIndicators \times \{TRUE\}$
         $\}(bool(hazardWarningSwitchOn = FALSE))$

       ...
end

We have applied the same modeling process to the Light Rotary Switch.

Splitting the event makes the proof obligations easier to discharge even if more proof obligations are generated.

4.2 Managing Priorities Between Requirements

Some requirements can be in conflict because they have common system states with different transitions. This is the case for Requirements ELS-16 and ELS-17. On one hand, ELS-16 states that if the key state is **inserted** then the low beam headlights are **off**. This is specified in Invariant inv4 of Machine M3:

$$ELS16 = TRUE \wedge ... \Rightarrow headingState[LowBeams] = 0$$

where Variable $ELS16$ is **TRUE** if the key state is **inserted**.

On the other hand, ELS-17 states that if the daytime running light is activated then the low beam headlights are activated after starting the engine and remain activated as long as the key is not removed, that is, either the key position is **inserted** or the ignition is **on**.

We have detected the conflict when we have animated the specification. The solution is to prioritize the requirements. After discussing with the case study authors, a priority for ELS-16 over ELS-17 has been set; this is specified in Invariant inv5 of Machine M3 that translates ELS-17:

$$(... \vee dayTimeLightCont = TRUE) \wedge ... \wedge ELS16 = FALSE \wedge ...$$
$$\Rightarrow headingState[LowBeams] = 100$$

where Variable $dayTimeLightCont$ is **true** if the daytime running light is activated.

4.3   Modeling Time Duration

In EVENT-B, a specification of requirements that involves time duration requires to explicitly model time. In this case study, time can trigger changes on the state of lights (*e.g.* Requirements ELS-18, 19, 24, ... specify time intervals where particular lights have to be activated or not). A variable *currentTime* has been introduced in Machine M1 to model the time progression together with Event progress that increments this variable by an arbitrary positive number (Action act2). Action act1 specifies the lights whose state can be modified by a time progress.

Event progress $\widehat{=}$
refines headLightSet
    any
        hl
        step
    where
        grd1: $hl \in \quad LowBeams \cup tailLamps \cup directionIndicators \cup$
                $\{corneringLightLeft, corneringLightRight\} \twoheadrightarrow LigntnessLevel$
        grd2: $step \in N1$
    then
        act1: $headingState := headingState \mathbin{\lhd\mkern-9mu-} hl$
        act2: $currentTime := currentTime + step$
        ...
    end

Event progress is refined in Machines M3, M4, M5 by detailing how each kind of lights is impacted. For instance, in M3, the exterior brightness (ELS-18) and the ambient light (ELS-19) imply to activate the low beam headlights for a given time interval.

4.4   Model Statistics

Table 2 describes the size of the model. Since RODIN does not use text files to store models, there are various ways of counting the lines of code (LOC) of a model. Moreover, code is inherited when refinement and event extension is used. Lines of code are computed using the CAMILLE editor representation of the EVENT-B model, which does not count inherited LOC through event extension and puts all variables on the same line. Total LOC, which includes inherited LOC, is provided within "( )", and computed using the pretty printer of the RODIN EVENT-B Machine Editor. Comments are excluded. Since we do not use data refinement (*i.e.*, no variable is replaced through refinement), we provide the total number of variables for each machine along with the number of new variables (*i.e.*, introduced in a refinement) enclosed by "( )". Invariants are specific to each machine. Since some events are renamed by refinement, we provide the total and new events introduced in each machine.

Table 2. Model size

Component	Size in LOC (extended)	Constants/ variables Total (New)	Axioms/ invariants New	Events Total (New)
C0	15	(17)	7	
C1	15	(17)	7	
C2	8	(2)	2	
C3	10	(2)	2	
C4	16	1	10	
M0	21 (28)	1 (1)	1	1
M1	215 (320)	15 (14)	13	12 (11)
M2	382 (691)	25 (10)	18	14 (2)
M3	908 (1619)	37 (12)	36	19 (5)
M4	885 (2377)	50 (13)	15	20 (1)
M5	416 (2694)	61 (11)	15	23 (3)
Total	2875		126	

5 Validation and Verification

To verify and validate the EVENT-B models presented in the previous sections, we have proceeded into three steps detailed hereafter.

5.1 Model Checking of the Specification

In this step, the PROB tool is used as a model checker in order to ensure that the specification is invariant violation-free, that is, there is no trivial scenario that violates the invariants. From a practical point of view, PROB can find a sequence of events that, starting from a valid initial state of the machine, leads to a state that violates its invariant. Such scenarios (or counterexamples) may result from a guard/action missing but also from an incorrect invariant. This step permits us to fix trivial bugs before the proof phase that can be very long and hard. It is worth noting that even if the tool does not find any invariant violation, it does not mean that the specification is correct. Indeed, there may be a scenario that the tool fails to find for different reasons like a timeout on the model checking process. In the present case study, the model checking step permits us to detect missing actions in particular those related to the variables representing the previous state of an element. Indeed, this makes the invariants depending on such variables violated as they should be verified only when the current and the previous values of these variables are different.

5.2   Validation with Scenarios

The goal of this phase is to be sure that the specification satisfies the requirements. To this aim, we used the animation capability of PROB and played the different scenarios provided with the case study. This step permits us to exhibit several flaws/ambiguities in the initial release of the description documents (see Sect. 6 for more details). As examples of such flaws, we can cite the lack of prioritization between some requirements like ELS-16 and ELS-17 that share the same activation conditions when the *daytime running light* option is activated with the ignition in the Off position and the driver turns the switch in the Auto position. To correct these flaws/ambiguities, we have discussed with the case study authors because we are not specialists of the domain. For the above particular example, a priority is given to ELS-16 over ELS-17. It is worth noting that such flaws/ambiguities can not be detected in the model checking phase because they make the guard of some events unsatisfied, thus the event is not enabled and the invariant is thus not violated. Let us note that we had some problems to animate the first version of our models where we have kept the event parameter *hl* as a partial function on the set of all the lights. Indeed in that case, PROB checks all the possible partial functions on these lights which leads to a timeout. To overcome this issue, we have replaced each partial function by a more restrictive total function on the right domain, that is, the lights whose state actually changes after the execution of the event.

5.3   Proof of the Specification

It is the last step, whose goal is to ensure the correctness of the specification by discharging proof obligations generated by RODIN. These proof obligations aim at proving invariant preservation by each event, but also to ensure that the guard of each refined event is stronger than that of the abstract event. These guard strengthening refinement proof obligations ensure that event parameters like *hl* mentioned above are properly refined. For instance, *hl* is defined as a partial function in the abstract event headLightSet; it is refined using total functions by giving its value for each refining event. So, we have to ensure that these values satisfy the initial guard. Figure 2 provides the proof statistics of the case study: 1643 proof obligations have been generated, of which 23% (385) were automatically proved by the various provers. The remaining proof obligations were discharged interactively since they needed the use of external provers like the Mono Lemma prover that has shown to be very useful for arithmetic formulas. In addition, we have added some theorems on min/max operators (a min/max of a finite set is an element of the set, etc).

Let us note that the results of this phase has especially impacted some modeling choices. For instance, to speed up the proof phase, we have included in the guards some properties tagged as theorems in order to prove them only once and reuse them in all the proofs that need them for that event. This is the case of Guards grd9, grd10 of insertKey in Machine M3 that state:

grd9: $lowBeamRight \in dom(hl) \Rightarrow hl(lowBeamRight) \in 0..100$

grd10: $lowBeamLeft \in dom(hl) \Rightarrow hl(lowBeamLeft) \in 0..100$

Element Name	Total	Auto	Manual	Reviewed	Undischarged
ELS_1112	1643	385	1258	0	0
C0	0	0	0	0	0
C1	0	0	0	0	0
C2	0	0	0	0	0
C3	0	0	0	0	0
C4	14	9	5	0	0
M0	2	1	1	0	0
M1	88	55	33	0	0
M2	206	25	181	0	0
M3	738	131	607	0	0
M4	402	128	274	0	0
M5	193	36	157	0	0

Fig. 2. RODIN proof statistics of the case study

6   Other Points

6.1   Feedback on the Requirements Document

The formal modeling of the requirements document [3] lead us to identify a number of ambiguities and some contradictions with the test scenarios provided. We have communicated these to the authors of the requirements document, and a number of revisions were produced, following our comments. Our comments induced 9 of the 17 versions produced after the publication of the initial version of the requirements document. These modifications impacted 18 of the 49 requirements of the Exterior Light System. A detailed list of these elements are described in the last version (*i.e.*, 1.17) of the requirements document. We have mainly rephrased some requirements for which the applicability conditions should hold at different time points. For instance, in requirement ELS-16, the condition "the switch in position Auto" should happen after the condition "the ignition is already Off". Moreover, we have defined priorities between requirements to make the specification deterministic: ELS-16 has priority over ELS-17, ELS-19 has priority over ELS-17, etc. We have also rephrased some sentences to clarify them. For instance in the first version of the document, the word "released" was used with the meaning "button pushed" in some places and with the meaning "button not pushed" in some others. To remove this ambiguity, we have replaced it with the terms "active" and "not active". Finally to make the modeling easier and after a discussion with the case study authors, the signal *pitmanArm* has been splitted into signals *pitmanArmForthBack* and *pitmanArmUpDown* with their corresponding positions (states) and the possible transitions between them.

6.2   Modeling Temporal Properties

Dealing with previous values to prove temporal properties turned out to be a significant burden. To improve/facilitate the specification of such kind of properties, which are probably very common in control systems, it would be interesting to study how they could be handled in RODIN or in some other plugin like the EVENT-B State machines plugin[1]. This plugin permits to generate EVENT-B events from a state machine including their guards that specify the requirements modeled by the state machine but without producing the related invariants. In that case, it becomes difficult to trace and justify the usefulness of the generated guards.

6.3   Identifying a Refinement Strategy

The crux in defining the structure of the EVENT-B model was to define the requirements elements to include at each refinement level. Recall that once a variable is introduced in a model, it cannot be modified by new events of subsequent refinements. Thus, when a variable is introduced, each event that needs to update it must be also introduced. In this case study, there are several dependencies between requirements elements. As many lights mutually rely on the same sensors and are correlated in terms of behavior, we have defined a single event, in the first machine, to model the light state changes and refined it according to the different actuators/sensors. But, we think that it would be interesting to look deeper into the existing structuring approaches for EVENT-B: decomposition [12] or modularization [4], in order to structure the specification into smaller logical units to make the proofs easier. A refactoring tool based on the read/update dependencies between events and state variables would be nice. It could help in finding an optimal decomposition based on the connected components of a dependency graph for a given machine. Building such a graph from the requirements is not easy, as one typically needs to formalize the requirements to precisely understand which variables are needed and where. So, the specifier typically finds the ideal refinement structure only after creating a potentially non optimal refinement structure. Often a lot of effort has been invested in creating this first model, and there is no resource left to do a refactoring to obtain a better model. By better, we mean a model whose refinement decomposition would yield easier proofs for the same set of properties.

7   Conclusion

We have presented an EVENT-B model for the ELS case study. Our model takes into account all of the requirements. The model was verified by proving a large number of properties (98 invariants) and by simulation using PROB. Temporal properties involving two consecutive states were proved using variables storing previous state values. Due to the model size (61 state variables), PROB was

---

[1] http://wiki.event-b.org/index.php/Event-B_Statemachines.

unable to verify invariant or temporal properties. The proof effort was quite significant: 1258 proofs obligation (76%) had to be manually discharged. The last EVENT-B machine is quite large (2 694 LOC), which denotes that the case study was an interesting modeling and verification challenge. The RODIN provers were less efficient than in previous ABZ case studies, where the manual proofs ratio was closer to 30% [8, 9].

The formalization lead us to identify several small ambiguities in the requirements. They have been discussed with the case study authors as they were discovered, which lead to 9 out of the 17 revisions of the case study text that were published during the modeling process. This shows that formalization is an effective technique to discover defects early in the software development process. It is well-known in the software engineering literature that the earlier a defect is found, the cheaper it is to fix it.

Determining the best refinement strategy remains a challenge in EVENT-B. We fell short of time to try out the model decomposition plugins available in RODIN. They might have been useful in decomposing the specification into smaller, more manageable parts. This case study is of a different nature than the previous ones in the ABZ conference series (*i.e.*, 2014 Landing gear, 2016 Hemodialysis, 2018 ERTMS). Its elements are more tightly coupled, which made it more difficult to find an appropriate refinement strategy. It contains more properties to prove than the previous ones, but they are more localized properties (*i.e.*, each property referring to a small number of events on at most two consecutive states) that do not depend on the relationship between monitored variables and controlled variables. However, we really think that the EVENT-B method must include modularization clauses as native structuring mechanisms like those of the B method that permit to have a modular specification since the first phases of the development. This will make EVENT-B more suitable for the development of big and complex systems. For comparison, in the ERTMS case study, we had to build a relationship between the real (actual) positions of the trains and the controller view of the train positions to prove safety properties. There were no such issues in the ELS case study.

Acknowledgments. The authors would like to thank the case study authors, and Frank Houdek in particular, for his responsiveness and useful feedback during the modeling process when questions were raised or when ambiguities were found. The authors would also like to thank Michael Leuschel for his quick feedback on using PROB for this large case study.

References

1. Abrial, J.: Modeling in Event-B. Cambridge University Press, Cambridge (2010)
2. Back, R.J.R., Sere, K.: Stepwise refinement of action systems. In: van de Snepscheut, J.L.A. (ed.) MPC 1989. LNCS, vol. 375, pp. 115–138. Springer, Heidelberg (1989). https://doi.org/10.1007/3-540-51305-1_7
3. Houdek, F., Raschke, A.: Adaptive exterior light and speed control system, November 2019. https://abz2020.uni-ulm.de/case-study#Specification-Document

4. Iliasov, A., et al.: Supporting reuse in Event B development: modularisation approach. In: Frappier, M., Glässer, U., Khurshid, S., Laleau, R., Reeves, S. (eds.) ABZ 2010. LNCS, vol. 5977, pp. 174–188. Springer, Heidelberg (2010). https://doi.org/10.1007/978-3-642-11811-1_14

5. Leuschel, M., Bendisposto, J., Dobrikov, I., Krings, S., Plagge, D.: From animation to data validation: the ProB constraint solver 10 years on. In: Boulanger, J.L. (ed.) Formal Methods Applied to Complex Systems: Implementation of the B Method, chap. 14, pp. 427–446. Wiley ISTE, Hoboken (2014)

6. Mammar, A., Frappier, M., Laleau, R.: An Event-B model of an automotive adaptive exterior light system, January 2020. http://www-public.imtbs-tsp.eu/~mammar_a/LightControlSystem.html and http://info.usherbrooke.ca/mfrappier/abz2020-ELS-Case-Study/

7. Mammar, A., Frappier, M.: Proof-based verification approaches for dynamic properties: application to the information system domain. Formal Aspects Comput. **27**(2), 335–374 (2014). https://doi.org/10.1007/s00165-014-0323-x

8. Mammar, A., Frappier, M., Tueno Fotso, S.J., Laleau, R.: An EVENT-B model of the hybrid ERTMS/ETCS level 3 standard. In: Butler, M., Raschke, A., Hoang, T.S., Reichl, K. (eds.) ABZ 2018. LNCS, vol. 10817, pp. 353–366. Springer, Cham (2018). https://doi.org/10.1007/978-3-319-91271-4_24

9. Mammar, A., Laleau, R.: Modeling a landing gear system in Event-B. In: Boniol, F., Wiels, V., Ait Ameur, Y., Schewe, K.-D. (eds.) ABZ 2014. CCIS, vol. 433, pp. 80–94. Springer, Cham (2014). https://doi.org/10.1007/978-3-319-07512-9_6

10. Event-B Consortium. http://www.event-b.org/

11. Parnas, D.L., Madey, J.: Functional documents for computer systems. Sci. Comput. Program. **25**(1), 41–61 (1995)

12. Silva, R., Pascal, C., Hoang, T.S., Butler, M.J.: Decomposition tool for Event-B. Softw. Pract. Exper. **41**(2), 199–208 (2011)

13. Snook, C., Butler, M.: UML-B: a plug-in for the Event-B tool set. In: Börger, E., Butler, M., Bowen, J.P., Boca, P. (eds.) ABZ 2008. LNCS, vol. 5238, p. 344. Springer, Heidelberg (2008). https://doi.org/10.1007/978-3-540-87603-8_32

Modeling of a Speed Control System Using Event-B

Amel Mammar[1]([✉])[iD] and Marc Frappier[2][iD]

[1] SAMOVAR, Institut Polytechnique de Paris, Télécom SudParis, Évry, France
amel.mammar@telecom-sudparis.eu
[2] Laboratoire GRIF, Département d'informatique, Faculté des sciences,
Université de Sherbrooke, Québec, Canada
marc.frappier@usherbrooke.ca

Abstract. The present paper presents our proposal of an EVENT-B model of a speed control system, a part of the case study provided in the ABZ2020 conference. The case study describes how the system regulates the current speed of a car according to a set criteria like the speed desired by the driver, the position of a possible preceding vehicle but also a given speed limit that the driver must not exceed. For that purpose, this controller reads different information form the available sensors (key state, desired speed, etc.) and takes the adequate actions by acting on the actuators of the car's speed according to the read information. To formally model this system, we adopt a stepwise refinement approach with the EVENT-B method. We consider most features of the case study, all proof obligations have been discharged using the Rodin provers. Our model has been validated using ProB by applying the different provided scenarios. This validation has permitted us to point out and correct some mistakes, ambiguities and oversights contained in the first versions of the case study.

Keywords: Speed control system · EVENT-B method · Refinement · Verification

1 Introduction

The case study, proposed in the context of the ABZ2020 conference, is composed of two parts: *Adaptive Exterior Light and Speed Control Systems*. Since the whole case study is quite lengthy/complex and the two parts are only loosely coupled as stated in the description document, we chose to handle each part in a separate paper. The present paper deals with the speed control system whereas a companion paper considers the adaptive exterior light system [7].

The goal of the speed control system is to regulate the current speed of a car according to a set of criteria like the speed desired by the driver, the position of

This work was supported in part by NSERC (Natural Sciences and Engineering Research Council of Canada).

A. Raschke et al. (Eds.): ABZ 2020, LNCS 12071, pp. 367–381, 2020.
https://doi.org/10.1007/978-3-030-48077-6_29

a possible preceding vehicle but also a given speed limit that the driver must not exceed. The system can behave according to two options: the first one regulates the speed independently on the any preceding vehicle, the second component takes into account the position of a possible preceding vehicle by maintaining a safety distance. The driver has the possibility to choose which option to activate at a given moment. Like a controller, in both options, the system reads different informations from the available sensors (key state, desired speed, the preceding vehicle position, etc.) and takes the adequate actions by acting on the actuators of the car's speed according to the read information.

The present paper describes the formal modeling of the speed control system using the EVENT-B method and its refinement technique that permits to master the complexity of a system by gradually introducing its different elements/characteristics. Proposed by Abrial as a successor of the B method [1], the EVENT-B method [2] permits to model discrete systems using mathematical notations. An EVENT-B specification is made of two elements: *context* and *machine*. A context describes the static part of an EVENT-B specification; it consists of constants and sets (user-defined types) together with axioms that specify their properties. The dynamic part is included in a machine that defines variables and a set of events. The possible values that the variables can hold are specified by an invariant using a first-order formula on the state variables. The different machines composing an EVENT-B specification are related with a refinement relation whereas the contexts are linked with an extension link (extends). Each refinement adds new information to a model; these could be new state variables, new events or new properties. EVENT-B refinement allows for guard strengthening, nondeterminism reduction, and new events introduction. New events of a model M' that refines a model M are considered to refine a *skip* event of M, hence they cannot modify a variable of M. Therefore, all events that need to modify a variable v must be defined in the same model where v is first introduced. The correctness of an EVENT-B model is ensured by proof obligations that verify that the invariant is preserved by each event and that the refinement preserves the properties of the system.

The development of our EVENT-B models has been done under the Rodin platform [3] that provides editors, provers and several other plugins for various tasks like animation and model checking with PROB [5]. We use PROB in order to animate the built models with two purposes: exhibiting the problematic scenarios that violate the invariant prior to the hard/long proof phase, but also validating the specification by playing the provided scenarios in order to be sure that we have specified the right system.

The rest of this paper is structured as follows. Section 2 describes our modelling strategy. Section 3 describes our model in more details. Section 4 describes the validation and verification of our model. Section 5 identifies the weaknesses of the requirements document provided for the case study, and the adequacy of the EVENT-B method for constructing a model of this case study. We conclude in Sect. 6.

2   Modelling Strategy

The speed control system subject of this paper can be seen as a control system that interacts with its environment through a set of sensors, which provide it with information about the state of the physical elements, and a set of actuators that are used to transmit the adequate orders to these elements. In this paper, we use the concepts described in [10]. A sensor measures the value of some environment elements m, called a *monitored* variable (*e.g.*, the state of the ignition key), and provides this measure (*e.g.*, whether the key is inserted or not) to the software controller as an *input* variable i. The software controller can influence the environment by sending commands, called *output* variable o to actuators. An actuator influences the value of some characteristics of the environment, call a *controlled* variable c. Variables m and c are called *environment variables*. Variables i and o are called *controller variables*. Finally, a controller has its own internal state variables to perform computations. In this case study, we use EVENT-B state variables to represent both environment and controller variables. We do not model sensor or actuator failures.

A well-known architecture of a control system is a control loop that reads all input variables at once, at a given moment, and then computes all output variables in the same iteration. But, it can be also viewed as a continuous system that can be interrupted by any change in the environment represented by a new value sent by a sensor. In this paper, we see the controller as a distributed system; each sub-system is associated to a given sensor. In that case, the system reacts to each single modification of the sensor. This approach can be seen as a more abstract approach, as it is common in the EVENT-B style of system modeling. We define one event for each input variable change, which allows for a more modular specification that is easier to prove. This is closer to an interrupt-driven control system. Our EVENT-B abstraction is also a reasonable abstraction for a control loop, considering that in most cases, a single input variable changes between two control loop iterations. The control loop can be derived from our specification by merging all events and defining priorities between events.

3   Model Details

This section briefly describes the main modeling elements that characterize our specification. The complete archive of the EVENT-B project is available in [6]. Let us note that the development of our model (EVENT-B components, proofs, animation, etc.) took about two months including the different exchanges with the authors of the case study. Our model contains 4 contexts and 4 machines/refinements. Table 1 relates the components of our model with the requirements listed in [4]. As one can remark, some requirements are modelled as invariants whereas others are dealt with in the adequate events. We chose to do not model some requirements as invariants because this would make the modeling and the proof activities more complex and difficult. Requirement SCS-41 for example: "... *the self-test of the radar system is restarted every 10 min*" is modeled by a variable *nextTest* that is set to the current time plus 10 min in the events that represent the movement of the key and the progression of the time because the self-test of the radar system should be performed at the start (the key is in the ignition position) and then every 10 min. Modeling this requirement as an invariant would require the introduction of two extra variables to store the moments of two consecutive self-test activities, then we have to state that 10 min should be elapsed between these two moments.

Machine $M0$ models the current speed of the studied car independently from any preceding vehicle and also without giving any condition on its evolution. This machine defines the following unique invariant:

$$currentSpeed \in rangeSpeed$$

where *rangeSpeed* denotes a constant defined in the context $C0$ to set the range values for the speed ($rangeSpeed = 0..5000$). Machine $M0$ defines a unique event *updateVehicleSpeed* to set the current speed of the car as follows:

Event updateVehicleSpeed $\widehat{=}$
   any
      val
   where
      grd1: $val \in rangeSpeed$
   then
      act1: $currentSpeed := val$
   end

Machine $M1$ introduces the physical elements that are manipulated by the driver and that have an impact on the current speed of the car. These elements include gas/brake pedal, key, cruise control lever, etc. Machine $M1$ describes how the position of each of these elements evolves depending on its current position. In this same machine, we also introduce the event **progress** that makes the current time keep progressing. Machine $M2$ models the desired speed together with the activation of the normal/adaptive cruise control and also the traffic sign detection that has an impact on the value of the desired speed according to the requirements (SCS-36,SCS-39). It is worth noting that some events, like that related to the traffic sign detection, are introduced in $M1$ even if this aspect is really dealt with in Machine $M2$. Indeed, these events need to modify some variables that are introduced in $M1$ and, as noted before, a new event cannot modify a variable defined in a previous refinement level. Machine $M3$ specifies the different aspects that depend on or impact the desired/current speed like speed-dependent safety distance that also depends on the speed of the preceding vehicle but also the faults that can happen on the radar system. The main elements of these EVENT-B components are described hereafter.

3.1   Machine M1: Physical Elements

This machine refines Machine M0 by introducing the different elements that impact the current speed of the car. This includes the physical elements that

Table 1. Cross-reference between the components of our model and the requirements of [4]

Requirements [4]	Component	Invariant/event
SCS-1, SCS-31	M2	$inv4$
SCS-2, SCS-31	M2	$inv5$
SCS-3, SCS-12, SCS-13, SCS-16, SCS-17, SCS-31	M2	$inv6$ and $inv7$
SCS-4, SCS-19, SCS-31	M2	$inv8$
SCS-5, SCS-19, SCS-31	M2	$inv9$
SCS-6, SCS-19, SCS-31	M2	$inv10$ and $inv11$
SCS-7, SCS-19, SCS-31	M2	$inv12$
SCS-8, SCS-19, SCS-31	M2	$inv13$
SCS-9, SCS-31	M2	$inv14$
SCS-10, SCS-31	M2	$inv15$
SCS-11, SCS-31	M2	$inv16$
SCS-14		Not covered since no information is given on how the system reaches/ maintain the desired speed
SCS-15	M3	$inv13$
SCS-18	M3	$inv24$, $inv25$, $inv26$
SCS-20	M3	$inv12$
SCS-21		Not covered
SCS-22	M3	$inv11$
SCS-23	M3	$inv14$, $inv15$ and $inv16$
SCS-24	M3	$inv17$
SCS-25	M3	$inv19$
SCS-26	M3	$inv20$
SCS-27-SCS-28		Not covered
SCS-29	M3	Event moveSpeedLimiterSwitch
SCS-30		Not covered since it is related to the interface appearance
SCS-32, SCS-33, SCS-34	M2	$inv21$
SCS-35	M2	$inv26$ and moveSpeedLimiterSwitch
SCS-36, SCS-37, SCS-38, SCS-39	M2	$inv24$
SCS-40 and SCS-41	M2	Event moveKey and progress
SCS-42	M3	$inv13$
SCS-43		Not covered since the light system is not included

the driver manipulates, the radar system that gives the distance to the nearest obstacle but also the time progression since it makes some variables evolve like the desired speed. For that purpose, several variables/invariants are introduced to model how the position of the physical elements evolves depending on its current position. In this paper, we give details about the radar the system, the time progression and also the cruise control lever.

The state of the radar system is modelled by a Boolean variable $rangeRadarState$. This variable is initialized to FALSE since the ignition is Off at the beginning then its state is updated each 10 min. Therefore, we define a variable $nextTest$ to store the moment of the next radar system self-test. These variables are defined by the following invariants:

$$keyState = KeyInIgnitionOnPosition \wedge$$
$$keyStateP \neq KeyInIgnitionOnPosition$$
$$\Longrightarrow$$
$$nextTest = currentTime + 6000$$

where $KeyState$ is a variable representing the position of the key ($\{NoKeyInserted, KeyInserted, KeyInIgnitionOnPosition\}$). This invariant expresses that the state of the radar system is checked 10 min after the state ($keyState = KeyInIgnitionOnPosition$). Let us remark the value of 6000 is equal to (10×600) since we choose a progression time step of a tenth of a second because some data in the case study are with 0.1 precision as depicted by the following **progress** event that models the time progression:

Event progress $\widehat{=}$
refines updateVehiculeSpeed
    any
        val
        radstate
    where
        grd1: $keyState \neq KeyInIgnitionOnPosition \vee$
            $nextTest \neq currentTime + 1 \implies radstate = rangeRadarState$
        grd2: $keyState = KeyInIgnitionOnPosition \wedge$
            $nextTest = currentTime + 1 \implies radstate \in BOOL$
        ....
    then
        act1: $currentTime := currentTime + 1$
        act2: $rangeRadarState := adstate$
        act3: $nextTest := \{TRUE \mapsto 6000, FALSE \mapsto nextTest\}$
                  ($\textbf{bool}(keyState = KeyInIgnitionOnPosition \wedge$
                                $nextTest = currentTime + 1))$
        ...
    end

Guard $grd1$ specifies that when the time progresses to the next self-test moment ($nextTest = currentTime + 1$) and the stating of the system ($keyState = KeyInIgnitionOnPosition$), the state of the radar system is chosen randomly ($rdstate \in BOOL$) otherwise its state remains the same ($radstate = rangeRadarState$ in $grd2$).

Similarly, cruise control lever is modeled by the variable $SCSLeverUD$ and its typing invariant: $SCSLeverUD \in SCSLeverPositions$ where $SCSLeverPositions$ is a given set defined in Context $C1$ seen by M_1:

$partition(SCSLeverPositions, Upward, Downward,$
$$\{Backward\}, \{Forward\}, \{Neutral\})$$
$partition(Upward, \{Upward5\}, \{Upward7\})$
$partition(Downward, \{Downward5\}, \{Downward7\})$

For each of these elements, invariants are defined in Machine $M1$ to specify the authorized position changes together with the event that models them. The following invariant states that the cruise control level cannot directly move from an $Upward$ position to a $Downward$ position bypassing the $Neutral$ position. As we can remark, the above invariant uses an extra variable $SCSLeverUDP$ to model the previous position of the cruise control level. In the next section, we show that this kind of variables is also relevant for modeling some requirements that need to make reference to the current and previous states of the system.

$SCSLeverUDP \neq Neutral \implies$
$\quad SCSLeverUD = SCSLeverUDP$
$\qquad \vee$
$\quad (SCSLeverUDP \in Upward \wedge SCSLeverUD \in Upward)$
$\qquad \vee$
$\quad (SCSLeverUDP \in Downward \wedge SCSLeverUD \in Downward)$
$\qquad \vee$
$\quad SCSLeverUD = Neutral$

Machine $M1$ defines event moveSCSLeverUD that models the cruise control level movements where $grd2$ permits to make the invariant preserved after the observation of this event:

Event moveSCSLeverUD $\widehat{=}$
   any
      valSCS
   where
      grd1: $valSCS \in Upward \cup Downward \cup \{Neutral\}$
      grd2: $SCSLeverUD \neq Neutral \implies$
       $(SCSLeverUD \in Upward \wedge valSCS \in Upward)$
       $\vee$
       $(SCSLeverUD \in Downward \wedge valSCS \in Downward)$
       $\vee$
       $(valSCS = Neutral)$
   then
      act1: $SCSLeverUD := valSCS$
      act2: $SCSLeverUDP := SCSLeverUD$
      ...
  end

3.2 Machine M2: Desired Speed

This machine describes how the desired speed evolves according to the requirements (SCS-1 to SCS-12) by moving the cruise control level into different positions. We also model the activation of the normal/adaptive cruise control as described in the document. In addition, we specify the speed limit requirements (SCS-29 to SCS-34) because the calculation of the current speed must respect such a limit.

Mainly, this machine introduces some additional variables to model the desired speed (*desiredSpeed*) and the normal/adaptive cruise control (*normContr* and *adapContr*) with their associated variables to represent their previous values. For instance, the following invariant defines the activation of the normal cruise control:

$$normContr = TRUE$$
$$\Leftrightarrow$$
$$((SCSLeverFB = Forward \wedge SCSLeverFBP \neq Forward \wedge$$
$$(currentSpeed \geq 2000 \vee desiredSpeed \neq 0))$$
$$\vee$$
$$(normContrP = TRUE \wedge SCSLeverFB \neq Backward)) \wedge$$
$$cruiseControlMode = 1 \wedge brakePedal = 0$$

The invariant states that, if the normal mode is selected for the cruise control and the brake pedal is not activated, the normal cruise control is activated the first time when the cruise control level moves to the forward position while the current speed is greater than 200 km/h and or the desired speed is not null and remains activated as long as the cruise control level is not put in the Backward position.

To model the desired speed whose evolution depends on the time, we store the last time (*lastTimeSCSLeverUD*) when the cruise control level has been in the Up/down positions. Thus requirements SCS-4 and SCS-7 are modeled as follows. Requirement SCS-4 specifies that, while the cruise control is activated, the desired speed increases by 1 the first time the cruise control level is put in position *Upward5* whereas Requirement SCS-7 states that the desired speed continues to increase by 1 by each second as long as the cruise control level stays in that position for more than 2 s. Variable *lastdesiredSpeed* represents the desired speed when the lever has been moved into a given position.

$$SCSLeverUDP \neq Upward5 \wedge SCSLeverUD = Upward5$$
$$\wedge$$
$$(adapContrP = TRUE \vee normContrP = TRUE)$$
$$\implies$$
$$desiredSpeed = min(\{200, desiredSpeedP + 1\})$$

and

$$(normContr = TRUE \vee adapContr = TRUE) \wedge SCSLeverUD = Upward5$$
$$\wedge$$
$$currentTime - lastTimeSCSLeverUD \geq 20$$
$$\implies$$
$$desiredSpeed =$$
$$min(\{200, lastdesiredSpeed + (currentTime - lastTimeSCSLeverUD - 10) \div 10\})$$

Let us give more explanation about the last invariant. Expression $(currentTime - lastTimeSCSLeverUD - 10)$ permits to update the desired speed immediately after 2 s, this is why we subtract 10 units of time and not 20. As stated before, as we chose a progression step of tenth of a second, we must divide by 10 each data related to the time. To make these invariants preserved, we have refined the moveSCSLeverUD event according to Requirement SCS-4 but also the progress event with respect to Requirement SCS-7. Event progress for instance is refined by adding the following guard that calculates the new desired speed:

$$(normContr = TRUE \lor adapContr = TRUE) \land SCSLeverUD \neq Neutral$$
$$\Longrightarrow$$

$despeed=$
$\quad \{TRUE \mapsto$
$\qquad \{TRUE \mapsto$
$\qquad\quad \{TRUE \mapsto min(\{200, lastdesiredSpeed+$
$\qquad\qquad ((currentTime + 1 - lastTimeSCSLeverUD - 10) \div 10)\}), //(*\text{case } 2*)$
$\qquad\quad FALSE \mapsto min(\{200, lastdesiredSpeed+$
$\qquad\qquad ((currentTime + 1 - lastTimeSCSLeverUD) \div 2)\}) //(*\text{case } 3*)$
$\qquad\quad \} (\textbf{bool}(SCSLeverUD = Upward5)),$
$\qquad FALSE \mapsto$
$\qquad\quad \{TRUE \mapsto max(\{10, lastdesiredSpeed-$
$\qquad\qquad (currentTime + 1 - lastTimeSCSLeverUD - 10) \div 10\}), //(*\text{case } 4*)$
$\qquad\quad FALSE \mapsto max(\{10, lastdesiredSpeed-$
$\qquad\qquad ((currentTime + 1 - lastTimeSCSLeverUD) \div 2)\})//(*\text{case } 5*)$
$\qquad\quad \}(\textbf{bool}(SCSLeverUD = Downward5))$
$\qquad \}(\textbf{bool}(SCSLeverUD \in Upward)),$
$\quad FALSE \mapsto desiredSpeed //(*\text{case } 1*)$
$\quad \}(\textbf{bool}(currentTime + 1 - lastTimeSCSLeverUD \geq 20))$

The above guard distinguishes different cases according to the position of the control lever and the time elapsed since its last position change($currentTime + 1 - lastTimeSCSLeverUD \geq 20$). The term ($currentTime + 1$) denotes the after-value of $currentTime$ when Event progress is observed. The following cases have been distinguished:

1. if the time elapsed from the last movement of the lever is less than 2 s then, the desired speed does not change (case 1), otherwise
2. if the lever is at the $Upward5$ position, the desired speed increases by 1 every second (10 × tenth of a second): SCS-7, case 2. otherwise the lever is in the $Upward7$ position and the desired speed increases to the next ten's place after each 2 s: SCS-8, case 3.
3. if the lever is at the $Downward5$ position, the desired speed decreases by 1 every second (10 × tenth of a second): SCS-9, case 4. otherwise the lever is in the $Upward7$ position and the desired speed increases to the next ten's place after each 2 s: SCS-10, case 5.

Let us note that the EVENT-B method and its underlying language is not well-adapted to model the evolution of the speed vehicle according to its acceleration/speed and the time passing. Indeed, since the language does not support real numbers, we model the current speed as an integer amount that evolves according to the usual equation $(V = \gamma \times t + V_p)$ where the γ represents the acceleration/deceleration of the vehicle, $(t = 1)$ the time progression and V_p the previous speed. As our time progression is by a tenth of a second, the progression of the speed is very small, that is, less than one kilometer. This progression can not be taken into account using the B language. To overcome such a limit, we proceed as follows. We do not include the increasing/decreasing of the current speed in the event that makes the time progress but we introduce a new event setSpeed that sets the current speed to a given value. This also permits to play and produce the scenarios provided in the case study. Another alternative to overcome the lack of reals in the EVENT-B language is to define or reuse an existing theory plugin that models them [11]. However, this will make the development and the proofs more complex since the interactive prover of Rodin does not adequately support such a concept, that it a proof that uses a theory can not be saved.

3.3   Machine M3: Other Elements

In this level, we model the different aspects that depend on or impact the desired/current speed, like speed-dependent safety distance and the speed of the preceding vehicle. Moreover, we model the faults that can happen on the radar system. Machine $M3$ introduces two new events turnHead and VehicHeadDetect to model respectively the selection of a safety level by turning the cruise control lever head and the detection of a preceding vehicle by catching its speed that is relevant for determining the speed-dependent safety distance and also to make the system decelerates if it is necessary. Event VehicHeadDetect for instance is specified as follows:

Event VehicHeadDetect $\hat{=}$
  any
      val stv brk secdis speh
  where
      grd1: $val \in rangeRadarSensorValues$
      grd2: $rangeRadarState = FALSE \Leftrightarrow val = 255$
      grd3: $speh \in rangeSpeed$
      grd4: $speh \leq 200 \wedge speedOfHead > speh \wedge speh \neq 0 \wedge$
        $adapContr = TRUE \wedge val \notin \{0, 255\}$
        $\Longrightarrow$
        $secdis = 25 \times currentSpeed \div 360$
      grd5: $speh = 0 \wedge currentSpeed = 0 \wedge adapContr = TRUE \wedge$
        $val \notin \{0, 255\}$
        $\Longrightarrow$
        $secdis = 2$
      grd6: $speedOfHead < speh \wedge speh \neq 0 \wedge speh \leq 200 \wedge$
        $adapContr = TRUE \wedge val \notin \{0, 255\}$
        $\Longrightarrow$
        $secdis = 30 \times currentSpeed \div 360$

grd7: $speh > 200 \wedge adapContr = TRUE \implies secdis = safetyDistance \times currentSpeed \div 360$

grd8: ...

then

act1: $rangeRadarSensor := val$

act2: $speedOfHead := speh$

act3: securedistanceToHead:= secdis

act4: ...

end

Event parameter *val* represents the distance between the studied car and a possible preceding vehicle as provided by the radar. Guard *grd2* states that such a value should be equal to 255 if the radar system is not ready. Guards *grd4-grd7* permit to calculate the new value for the speed-dependent safety distance according to the requirements SCS-23 and SCS-24 with the event parameter *speh* denoting the speed of the preceding vehicle.

Already existing events of $M2$ are refined in $M3$ in similar way by calculating the value of the different variables. For instance, the desired speed should be updated when a traffic sign is detected, the speed-dependent safety distance is updated when the current speed is modified or the speed of a preceding vehicle changes. More details can be found in [6].

4   Validation and Verification

To ensure the correctness and validate the built EVENT-B models, we have proceeded into three steps detailed hereafter.

4.1   Model Checking of the Specification

We used the PROB tool as a model checker in order to ensure that all the invariants of each machine are preserved after the observation of each event, that is, there is no sequence of events that makes an invariant not satisfied. Basically, when an invariant becomes violated, PROB exhibits such a sequence of events that, starting from a valid initial state of the machine, leading to a state that violates the related invariant. Such specification errors can be due to a guard/action missing, to an incorrect specification of the invariant but sometimes also to an incorrect property, that is the system really does not satisfy the property. Let us note that even if no invariant violation is found by the tool, there may still exist scenarios that violate the invariant that the tool cannot find due to their complexity or/and the timeout on the model checking process. This is why a proof phase should be performed to ensure that the specification is invariant-violation free.

4.2   Validation with Scenarios

This step aims at verifying that we have built the right model whose behaviors conform to the desired ones as described by the scenarios of the specification document. For that purpose, the animation capability of PROB is used to play the different scenarios provided in the case study. This step allows us to point out some flaws/ambiguities in the initial release of the description document. For instance, the initial examples provided to illustrate the requirements SCS-5-SCS-9 were incorrect with respect to the requirements. In addition, in some place like SCS-7-SCS-9, the term "target speed" is used instead of "desired speed", etc. All these aspects have been discussed with the case study authors because we are not specialists of the domain. Let us note that we have faced some difficulties to play the provided scenarios since no information is provided on how the controller calculates the acceleration at each step. So, we have made our best to "simulate" these values without any representation about their suitability, reliability.

4.3   Proof of the Specification

This last phase aims at ensuring the correctness of the specification by discharging all the proof obligations generated by RODIN to prove that the invariants are preserved by each event, but also that the guard of each refined event is stronger than that of the abstract one. Figure 1 provides the proof statistics of the case study: 579 proof obligations have been generated, of which 60% (345) were automatically proved by the various provers. The remaining proof obligations were discharged interactively since they needed the use of external provers like the Mono Lemma prover that has shown to be very useful for arithmetic formulas even if we had to add some theorems on min/max operators (a min/max of a finite set is an element of the set, etc) but also on the transitivity property of the comparison operator ($\geq$, $\leq$, etc.).

5   Other Points

This section reports on some points about the choices made during the Event-B modeling of the speed control system.

5.1   Feedback on the Specification Document

The formal modeling of the specification document [4] lead us to question ourselves about the semantics of some requirement and identify a number of ambiguities and some contradictions with the test scenarios provided. Being not specialist of the domain, we have communicated these to the authors of the requirements document, and a number of revisions were produced, following our comments. Our discussion and exchange lead to the modification/revision of a set of requirements to make them clearer and consistent. A detailed list of these elements are described in the last version (*i.e.*, 1.17) of the requirements document:

Element Name	Total	Auto	Manual	Reviewed	Undischarged
SCS_012020	579	345	234	0	0
C0	0	0	0	0	0
C1	0	0	0	0	0
C2	18	15	3	0	0
C3	0	0	0	0	0
M0	2	2	0	0	0
M1	78	72	6	0	0
M2	215	104	111	0	0
M3	266	152	114	0	0

Fig. 1. RODIN proof statistics of the case study

1. Correction of the examples in SCS-7, SCS-8 and SCS-9 since the values do not respect the requirements.
2. Modification of signal description *setVehicleSpeed* to make its meaning clearer.
3. Replacing 'target speed' by 'desired speed' in requirements SCS-7 and SCS-8.
4. Adjustment of the maximum acceleration and deceleration values in SCS-20, SCS-22.
5. Stating that SCS-23 applies when the speed is 20 km/h or below.
6. Clarification of priority between adaptive cruise control and emergency braking assistant in case of brake activation in SCS-28
7. the signal *SCSLever* has been splitted into signals *SCSLeverForthBack* and *SCSLeverUpDown* with their corresponding positions (states) and the possible transitions between them.

   As already well-known, the use of a formal method does not only permit to built a correct system but it also allows to make the requirement document clearer and precise by removing ambuities and errors.

5.2 Modeling Temporal Properties

As stated before, a number of requirements refer to the current and previous state of an element. In order to be able to verify these requirements using a proof strategy, we modeled them as invariants by introducing two variables for each element to store their current and previous values. The obtained specification is quite cumbersome especially that we have to add for each event that does not modify a variable that its previous value is equal to its current value. We think that it would be interesting to investigate existing tools/approaches that could help us specify this kind of properties in a simpler manner. An example of such tools is the EVENT-B State machines plugin[1] that produces EVENT-B

---

[1] http://wiki.event-b.org/index.php/Event-B_Statemachines.

events from a state machine including their guards that specify the requirements modeled by the state machine but without producing the related invariants. This plugin makes difficult to trace and justify the usefulness of the generated guards.

6 Conclusion

This paper presents a formal modeling proposal of a speed control system using the EVENT-B method. We have modelled most of requirements that permits us to point out some ambiguities in the requirements that we have discussed and clarified with the case study authors by rephrasing them. These ambiguities have been discovered during during different development phases: formalization, proof and validation using the provided scenarios. This experience has affirmed that the formal modeling of a system helps the software users detect error in early development phase that makes its correction cheaper.

The main difficulty when modeling the speed control system is to determine the order in which elements should be introduced during the refinement especially that many elements are interdependent. Due to time constraints, we were unfortunately not able to explore the different decomposition plugins of RODIN that might produce smaller specification parts that would be easier to understand and maintain. We plan to explore some decomposition techniques as future work even if we really think that the EVENT-B method should include modularization clauses as native structuring mechanisms like those of the B method that permit to have a modular specification since the early development phases to make EVENT-B method more usable for the development of big and complex systems. Another point concerns the PROB plugin under Rodin that unfortunately does not permit to store an already played scenario, so we are obliged to manually replay each scenario; this is a very time-consuming for long traces.

The work presented in this paper can also be extended by considering the remaining requirements that need more clarifications. Requirement SCS-21 for instance needs more information on how the system can deduce that deceleration of $3\,\mathrm{m/s^2}$ is insufficient to prevent a collision without having any information about the acceleration of the preceding vehicle. Also, we think that more information should be provided on the internal variables like *setVehicleSpeed* that represents the automatic acceleration of the system in order to able to build a more complete system. Finally through the different case studies proposed in the ABZ conference [8,9], we are now convinced of the need to improve the EVENT-B language to make it supports the real numbers as basic types. Its prover should be also extended to include more rules on arithmetic and set theories.

Acknowledgements. The authors would like to thank the case study authors, and Frank Houdek in particular, for his responsiveness and useful feedback during the modeling process when questions were raised or when ambiguities were found. The authors would also like to thank Michael Leuschel for his quick feedback on using PROB for this large case study.

References

1. Abrial, J.: The B-Book - Assigning Programs to Meanings. Cambridge University Press, Cambridge (1996)
2. Abrial, J.: Modeling in Event-B. Cambridge University Press, Cambridge (2010)
3. Event-B Consortium. http://www.event-b.org/
4. Houdek, F., Raschke, A.: Adaptive exterior light and speed control system, November 2019. https://abz2020.uni-ulm.de/case-study#Specification-Document
5. Leuschel, M., Bendisposto, J., Dobrikov, I., Krings, S., Plagge, D.: From animation to data validation: the prob constraint solver 10 years on. In: Boulanger, J.L. (ed.) Formal Methods Applied to Complex Systems: Implementation of the B Method, Chap. 14, pp. 427–446. Wiley ISTE, Hoboken (2014)
6. Mammar, A., Frappier, M.: Modeling of a Speed Control System using Event-B, January 2020. http://www-public.imtbs-tsp.eu/~mammar_a/SpeedControl.html
7. Mammar, A., Frappier, M., Laleau, R.: An Event-B Model of an Automotive Adaptive Exterior Light System, January 2020. http://www-public.imtbs-tsp.eu/~mammar_a/LightControlSystem.html and http://info.usherbrooke.ca/mfrappier/abz2020-ELS-Case-Study/
8. Mammar, A., Frappier, M., Tueno Fotso, S.J., Laleau, R.: An EVENT-B model of the hybrid ERTMS/ETCS level 3 standard. In: Butler, M., Raschke, A., Hoang, T.S., Reichl, K. (eds.) ABZ 2018. LNCS, vol. 10817, pp. 353–366. Springer, Cham (2018). https://doi.org/10.1007/978-3-319-91271-4_24
9. Mammar, A., Laleau, R.: Modeling a landing gear system in Event-B. In: Boniol, F., Wiels, V., Ait Ameur, Y., Schewe, K.-D. (eds.) ABZ 2014. CCIS, vol. 433, pp. 80–94. Springer, Cham (2014). https://doi.org/10.1007/978-3-319-07512-9_6
10. Parnas, D.L., Madey, J.: Functional documents for computer systems. Sci. Comput. Program. **25**(1), 41–61 (1995)
11. Su, W., Abrial, J.R., Zhu, H.: Formalizing hybrid systems with Event-B and the Rodin platform. Sci. Comput. Program. **94**, 164–202 (2014)

A Verified Low-Level Implementation of the Adaptive Exterior Light and Speed Control System

Sebastian Krings[1]([✉]) [iD], Philipp Körner[2] [iD], Jannik Dunkelau[2] [iD], and Chris Rutenkolk[2] [iD]

[1] Institute for Information Security, Niederrhein University of Applied Sciences, Mönchengladbach, Germany
`sebastian@krin.gs`
[2] Institut für Informatik, Heinrich-Heine-Universität, Universitätsstr. 1, 40225 Düsseldorf, Germany
{`p.koerner,jannik.dunkelau,chris.rutenkolk`}`@hhu.de`

Abstract. In this article, we present an approach to the ABZ 2020 case study, that differs from the ones usually presented at ABZ: Rather than using a (correct-by-construction) approach following a formal method, we use MISRA C for a low-level implementation instead. We strictly adhere to test-driven development for validation, and only afterwards apply model checking using CBMC for verification. In consequence, our realization of the ABZ case study can serve as a baseline reference for comparison, allowing to assess the benefit provided by the various formal modeling languages, methods and tools.

1 Introduction

The ABZ 2020 Case Study [18] describes two assistants commonly found in modern cars. The overall system consists of two loosely coupled components, namely an adaptive exterior light system (ELS) and a speed control system (SCS). The ELS controls head- and taillights, setting their brightness depending on the surroundings and user preference. At the same time, the SCS controls the vehicle's speed, again by taking into account the environment as well as parameters given by the driver. Obviously, both are safety critical components, rendering safety and security a development priority.

Used Methods and Tools. In this article, we present our implementation of the ABZ 2020 Case Study. Our approach differs from the ones usually followed by the ABZ community: we do not employ a fully formal development method. Instead, we attempted an approach closer to what might happen in industries, where formal methods are not common yet. To do so, we implemented both the ELS and the SCS directly in (MISRA) C, following a test-driven development workflow. Only afterwards, we performed formal verification attempts directly on the C code, using the CBMC model checker [11]. Both MISRA C and CBMC

© Springer Nature Switzerland AG 2020
A. Raschke et al. (Eds.): ABZ 2020, LNCS 12071, pp. 382–397, 2020.
https://doi.org/10.1007/978-3-030-48077-6_30

will be introduced more thoroughly in Sects. 2.1 and 4.2 respectively. Test-driven development and mocking of test objects will be presented in Sect. 2.2.

Rationale. Often, formal methods practitioners claim to hold a high ground over "traditional" software development or at least that there rarely are disadvantages [7,14]. The argument seems convincing; yet, we are not aware of any (case) study comparing two teams working on the same project, one employing a formal approach and the other working "traditionally". For this case study, we aim at providing a baseline that can be compared to fully formal approaches or other approaches combining formal and informal verification, e.g., as suggested for spacecrafts [21]. We opted to postpone verification as much as possible, as one would expect a group focusing on embedded systems to work. This allows a fair evaluation of (dis-)advantages of the individual approaches. Our aim is to examine, whether a rigorous approach is beneficial in the context of the case study. If so, we hope to add to the body of evidence that formal methods actually *are* beneficial compared to "traditional" software development.

Distinctive Features. There are several features rendering our approach unique: Firstly, as the implementation is written in C, it could be directly deployed to an embedded system. Models written in formal specification languages would have to be refined to an implementation level before code can be generated. Furthermore, code generators usually are not proven and might introduce new errors. In cases where code generation is not easily applicable, side-by-side development of code is suggested. However, this approach is error-prone as well.

Secondly, the implementation is close to the actual hardware. Code that interacts with sensors or user input is separated, i.e., it could immediately be linked to actual hardware. Additionally, our implementation makes use of real threads, just as the sub-components of the system would run in parallel. We expect that most specifications using formal methods simply allow some non-determinism concerning the ordering of state transitions. This has some more consequences: Our implementation allows real-time simulation of the system, whereas state transitions using formal methods usually happen instantly and do not amount for any time elapsed during calculations. This also allows usage of our implementation for hardware-in-the-loop tests, which are common for automotive (cf. [13,20]) in order to test the entire system.

Thirdly, MISRA C is a language that stems from the automotive industry. It is a somewhat formal language, in the sense that certain rules are required to be followed. Yet, it is also relatively flexible, since other rules are only advisory.

2   Modeling Strategy and Implementation

In the following section, we will discuss how we approached the initial implementation in C, starting with details on the C dialect we use in Sect. 2.1. Afterwards, the general structure of our implementation is presented in Sect. 2.3, followed by a discussion of the limitations of our approach and implementation in Sect. 2.4.

For the sake of brevity, we will only show small code snippets in this paper. The full model is available at

https://github.com/wysiib/abz2020-case-study-in-c-public.

2.1 MISRA C

MISRA C is a set of development and style guidelines for C, introduced by MISRA, the Motor Industry Software Reliability Association. The standard [1] defines a subset of C meant to be used for safety critical systems, in particular in the automotive sector. In fact, both ISO 26262 [19] and the software specification by AUTOSAR [2] reference or suggest the usage of MISRA C for automotive applications.

The overall goal of MISRA C is to increase both safety and security by avoiding common pitfalls. Thus, the rules prohibit or discourage the use of unsafe constructs, try to avoid ambiguities, and so on. The MISRA C standard distinguishes between three kinds of rules: those that are mandatory, those that are required but could be ignored if a rationale is given and rules that are meant as advisory only. For instance, there is a required rule stating that any switch statement should have a default label and mandatory rule stating that any path through a non-void function should end in a return statement.

While of course all coding rules could be checked by hand in theory, we used cppcheck[1] to verify compliance of our code to most of the MISRA rules. However, given that not all rules can be statically checked the result is only an indication and some manual review is required as well.

Despite its prevalence in the automotive industry, MISRA C has been criticized regarding both efficiency and ease of use. In particular, the possibilities of false positives [17] and of introducing new errors by (unreflectingly) changing code to adhere to the rules [6] should be carefully considered. Both factors again allow for comparison to the formal development methods present at ABZ. Despite the criticism, MISRA C remains the de facto standard in the automotive industry and is used throughout all production code in this case study.

2.2 Test-Driven Development and Mocking

Test-driven Development is an approach to software development, that follows a certain development cycle: before implementing a new feature of fixing an issue, an appropriate test case is formulated and execute [4]. Naturally, the test fails, as no code implementing the scenario has been added yet.

Only afterwards, the code is extended and improved to make the test pass. As a result, a high confidence can be achieved. Furthermore, the test suite developed helps during refactoring later on.

To simplify formulating tests and to allow testing program parts in isolation, mocks can be used. A mock is an object or library that simulates the input and output behavior of program parts [4]. However, rather than implementing

---

[1] http://cppcheck.sourceforge.net.

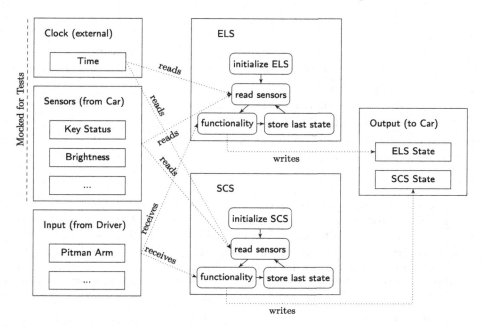

Fig. 1. System architecture and internal communication

the full functionality, a mock is usually much simpler than the code it replaces. For instance, mocks are often supposed to behave deterministically or even to provide constant outputs. For testing purposes, mocks often record the inputs to them and provide them to assertions.

2.3   Code Structure

The overall architecture of our implementation is depicted in Fig. 1. We follow a structure that is fairly similar to the one the specification provides. Since two subsystems are specified, the code is separated into two folders, one for the cruise control and the other for the light system. This is to help ensure that the systems are independent of each other. Shared type definitions, e.g., the pedal deflection, the sensor state enumeration, and shared sensors, are stored separately. An artificial time sensor was introduced for testing, but can easily be replaced by an actual clock.

Each of the subsystems is split into three header files and implementations. The first header file declares the accessible and shared sensors for the subsystem, and contains relevant type definitions. Another header file defines the user interface, e.g., how the pitman arm may be moved or what input the pedals for gas and brakes may yield. The last header file contains definitions for the actuators, i.e., what the system is allowed to do. Only the latter two header files are actually implemented, eventually resulting in three C files:

Table 1. Development time

Task	Time in hours
Basic implementation and code structure	2
ELS implementation, tests and scenarios	30
SCS implementation, tests and scenarios	22
Model checking	3
Refactoring and code cleanup	2
State visualization	6

- A state struct that contains all the data relevant to the subsystem.
- The user interface such that user input can be simulated. This changes some internal variables that keeps track of the state of the UI; in a deployed system, this can be replaced by additional sensors. The attributes correspond to the signals that the subsystem has to communicate.
- The realization of the state machine with several guarded state transitions. This is the actual implementation of the specified safety properties.

For the test cases, sensors are mocked. In order to get an actual executable, real sensors have to be linked during compilation. The time spend for development, validation and verification is given in Table 1.

2.4   Limitations

Due to time constraints, we opted not to implement every single requirement but tried to cover as much as possible. Aside from the emergency brake light, all requirements have been taken into account for the ELS. For the SCS, we implemented about two-thirds of the requirements, up to (including) SCS-28. While it would be nice to have a more complete implementation, we do not think that it would impact our gathered conclusions.

A feature of the requirements that is not addressed satisfyingly are timers. We are convinced that any modern CPU to be used in cars is fast enough to execute an iteration of the state machine withing a reasonable time frame. Thus, any real system realized following our approach should be able to guarantee execution within the smallest time resolution that is relevant to the subsystems and their respective requirements.

Yet, it is hard to give any real-time guarantees. The only evidence that can be given is to run the system often enough and measure whether execution is kept in the specified tolerances. However, this is still better than what we expect of more formal approaches, which usually do not account for wall time at all.

2.5   Formalization Approach

As mentioned earlier, we postponed actual verification work as much as possible. Instead, as our first step, we set up the validation sequences as unit tests first.

Then, in a test-driven development manner, we added to the implementation code by only considering the next assertion in a scenario. Once the test passed, we moved on to the next. In a second step, we added test cases that are directly related to one or sometimes several requirements.

Finally, we set up CBMC and tried to verify the properties described by the requirements. As stated, we use the same code for testing and formal verification, avoiding any translation between formal verification and testing environment as done for instance by Chen et al. [10] and others. However, both approached remain distinct rather than being combined into a single verification procedure [22].

As part of possible future work, we intend to use CBMC to try to provide real-time guarantees and to verify the correct behavior in presence of scheduling and limited by the actual specifications of an embedded device. Both could be verified by providing a Verilog model of the hardware, sensors and connections. Afterwards, co-verification of the implementation in C with the Verilog circuit model can be performed by CBMC [12]. Additionally, we would like to consider other tools that work directly on the C code, e.g., Symbiotic [9] or Klee [8].

3   Model Details

In the following, we will detail our implementation idioms we employed to ensure easier handling and verification of the involved state machine, and explore some crucial snippets of our code to show these idioms in practice. Contrary to the proposed outline, we will present key snippets as well in Sect. 4.

3.1   Idioms

Types. We opted to define all types as enumeration types. This is to be expected for some data types, which are true enumerations, such as:

```
typedef enum {Ready, Dirty, NotReady} sensorState;
```

Yet, we also defined integer types as enumerations, e.g.:

```
typedef enum {
    percentage_low = 0,
    percentage_high = 100
} percentage;
```

The reasons for this are twofold: first, we can easily identify thresholds and the value range for each type. While percentages are straightforward to everyone, e.g., the translation of the steering wheel angle into human-understandable semantics is hard. An excerpt of the corresponding type definition is as follows (analogously for turning the steering wheel to the right):

```
typedef enum {
    st_calibrating = 0,
    st_hard_left_max = 1,    /* 1.0 deg */ st_hard_left_min = 410,
    st_soft_left_max = 411, /* 0.1 deg */ st_soft_left_min = 510,
    st_neutral_maxl = 511, st_neutral = 512, ...
} steeringAngle;
```

Such a type definition renders it easier to identify, e.g., in what direction the steering wheel is turned and how far, i.e., To check if it is turned far to the left, `st_hard_left_max <= angle && angle <= st_hard_left_min` can be used.

C behavior is undefined if a value that is out of range of the corresponding enumeration is passed. Thus, our second intention was that model checking tools could easily deduce the actual value range and do not consider, e.g., the full range of 32-bit integers in their stead. This will be discussed further in Sect. 4.3.

Do Not Expose Mutability. It is easy to write broken code when using mutable structs, especially if they are used in order to communicate between threads. Instead, we pass *values* to and from interface functions. This means, that values are copies of the data which are not referenced from anywhere else in the program and the receiver may do however they please with it. An example is that the state from the light sub-system can be queried (for test cases). The returned value will never change unless the test case chooses to do so; no action in the ELS influences it. This also allows reading multiple output variables consistently.

On the other hand, *internal* variables that may change frequently, which are not meant to be read by anyone else, are declared as local (using the `static` keyword). They are always stored in the same "place" and may not be exposed; in particular, there are no getter functions for these variables.

3.2 Timers

When writing code that takes time into account, one is easily tempted to access the current time provided by the operating system. This is a bad idea when such time properties shall be tested: then, tests would have to be enriched with additional sleep statements in order to achieve proper timing for the situation under test.

Instead, we introduced an artificial sensor that may be accessed by both sub-systems. The sensor reports the current time in milliseconds, comparable to a common unix timestamp. During test cases, this sensor is mocked and some artificial time is provided. The code does not know anything about time, but just reads a sensor returning an integer value.

The implementation only assumes that one cannot go back in time, no further assumptions regarding the progression of time are made. In consequence, the step functions can simply be called in a continuous loop, independent of the computing speed and time needed for a single iteration. On fast hardware, there might even be several executions within the same timestamp (e.g., if the resolution is milliseconds) or timestamps might pass without an execution following (e.g., when using nanoseconds). Mocking the sensor also has the advantage that test scenarios, that would take several minutes of wall time, can be executed in milliseconds instead.

If the entire piece of software was to be shipped, it would be trivial to swap out the sensor: One only has to link an implementation that provides the real time, which may be the provided by the operating system.

4   Validation and Verification

We tried to validate our implementation throughout the whole development process by using test-driven development, as we will discuss in Sect. 4.1. In addition, we used the CBMC model checker to fully verify different properties of our implementation directly on the C code as we will describe in Sect. 4.2.

4.1   Test-Driven Development Using Cmockery

We used test-driven development based on the provided scenarios. For this, we rely on Google's cmockery library[2], which provides a unit testing framework and allows mocking functions. Since we did not want to execute all tests in real-time, we mocked functions that extract sensor data as well as the current time in our test cases.

We used two different kinds of test cases for a first quick validation:

– The provided scenarios were automatized and used as integration tests.
– In addition, we implemented unit tests for all requirements given in the specification document. Of course, each unit test only covers a minimal scenario that shows how the requirement is supposed to be understood and automatizes the verification of that single scenario.

A snippet taken from the test case of the requirement ELS-3 is show in Listing 1. The system is initialized to belong to an EU-based car with left-hand drive and without any extras such as ambient light. Initialization and assertions regarding the correctness of the initial state are not shown in the snippet. Afterwards, in lines 2 to 9, we update the sensors to the values they should hold at the start of the test scenario and the code setting up the mocked functions is called. In particular, we set the time sensor that is used to simulate the actual clock as described in Sect. 3.2. Overall, the test setup phase ensures that our artificial sensors inside the mock report the required values if and when the system reads them.

Line 10 shows the difference between sensors and driver interaction: While sensors have to be mocked in order to simulate an actual system, user input is given directly. This corresponds to what will happen in an actual car: the system has to react to user input immediately and at any time, while it can read sensor data arbitrarily.

---

[2] https://github.com/google/cmockery.

Listing 1. Test of Requirement ELS-3

```
1   // ignition: key inserted + ignition on
2   sensor = update_sensors(sensor, sensorTime, 1000);
3   sensor = update_sensors(sensor, sensorBrightnessSensor, 500);
4   sensor = update_sensors(sensor, sensorKeyState, KeyInIgnitionOnPosition);
5   sensor = update_sensors(sensor, sensorEngineOn, 1);
6
7   mock_and_execute(sensor_states);
8
9   sensor = update_sensors(sensor, sensorTime, 2000);
10  pitman_vertical(pa_Downward5);
11  mock_and_execute(sensor_states);
12
13  assert_partial_state(blinkLeft, 100, blinkRight, 0);
14  pitman_vertical(pa_ud_Neutral);
15  sensor = update_sensors(sensor, sensorTime, 2000);
16  mock_and_execute(sensor);
17
18  pitman_vertical(pa_Upward7);
19
20  progress_time_partial(2000, 2499, blinkLeft, 100, blinkRight, 0);
21  progress_time_partial(2500, 2999, blinkLeft, 0, blinkRight, 0);
22
23  int i;
24  for (i = 3; i < 6; i++) {
25      progress_time_partial(i * 1000,        i * 1000 + 499,
26                            blinkLeft, 0, blinkRight, 100);
27      progress_time_partial(i * 1000 + 500, i * 1000 + 999,
28                            blinkLeft, 0, blinkRight, 0);
29  }
```

Line 13 asserts that the left blinker is on 100% and the right one is on 0% once the step function was executed after the user input was given. We use assert_partial_state, since we only make an assertion regarding the two variables blinkLeft and blinkRight, rather than making an assertion over all state variables.

Finally, Lines 20–21 as well as 25–28 assert that for each millisecond in the time interval, the provided values remain the same, i.e., that the step function does not change output values during that time frame.

As can be seen, we have implemented different C macros to simplify test case development:

- assert(_partial)_state which checks if the internal states of ELS and SCS correspond to given assertions. The assertions can specify the state both partially, as done in the listing, and fully.
- progress_time(_partial) combines assertions on the state with a progression of time as reported by the time sensor.

Validation Results. As expected, using unit and integration testing as parts of a test-driven development workflow helped us during the initial development. Using test-driven development provided the usual benefits:

- having to formulate test cases helped us gain an understanding of the requirements and how they are supposed to work,
- refactoring was made easier and more secure, and
- the implementation was closer to the actual specification from the start.

The fact that we are working with an actual implementation made test-driven development come naturally. However, different ways of combining formal methods with test-driven development have been discussed [3] as well. In addition, developing specifications using continuous testing has been suggested for former ABZ case studies in the context of the B method [15,16].

Influences on Code. Using the macros above, our initial design of splitting sensors, user input and actuators did not have to be adapted further to be testable. Yet, it created a vast amount of code entirely dedicated to testing. Of 5223 source code lines (which also contain a Makefile and code for state (graph) visualization), 3786 lines are test code. Comments and blank lines are already excluded.

4.2   Model Checking Using CBMC

As stated above we used CBMC [11] to verify properties of our implementation directly on the MISRA C code. CBMC is a model checker for programs written in C. It uses bounded model checking [5] to verify a default set of properties, mostly related to common programming errors, such as: memory safety, including bounds checks and pointer safety, occurrence and treatment of exceptions, and presence of undefined behavior due to C quirks.

Additionally, it can be used to verify user-given assertions stated as C-style assertions using the macros in `assert.h`. Depending on where they are placed in the code, they correspond to different kinds of properties commonly used in state-based formal methods:

- If placed at the end of the loop implemented by the ELS and the SCS state machines depicted in Fig. 1, assertions correspond to safety invariants that have to hold in every state reachable by one of the subsystems.
- If placed anywhere inside the loop, assertions can be used as invariants on intermediate states.
- If placed outside the loop, we can check if properties hold after a certain number of iterations (controlled by CBMC's unrolling preferences).
- By using additional variables, we can communicate between states and implement a lightweight verification of temporal properties. Of course, this is not as powerful as LTL or CTL, as we have to rely on unrolling.

392   S. Krings et al.

Listing 2. Partial CBMC Output

```
State 59 file light/light-impl.c line 242 function light_do_step thread 0
------------------------------------------------------
  ks=/*enum*/NoKeyInserted (00000000000000000000000000000000)

State 63 file light/light-impl.c line 242 function light_do_step thread 0
------------------------------------------------------
  ks=/*enum*/KeyInIgnitionOnPosition (00000000000000000000000000000010)

State 65 file light/light-impl.c line 244 function light_do_step thread 0
------------------------------------------------------
  engine_on=FALSE (00000000)

State 69 file light/light-impl.c line 244 function light_do_step thread 0
------------------------------------------------------
  engine_on=TRUE (00000001)
```

4.3   Example: Verification of ELS-22

Requirement ELS-22 is a great example for an invariant. It states "Whenever the low or high beam headlights are activated, the tail lights are activated, too". For this, we can add an assertion such as:

```
assert(implies(get_light_state().lowBeamLeft > 0,
               get_light_state().tailLampLeft > 0 ||
               get_light_state().tailLampRight > 0));
```

The disjunction in the second part of the implication is important for American cars: as tail lamps are used for indicators, it is accepted behavior if one tail lamp is temporary deactivated during a flashing cycle. When running CBMC, it immediately came up with a counterexample. A snippet can be found in Listing 2.

The counterexample shows how the two system variables ks, i.e., the key state, and engine_on, i.e., the engine's ignition state, change while our main step function light_do_step is executed.

The main issue with such a counterexample is that each variable assignment, function call and return from a function introduces a new state. While this representation mimics the internal workings of the C code, it does not correspond to the mental model: comparable to common state-based formal methods, we regarded a state change to include multiple variables at once.

Hence, as we were only interested in comparing state variables per full iteration of light_do_step, the output was barely readable to us (the counterexample consists of more than 200 lines).

CBMC can optionally reduce the output by removing assignments that are unrelated to the property. This did not work well for us, as the assignment of signals for the low beam headlights was removed as well. We ended up manually

Table 2. Example trace violating ELS-22.

State variable	Iteration 1	Iteration 2
key_state	NoKeyInserted	KeyInIgnitionOnPosition
engine_on	FALSE	TRUE
all_doors_closed	FALSE	TRUE
brightness	0	37539
speed	0	936
daytime_light_was_on	FALSE	TRUE
low_beam_left	0	100
low_beam_right	0	100
last_engine	FALSE	TRUE
last_key_state	NoKeyInserted	KeyInIgnitionOnPosition
last_all_door_closed	FALSE	TRUE

writing state variables in a spreadsheet to comprehend the scenario. A (condensed) version can be found in Table 2. Here, the state changes between two full iterations of our step function are shown, rather than changes of individual variables during the execution. This representation aligned better to our mental model of the implementation and was thus more helpful for debugging.

The error in our code was that, based on ELS-17, only the low beam headlights were activated due to activated daytime running light. This was not uncovered by the test scenarios, since daytime light was only tested by night, where, coincidentally, other triggers activated the tail lamps.

Verification Results. However, the assertion still failed to verify. Upon further analysis of the property, we discovered a conflict between ELS-22 and hazard blinking in Canadian and US cars. In those cases, hazard blinking deactivates both tails lights for the dark cycle, thus violating the property. We extended our assertion by checking our variable for blinking direction beforehand:

```
assert(implies(blinking_direction != hazard, /* old assertion */));
```

Afterwards, we were able to successfully verify the property using CBMC.

Influences on Code. At first glance, using CBMC only required to add assertions to the code. As assertions are often introduced as part of understanding certain scenarios, this does not change the modeling strategy itself. Yet, CBMC comes with a flaw: it is not able to detect integer ranges given by enumerations. This means it frequently finds errors with values for enumerations, that are out of scope. As a consequence, one has to add assumptions about value ranges to the code, which cannot be compiled to actual code. Another assumption that needs to be added is that consecutive timestamps cannot get smaller. Thus, for useful verification, some form of conditional compilation is required.

5   Specification Ambiguities and Flaws

During development, we identified several shortcomings or ambiguities within the specification. These issues were found during analysis of the requirements and during implementing test cases for test-driven development. As we only performed validation steps after implementation, the validation steps just uncovered shortcomings of our own implementation and non-compliances w.r.t. the specification. Due to page limitations, we will only present some of them:

ELS-37 is somewhat broken or at least highlights an incompleteness in the specification. For now, there is no way to discern whether an adaptive cruise control is part of the vehicle; from the specification, we had to assume that it is installed in every system. Then, according to SCS-1, there does not even have to be a desired speed. We think that, in order to make sense at all, it rather should be "is active" than "is part of the vehicle". Also, this is the only part of the specification that refers to an **advanced** cruise control.

ELS-42 does not specify what should happen in case of sub-voltage. The only given information is that the adaptive high beam headlight is not available. Should manual high beam headlight be triggered instead? Should the high beam remain dark? This remains absolutely unclear.

ELS-19 contains a contradiction: first, it states that ambient lighting *prolongs* already active low beam headlights. Later, it says that the headlamps "remain active or *are activated*". We think that some actions are reasonable to activate the headlight even if it was not on before (e.g., opening the doors). Others definitely should not activate the headlight (e.g., if the brightness falls below the specified threshold, as passing cars and the setting sun might trigger the brightness sensor). Also, it does not have any constraints regarding the light rotary switch: if the switch is in the "off" position, we think the ambient light should not activate at all. This requirement needs some serious polishing.

While `currentSpeed` is specified as a sensor in the ELS, it is not clear how the SCS accesses this value. No sensor is provided according to the specification, and only the brake pressure is mentioned as actuator but not the gas pedal. Thus, the SCS as specified appears to only be responsible for determining the desired speed but not for actually deploying it to the current speed? To our understanding, the measured current speed should be a sensor to the SCS, let alone for the possibility to ensure whether more acceleration is required to maintain it or not.

SCS-23 specifies a safety distance of $2.5\,\text{s} \cdot$ `currentSpeed` for the adaptive cruise control when the current speed is below $20\,\text{km/h}$. It further specifies an absolute distance of $2\,\text{m}$ if both vehicles are standing. Assuming `currentSpeed` $\in$ $]0, 2.88[$ however, the safety distance according to SCS-23 is below $2\,\text{m}$ and effectively approaches 0 the closer the vehicle gets to a standstill. But once a standstill is reached, the safety distance is set to $2\,\text{m}$ and thus is violated instantly. It remains unclear whether these $2\,\text{m}$ distance is meant as minimum or is intended to delay the reaction to eventual acceleration of the vehicle in front.

SCS-28 references a maximum deceleration value, which was only described for the adaptive cruise control in SCS-20 and SCS-21. We assume that it references the same maximum deceleration of $5\,\text{m/s}^2$. It further specifies the acoustic

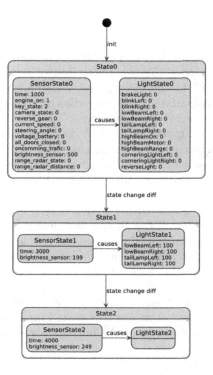

Fig. 2. Ad-hoc visualization

signal which is to be played if the time to reach a standstill with maximum deceleration $(5\,\mathrm{m/s}^2)$ is greater than the time until impact. This acoustic signal however may overlap with the signal specification given in SCS-21.

6   Conclusions

To summarize, we have implemented a low-level version of the ABZ 2020 case study in MISRA C, a language commonly used in the automotive industry. We relied both on common programming techniques such as test-driven development and formal verification using model checking. As we have not followed a fully formal development method, our implementation can serve as a baseline for comparison with the more formal approaches usually presented at ABZ.

We suspect that more rigorous approaches to software development will show both advantages and disadvantages to our approach. In particular, our approach stays close to the actual system and can easily be deployed to an actual car. Furthermore, our code can be used for simulation and hardware-in-the-loop tests.

However, we certainly missed the expressiveness and mathematical clarity that comes with more rigorous approaches. Compared to a formal method, we could only do very lightweight verification of temporal properties and would certainly have favored to be able to model check LTL or CTL properties. Thus, while we were able to verify our implementation to a certain degree, we suspect that a more thorough approach would be able to provide stronger guarantees.

In particular, our approach has only very limited support for verifying temporal properties (i.e., just by unrolling properties to a certain degree). Furthermore, we currently do not validate any properties on time constraints aside from simulating an external clock in the test cases.

That aside, all state properties given in the specification could in theory be verified using our approach even though we have not fully implemented all of them. Furthermore, given that we can place assertions everywhere in our C source code, we could reason about intermediate states as well.

Method and Tool Review. We are surprised how easy it was to implement the case study in C, especially as none of the authors is a professional C developer. While we were unsure during implementation, given our test harness and the results of CBMC, we now have more confidence in the correctness of our implementation.

CBMC was a great tool that found counterexamples, e.g., to the requirement ELS-22. Yet, we have to make the following observations: first, the output was barely readable, i.e., 52 state transitions represent two high level states after the initialization. As a result, we wrote our own state graph visualization tool based on plantuml[3] (cf. Fig. 2). Second, for the initial error, a simple assertion would already have tripped the test case.

The majority of our time, we spent implementing test cases for the individual requirements. Being aware of typical formal method workflows, we think that this must be done in every case study. Otherwise, without using animation to verify that the behavior is correct, one cannot have sufficient confidence in the model. This, combined with the tooling that is available for C code, makes us excited to see other case studies, and challenge them to name benefits of their individual approaches, as we now know the extent of access to (semi)-formal development the embedded software community has.

Nonetheless, we think these tools allow for interesting research for code generators: proven invariants on a high-level model could be compiled to C assertions. Then, they could be verified on the low-level code as well. It remains open how hard the translation process is and whether the power of these tools is sufficient.

References

1. MISRA C:2012 - Guidelines for the use of the C language in critical systems. MISRA (2013)
2. General Specification of Basic Software Modules. AUTOSAR, Munich (2019)
3. Baumeister, H.: Combining formal specifications with test driven development. In: Zannier, C., Erdogmus, H., Lindstrom, L. (eds.) XP/Agile Universe 2004. LNCS, vol. 3134, pp. 1–12. Springer, Heidelberg (2004). https://doi.org/10.1007/978-3-540-27777-4_1

---

[3] https://plantuml.com/.

4. Beck, K.: Test-Driven Development: By Example. Kent Beck Signature Book. Addison-Wesley, Boston (2003)
5. Biere, A., Cimatti, A., Clarke, E., Zhu, Y.: Symbolic model checking without BDDs. In: Cleaveland, W.R. (ed.) TACAS 1999. LNCS, vol. 1579, pp. 193–207. Springer, Heidelberg (1999). https://doi.org/10.1007/3-540-49059-0_14
6. Boogerd, C., Moonen, L.: Assessing the value of coding standards: an empirical study. In: Proceedings ICSM, pp. 277–286. IEEE (2008)
7. Bowen, J.P., Hinchey, M.G.: Seven more myths of formal methods. IEEE Softw. **12**(4), 34–41 (1995)
8. Cadar, C., Dunbar, D., Engler, D.R., et al.: KLEE: unassisted and automatic generation of high-coverage tests for complex systems programs. In: Proceedings OSDI, vol. 8, pp. 209–224. USENIX Association (2008)
9. Chalupa, M., Vitovská, M., Strejček, J.: SYMBIOTIC 5: boosted instrumentation. In: Beyer, D., Huisman, M. (eds.) TACAS 2018. LNCS, vol. 10806, pp. 442–446. Springer, Cham (2018). https://doi.org/10.1007/978-3-319-89963-3_29
10. Chen, M., Ravn, A.P., Wang, S., Yang, M., Zhan, N.: A two-way path between formal and informal design of embedded systems. In: Bowen, J.P., Zhu, H. (eds.) UTP 2016. LNCS, vol. 10134, pp. 65–92. Springer, Cham (2017). https://doi.org/10.1007/978-3-319-52228-9_4
11. Clarke, E., Kroening, D., Lerda, F.: A tool for checking ANSI-C programs. In: Jensen, K., Podelski, A. (eds.) TACAS 2004. LNCS, vol. 2988, pp. 168–176. Springer, Heidelberg (2004). https://doi.org/10.1007/978-3-540-24730-2_15
12. Clarke, E., Kroening, D., Yorav, K.: Behavioral consistency of C and Verilog programs using bounded model checking. In: Proceedings DAC, pp. 368–371. IEEE (2003)
13. Fathy, H.K., Filipi, Z.S., Hagena, J., Stein, J.L.: Review of hardware-in-the-loop simulation and its prospects in the automotive area. In: Modeling and Simulation for Military Applications, vol. 6228. SPIE (2006)
14. Hall, A.: Seven myths of formal methods. IEEE Softw. **7**(5), 11–19 (1990)
15. Hansen, D., Ladenberger, L., Wiegard, H., Bendisposto, J., Leuschel, M.: Validation of the ABZ landing gear system using ProB. In: Boniol, F., Wiels, V., Ait Ameur, Y., Schewe, K.-D. (eds.) ABZ 2014. CCIS, vol. 433, pp. 66–79. Springer, Cham (2014). https://doi.org/10.1007/978-3-319-07512-9_5
16. Hansen, D., et al.: Using a formal B model at runtime in a demonstration of the ETCS hybrid level 3 concept with real trains. In: Butler, M., Raschke, A., Hoang, T.S., Reichl, K. (eds.) ABZ 2018. LNCS, vol. 10817, pp. 292–306. Springer, Cham (2018). https://doi.org/10.1007/978-3-319-91271-4_20
17. Hatton, L.: Language subsetting in an industrial context: a comparison of MISRA C 1998 and MISRA C 2004. Inf. Softw. Technol. **49**(5), 475–482 (2007)
18. Houdek, F., Raschke, A.: Adaptive exterior light and speed control system (2020)
19. ISO: Road vehicles - functional safety (2011)
20. Short, M., Pont, M.J.: Assessment of high-integrity embedded automotive control systems using hardware in the loop simulation. J. Syst. Softw. **81**(7), 1163–1183 (2008)
21. Yang, M., Zhan, N.: Combining formal and informal methods in the design of spacecrafts. In: Liu, Z., Zhang, Z. (eds.) SETSS 2014. LNCS, vol. 9506, pp. 290–323. Springer, Cham (2016). https://doi.org/10.1007/978-3-319-29628-9_6
22. Yuan, J., Shen, J., Abraham, J., Aziz, A.: On combining formal and informal verification. In: Grumberg, O. (ed.) CAV 1997. LNCS, vol. 1254, pp. 376–387. Springer, Heidelberg (1997). https://doi.org/10.1007/3-540-63166-6_37

Short Articles of the PhD-Symposium
(Work in Progress)

A Correct by Construction Approach for the Modeling and the Verification of Cyber-Physical Systems in Event-B

Meryem Afendi[(✉)]

Université Paris-Est Créteil, LACL, 94010 Creteil, France
meryem.afendi@u-pec.fr

Abstract. Cyber-Physical Systems (CPSs) connect the real world to software systems through a network of sensors and actuators: physical and discrete components interact in complex ways by involving different spatial and temporal scales. One of the most common architectures for CPSs is a discrete software controller which interacts with its physical environment in a closed-loop schema where input from sensors is processed and output is generated and communicated to actuators. We are concerned with the construction and verification of the correctness of such discrete controller using a correct by construction approach, which requires correct integration of discrete and continuous models.

1 Introduction

In CPSs [1], the measurement of continuous behaviors is performed by sensors. Ideally sensors have a continuous access to these measurements, which can be captured by an abstract model of CPSs, called *Event-Triggered* system by Kopetz in [2]. However, implementing such models is difficult in practice. Therefore, an approach to develop CPSs is to introduce a more realistic model called Time-Triggered system where sensors take periodic measurements [2]. Contrary to *Event-Triggered* models, properties on *Time-Triggered* models are difficult to verify since their controllers are discrete. Platzer et al. [3,4] use this approach to model hybrid systems, which represent the most common mathematical models for CPSs. They have proved that a *Time-Triggered* model is a refinement of an *Event-Triggered* model, by using hybrid programs (HPs) [3] and an extension of the differential dynamic logic (d$\mathcal{L}$), called the differential refinement logic (dR$\mathcal{L}$). In addition to d$\mathcal{L}$ formulas, dR$\mathcal{L}$ introduces formulas of the form $\alpha \leq \beta$, α refines β, with α and β denoting HPs. However dR$\mathcal{L}$ is not supported by any prover and dR$\mathcal{L}$ formulas can only be manually proved, which heavily restricts its use, especially in an industrial context. This is why we propose the development of an approach that takes advantage of dR$\mathcal{L}$ and applies the reasoning of HPs in Event-B [5], a formal method based on set theory, first-order logic and predicate logic.

© Springer Nature Switzerland AG 2020
A. Raschke et al. (Eds.): ABZ 2020, LNCS 12071, pp. 401–404, 2020.
https://doi.org/10.1007/978-3-030-48077-6_31

2   State of the Art

The development of techniques and tools to effectively design hybrid systems has drawn the attention of many researchers [6–10]. Traditional approaches are based on simulation tools like Matlab/Simulink [11]. Since these tools are time-consuming and produce results tainted with uncertainty, these approaches can be very expensive and difficult to apply. To overcome these limitations, several formal approaches, which can be grouped into two categories, have been proposed: *model-checking-based* approaches [6,7] and *proof-based* approaches [3,8–10], have been proposed. Proof-based approaches use deductive verification to prove the properties of hybrid systems. One of the strong points of these approaches is that they support the description of any kind of hybrid systems. However, they require significant effort and a high expertise in modelling and proof phases. A. Platzer introduced in [3] a first-order dynamic logic in the domain of real ($\mathbb{R}$), called d$\mathcal{L}$, to specify hybrid systems and verify their correctness using its associated proof calculus. The major advantage of this logic is its ability to handle differential equations, even those with non-polynomial solutions. Moreover, d$\mathcal{L}$ and its associated proof calculus are supported by two automatic formal verification tools, KeYmaera and its successor KeYmaera X [12]. In [10], Dupont et al. introduced a proof-based approach to model and prove hybrid systems in Event-B. They use the Theory plug-in of Event-B to define theories that handle continuous aspects of hybrid systems such as differential equations. The behavior of CPSs is specified by three Event-B models: *System* model that is used to describe the continuous evolution of the physical part, *State-System* model that refines the previous model by adding the evolution of the discrete part (the controller) and *Controlled-System* model that refines *State-System* model by adding the interaction between the physical part and the discrete part.

3   Proposed Approach

Our work is supported in part by the DISCONT project [13] funded by the French National Research Agency (ANR). The objective of this project is to elaborate a correct-by-construction method, based on Event-B, to specify hybrid systems models. Two approaches are considered. The first one, developed by Dupont et al. [10], is based on a translation of hybrid automata in Event-B. In our approach we propose to model the high-level structure of hybrid programs in Event-B, and more precisely the generic templates defined for modelling *Event* and *Time-Triggered* systems in dR$\mathcal{L}$ to take advantages of its proof obligations. To handle continuous aspects of hybrid systems in Event-B, we use the abstract model *System* of Dupont et al. as a starting model, as well as their theories developed to handle differential equations and continuous functions. To validate our approach, we chose the *Stop Sign* case study [14] which deals with a stop sign controller whose objective is to ensure the stopping of a car before a stop signal SP. In this case study, the differential equation that represents the evolution of the physical part is linear and can be easily solved. To handle more difficult differential equations we plan to integrate an external mathematical tool as a back-end tool in Event-B associated tools.

4   Current Results and Future Work

So far, our approach has focused on the most abstract level of hybrid systems design where all transitions are instantaneous, that is, duration issues are not considered. It consists of three models (see Fig. 1). *Event-Triggered* model (*EventTriggered_Ctx* and *EventTriggered_M* of Fig. 1) is a generic abstract model designed to specify and prove systems with con-

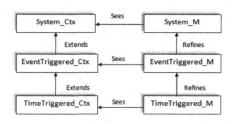

Fig. 1. Approach schema.

tinuous controllers in Event-B. It defines a formula $safe$ used in dR$\mathcal{L}$ to represent the system's safety envelope. To model the alternation between the controller and the physical part as represented in HPs $((ctrl; plant)^*$, where $ctrl$ denotes the discrete evolution, followed by the continuous evolution $plant$), *Event-Triggered* model defines a variable $exec$ that can take two values $ctrl$ and $plant$. In Event-B, the time must be explicitly handled. To be sure that this explicit time will only be updated after the execution of the controller, we added another value, prg, to $exec$. Therefore, our model follows the following structure: $init; (ctrl; plant; prg)^*$.

To model *Time-Triggered* systems in Event-B, we designed a new model, named *Time-Triggered* (*TimeTriggered_Ctx* and *TimeTriggered_M* of Fig. 1), that refines the previous one. The sensors of such systems take periodic measurements of physical state variables and the longest time between these sensors updates is bounded by a symbolic duration ϵ. For this purpose, *Time-Triggered* model defines a variable d to know whether the duration ϵ is reached or not. Since the controller of such systems must make a choice that will be safe for up to ϵ time, we replaced the formula $safe$ by a new safety envelope named $safeEpsilon$ that depends on both the current discrete state and the time duration ϵ, in addition to the current physical state. Finally, to prove that *TimeTriggered_M* refines *EventTriggered_M*, we establish that, during a control period ϵ, $safeEpsilon$ implies $safe$. Moreover, we must guarantee that the continuous controller is able to execute exactly when $safe$ is no longer satisfied. In addition, in the *Time-Triggered* model, we must also guarantee that the system will not exceed the domain of the safety envelope within time ϵ.

As future work, we plan to integrate Mathematica [15], a symbolic mathematical computation system that resolves differential equations, as a back-end tool in the Rodin platform to resolve differential equations. We also plan to define a refinement of the *Time-Triggered* model to take into account duration between the sending of continuous measurements by sensors and their processing by the controller as well as duration between the sending of actions by the controller and their execution in order to get a model that corresponds better to real CPSs, and consequently implements this concrete model.

References

1. Lee, E.A.: Cyber physical systems: design challenges. In: 2008 11th IEEE International Symposium on Object and Component-Oriented Real-Time Distributed Computing (ISORC), pp. 363–369. IEEE (2008)
2. Kopetz, H.: Event-triggered versus time-triggered real-time systems. In: Karshmer, A., Nehmer, J. (eds.) Operating Systems of the 90s and Beyond. LNCS, vol. 563, pp. 86–101. Springer, Heidelberg (1991). https://doi.org/10.1007/BFb0024530
3. Platzer, A.: Differential dynamic logic for hybrid systems. J. Autom. Reason. **41**(2), 143–189 (2008)
4. Loos, S.M., Platzer, A.: Differential refinement logic. In: 2016 31st Annual ACM/IEEE Symposium on Logic in Computer Science (LICS), pp. 1–10. IEEE (2016)
5. Abrial, J.-R.: Modeling in Event-B: System and Software Engineering. Cambridge University Press, Cambridge (2010)
6. Henzinger, T.A., Ho, P.-H., Wong-Toi, H.: HyTech: a model checker for hybrid systems. In: Grumberg, O. (ed.) CAV 1997. LNCS, vol. 1254, pp. 460–463. Springer, Heidelberg (1997). https://doi.org/10.1007/3-540-63166-6_48
7. Frehse, G.: PHAVer: algorithmic verification of hybrid systems past HyTech. In: Morari, M., Thiele, L. (eds.) HSCC 2005. LNCS, vol. 3414, pp. 258–273. Springer, Heidelberg (2005). https://doi.org/10.1007/978-3-540-31954-2_17
8. Butler, M., Maamria, I.: Practical theory extension in Event-B. In: Liu, Z., Woodcock, J., Zhu, H. (eds.) Theories of Programming and Formal Methods. LNCS, vol. 8051, pp. 67–81. Springer, Heidelberg (2013). https://doi.org/10.1007/978-3-642-39698-4_5
9. Banach, R., Butler, M., Qin, S., Verma, N., Zhu, H.: Core hybrid Event-B I: single hybrid Event-B machines. Sci. Comput. Program. **105**, 92–123 (2015)
10. Dupont, G., Aït-Ameur, Y., Pantel, M., Singh, N.K.: Proof-based approach to hybrid systems development: dynamic logic and Event-B. In: Butler, M., Raschke, A., Hoang, T.S., Reichl, K. (eds.) ABZ 2018. LNCS, vol. 10817, pp. 155–170. Springer, Cham (2018). https://doi.org/10.1007/978-3-319-91271-4_11
11. Sanfelice, R., Copp, D., Nanez, P.: A toolbox for simulation of hybrid systems in Matlab/Simulink: Hybrid equations (HyEQ) toolbox. In: Proceedings of the 16th International Conference on Hybrid Systems: Computation and Control, pp. 101–106 (2013)
12. Fulton, N., Mitsch, S., Quesel, J.-D., Völp, M., Platzer, A.: KeYmaera X: an axiomatic tactical theorem prover for hybrid systems. In: Felty, A.P., Middeldorp, A. (eds.) CADE 2015. LNCS (LNAI), vol. 9195, pp. 527–538. Springer, Cham (2015). https://doi.org/10.1007/978-3-319-21401-6_36
13. ANR-17-CE25-0005: DISCONT ANR project (2017). https://discont.loria.fr
14. Quesel, J.-D., Mitsch, S., Loos, S., Aréchiga, N., Platzer, A.: How to model and prove hybrid systems with KeYmaera: a tutorial on safety. Int. J. Softw. Tools Technol. Transfer **18**(1), 67–91 (2016)
15. Wolfram, S.: The Mathematica, 5th edn. Wolfram Media, Champaign (2003)

Improving Trustworthiness of Self-driving Systems

Fahad Alotaibi$^{(\boxtimes)}$ (ID)

ECS, University of Southampton, Southampton, UK
faa2n19@soton.ac.uk

1 Introduction

Self-Driving Vehicles (SDVs) are considered to be safety-critical system. They may jeopardize the lives of passengers in the vehicle and people in the street, or damaging public property such as the transportation infrastructure. According to the National Transportation Safety Board report [1] of an Uber self-driving crash, the accident was caused by the internal components of SDVs when the AI module failed to detect a victim. The autonomous system was implemented to give a human driver control of a vehicle on the unmanaged areas; however, the driver was distracted and did not react within the appropriate time.

In order to ensure the safety of SDVs, properties of the system must be demonstrated, especially the interactions between autonomous system and human driver in performing driving tasks. Event-B would help to emphasize the properties of the system and ensure a safe transition among multiple components of that system. Event-B uses a variety of extensive tools to support both theorem proving and model checking. According to a survey of formal verification tools [2], Event-B and its toolset (*Rodin*) provide a useful technique to support the goals of a *correct-by-construction* design. Therefore, Event-B and its extensive tools can be used to address issues related to SDVs.

2 Problems, Aims, and Objectives of the Research

Problems: Developers of SDVs are faced with one of the main challenges, specifically, ***establishing techniques for verifying safety properties***. There are three problems related to SDVs from the safety engineering perspective. The challenge in ensuring safety in the SDV starts from (1) *the complexity of the autonomous system*, (2) *the interactions between autonomous functions and the human driver*, and (3) *the ambiguous safety constraints*. **Firstly**, the complexity of SDVs system is determined by the automation level and its autonomous functions. According to SAE International [3], the levels of autonomy are classified from 0–5. The automation levels 1 to 4 (semi-automation) involve a human driver within the driving tasks, while automation level 5 (full automation) does not engage a human driver within the driving tasks. **Secondly**, based on the level of autonomy, the human driver and autonomous system may work together to

© Springer Nature Switzerland AG 2020
A. Raschke et al. (Eds.): ABZ 2020, LNCS 12071, pp. 405–408, 2020.
https://doi.org/10.1007/978-3-030-48077-6_32

perform driving tasks. Therefore, ensuring a secure transition of vehicle control between a human driver and the autonomous system is an important aspect. **Thirdly**, due to the variety of scenarios and faults that might occur in the SDV system, gathering the safety constraints to ensure the functionality of the autonomous system is a challenging task.

Aim and Objectives: The main aim of this research is ***to improve the trust-worthiness of SDVs by proposing a new methodology***. This aim can be split into a set of objectives as follows. **The first objective** is to reconsider and analyse the taxonomy requirements [3] in order to identify the safety requirements for the autonomous functions of SDVs on different levels of automation. **The second objective** focuses on the interaction among different components of an SDV at the system level in order to find the potential relationships among these components, especially when a human driver component and autonomous controller component frequently take control of the vehicle. **The third objective** deals with the input and output of autonomous functions and their relationships in order to identify the safety constraints for the autonomous functions.

3   The Current Development and Related Works

3.1   The General Approaches of Using Formal Methods Within the Autonomous System

Due to the complexity of the autonomous system architecture, there are many approaches that have been developed to verify and validate a system from the low- or high-level perspective. At the high level, the concept of a rational agent is used to focus on the autonomous controller, who is responsible for making decisions, and simplify the complexity of SDV system. A rational agent is a software that can perceive its environment via sensors and can explain its intentions [4]. The set of rules can be defined and formally verified into a rational agent entity [4]. In order to use the concept of the rational agent, the logical requirements (rules) must be defined. Consequently, formal methods such as LTL (*linear temporal logic*) can be used to verify the rational agent software.

There are some work in the literature that address the low-level issues of the SDV. The implementation of an SDV mainly relies on machine learning (ML) and its deep neural network (DNN). Although ML algorithms would perform with high accuracy in the image classification task, the work of ML might be affected by the input perturbations of image and lead to an incorrect classification. The input perturbations can be anything such as a 'shadow' and 'weather' which can affect the functionality of the SDV. This kind of manipulation is known as 'adversarial perturbations' [5]. Therefore, Huang et al. [5] proposed an automated verification framework for proving the adversarial robustness of the DNN. This approach is based on applying constraints through the layers of the DNN in order to prevent any misclassifications that might be caused by the adversarial examples. These constraints bound the regions of inputs for all points that are related to the same classification result. *Satisfiability Modulo*

Theory (SMT) is used to develop this verification framework. However, in order to apply this type of approach, the diameter of each region that belongs to a specific classification result must be known, and also the potential adversarial examples must be identified as well.

3.2 The Approaches of Using Event-B for Autonomous System

Constructing the Event-B Models at the System Level for Ensuring Interactions Between the Human Driver and Autonomous Systems: The inspired details of this approach are obtained from the cookbook [6] for the modelling and refinement of control systems. The guidelines mentioned in the cookbook suggest that the phenomenon of a system can be divided into two categories: 1) *variables that identify between environment and controller*, 2) *variables that represent the interaction between human operators and the environment*. There were two contributions related to this approach for modelling the functionality of the autonomous controller: a cruise control system, and lane departure warning system (LDWS) [7]. These autonomous functions belong to a lower level of automation (Level-1). However, reconsidering the ideas that were mentioned in [6,7] might help in either analysing the features of taxonomy requirements or modelling forward to the next automation level.

Constructing the Event-B Models Based on the Safety Constraints of an Autonomous Function: The SDV must implement fail-safe mechanisms, often known as the 'policing function' [8]. The concept of fail-safe mechanisms focuses on the functional requirements and is part of the system requirements. The policing function can check output values of autonomous functions such as an ML model at runtime. The important step of using a validation technique such as a policing function is to demonstrate the safety constraints for the autonomous functions, and are used either to validate the result of autonomous functions or detect failures in the runtime. According to Hoang et al. [8], the concept of metamorphic relationships that aim to discover an expected relationship between inputs and outputs can be used to identify the safety constraints which can be used to build a validation model.

4   Proposed Approach and Future Work

Finding a novel method to extract and identify either the safety constraints or the validation requirements for an autonomous function would be an important task. In order to achieve that, there are three layers that would be used to simplify the complexity of the SDV system as follows: *the specification of the features layer, the decision mechanism layer*, and *the actuation layer*. The aim of specification layer is to specify features that would be considered for making a driving decision by modifying the vehicle control variables at the actuation layer. Due to the driving decision might be made by the human-driver or autonomous controller, the local and global features are introduced. The local features focus on the autonomous functions and its safety constraints, while the global features consider the entire system and try to hold features that might be used when an autonomous function can not perform a driving task.

The local features can be identified from the in-deep knowledge about the expected output of autonomous function. For example, the centring lane lines function tries to keep the vehicle in the road lanes by identifying the lane boundaries and modifying the vehicle control variables. Therefore, '*left and right lane boundary*' and '*Yaw angel*' would be the local features which be controlled and monitored by the local monitor function in order to validate the work of the autonomous function. The local monitor function involves the constraints and procedures that can be used when an autonomous function cannot work as expected. For example, when an autonomous function cannot detect the lane lines, the local monitor function may need to notify the SDV system to use a global feature.

The global features such as '*Driver monitored feature*' and '*emergency stop*' might be applied to ensure the safety of system and avoid any potential mistakes of autonomous function. According to the taxonomy requirements (SAE) [3], the automation levels 1–4 require a human-driver in the loop of automation system. Therefore, it is necessary to implement a system for measuring the awareness level of a human-driver by installing a camera inside a vehicle and monitoring the eyes of driver. The global monitor function would hold global features for establishing a safe transition to the human-driver when an autonomous controller fails to preform the driving task.

Finally, Event-B models can be constructed based on the local and global monitor functions in order to emphasize the main properties of the SDV system. The next step is to apply the proposed approach to a practical case study. We will extend the work of the LDWS [7] to move forward into the next automation level (Level-2) where *a monitored human driver feature* is required.

References

1. National Transportation Safety Board (NTSB) (2018). Preliminary Report HWY18MH010
2. Armstrong, R.C., Punnoose, R.J., Wong, M.H., Mayo, J.R.: Survey of Existing Tools for Formal Verification (2014). https://doi.org/10.2172/1166644
3. SAE J3016: Taxonomy and definitions for terms related to on-road motor vehicle automated driving systems. Revision September 2016, SAE International
4. Fisher, M., Dennis, L., Webster, M.: Verifying autonomous systems. Commun. ACM **56**(9), 84–93 (2013). https://doi.org/10.1145/2494558
5. Huang, X., Kwiatkowska, M., Wang, S., Wu, M.: Safety verification of deep neural networks. Technical report (2016). http://arxiv.org/abs/1610.06940
6. Butler, M.: Modelling Guidelines for Discrete Control Systems, Deploy Deliverable D15, D6.1 Advances in Methods Public Document, Chapter 8 (2009)
7. Yeganefard, S., Butler, M.: Structuring functional requirements of control systems to facilitate refinement-based formalisation. ECEASST **46**, 8–11 (2011). https://eprints.soton.ac.uk/337259/1/695-2096-1-PB.pdf
8. Hoang, T.S., Sato, N., Myosin, T., Butler, M., Nakagawa, Y., Ogawa, H.: Policing functions for machine learning systems (2018)

A Formal Approach for the Modeling of High-Level Architectures Aligned with System Requirements

Racem Bougacha[✉]

Institut de Recherche Technologique Railenium, 59300 Famars, France
racem.bougacha@railenium.eu

1 Problem Statement and Motivations

IRT Railenium (http://railenium.eu/fr/) is a test and applied research center for the rail industry in France. One of its three R&D and innovation programs aims in particular to provide the technological tools and bricks necessary for the development of the Autonomous Train. This Autonomous Train program will thus address signaling, control-command, driving and railway operating systems. The Autonomous Freight Train project under the Autonomous Train program with the cooperation of several partners ("SNCF", "ALSTOM", "Hitachi Rail STS France", "ALTRAN" and "APSYS") targets performance improvements of the system thanks to the implementation of autonomy in railways operations. This system is classified as Cyber-Physical System (CPS) and depends more and more on effective solutions that can address heterogeneity and interplay of physical and software elements. In particular, modeling languages used for specifying CPS should incorporate, in a consistent manner, the essential concepts from multiple engineering disciplines that take part in the design of such systems.

In response to this need, our work addresses (1) the lack of a common modeling language between these disciplines, which can hamper reasoning about system properties. This issue is stemming from, on the one hand, several modeling languages specified for designing CPS that have been standardized but none of them provide the full range required to deal effectively with the heterogeneity of CPS elements. On the other hand, complex systems such as autonomous freight trains may include many concerns from different modeling formalisms.

A second issue addressed in our work is (2) the non-existence of a holistic approach for designing Autonomous Freight Trains from a high-level architecture to a formal specification where it can be possible to check the compliance with system requirements. Therefore, this issue could be divided into two sub-issues. The first sub-issue is deduced from the fact that architecture models actually used are semi-formal and/or informal, so their specifications are still no valid because they are not proved. Thus, formal specifications are needed in order to guarantee the consistency and the completeness of architecture models specification.

© Springer Nature Switzerland AG 2020
A. Raschke et al. (Eds.): ABZ 2020, LNCS 12071, pp. 409–413, 2020.
https://doi.org/10.1007/978-3-030-48077-6_33

The quality of a system is the main measure of its success, that depends on the degree to which it fulfills its requirements. Requirements modelling is an important activity in the process of designing and managing high-level architectures of systems. From this context the second sub-issue is raised, i.e alignment links between requirements models, domain models and architecture models should be established. These semantic links can be the support to prove the compliance of an architecture specification with the expression of system requirements.

From this perspective, we formulate the following research questions:

- **RQ1:** Can we provide a modeling language for High-Level architectures of Autonomous Freight Train?
- **RQ2:** What are the criteria for defining such a language?
- **RQ3:** How can we provide a formal specification of High-Level Architectures to verify its consistency?
- **RQ4:** How can we establish and verify alignment links between High-Level Architectures and System Requirements?

2   Related Work

Several architecture modeling languages have been proposed to reason about heterogeneous properties of CPS, and assessing multidisciplinary design languages. The authors of paper [1] propose an approach to combine SysML [2] and AADL [3]. These formalisms are two modeling languages specified for designing Integrated Control Systems (ICSs). These languages have been standardized but none of them provides the full range required to deal effectively with more general kinds of complex systems like CPS. In fact, SysML supports requirements engineering, traceability, and modeling of diverse physical phenomena. On the other hand, AADL is oriented to model real-time embedded systems. It provides software-to-hardware bindings allowing analyses of different system properties such as performance, timing, etc. This combination consists in extending SysML using the UML extension mechanism **profiles** to cover all AADL concepts and is called **Extended SysML for Architecture Analysis Modeling (ExSAM)**. This approach allows to design high-level architectures of complex systems. However, in this approach only semi-formal graphical models can be produced and thus no formal verification can be carried out. Furthermore, they do not permit to specify alignment links that can be used to ensure the compliance of architecture models with system requirements.

Authors in [5] proposes a multi approach for real-time systems specification and design. The purpose of the work is to couple MARTE [4] with the Event-B method. Papers [6] and [7] consider a meaningful subset of AADL which allows to specify respectively a class of embedded control systems and AADL data port protocols, and assign this subset a semantics in terms of Event-B refinements and model decomposition. Despite these approaches provide formal specifications, they do not consider high-level architectures but propose a temporal model of tasks execution on GPU.

Requirements Engineering (RE) is defined as the branch of software engineering which is focused on real-world goals, requirements development and management. [8] presents a review on Goal-Oriented Requirements Engineering (GORE) methods. This kind of methods defines goals as objectives that the system under consideration should achieve. [9] and [10] present an approach which aims to combine requirements engineering methods with formal methods. The main idea is to specify goal models by using a SysML-based language [2] extended with concepts of the KAOS goal language [8], to specify domain models by using an ontology-based language, and then to map them into an Event-B specification. However, these methods are just concerned by requirements models and do not provide formalism to specify system architecture models.

3   The Proposed Approach and Methodology

We propose to define a method for modeling high-level architectures for autonomous freight trains that integrates concepts from the different engineering disciplines that take part in the design process of such systems. The method needs to be multiviews: graphical in order to be validated by all the stakeholders and formal in order to verify both architecture models and their compliance with system requirements. This is a three-steps method based on the MDE approach shown in Fig. 1. In the first step, we propose to combine three graphical modeling languages: SysML, AADL and MARTE, in order to provide the full range required to deal effectively with Autonomous Freight Train concerns such as timing properties, non-functional properties, software-to-hardware bindings, etc. The second step consists in formalizing the concepts of the graphical language with Event-B in order to obtain an Event-B specification of architecture models. Thus their correctness and consistency can be proved. We have chosen the Event-B method since it is a formal method, widely used in industry, based

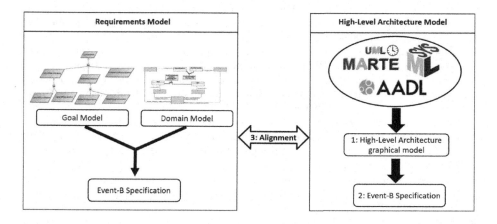

Fig. 1. The proposed approach

on refinement and decomposition mechanisms and supported by tools (provers, model-checkers, animators...). In the final step, we propose to specify alignment links between requirements models and high-level architecture models in order to prove that all the industrial stakeholders requirements are satisfied by high-level architecture models. We will reuse the SysML/KAOS method to produce requirements models (goal and domain models) and the corresponding Event-B specifications.

4   Current Assessment and Future Work

The current focus of this doctoral work, which is in its first year, is on the design of a graphical language for modeling high-level architectures. This first step attempts to answer RQ1 and RQ2 (step 1 of Fig. 1). The proposed modeling approach will be applied on industrial case studies. For instance we will consider the standard logical system architectures composed of three subsystems: on-board subsystem, trackside subsystem and communication subsystem, from the Autonomous Freight Train project. These case studies will be validated by the project partners (system, sub-system and requirements engineers) and verified by design models tools (for instance AADL verification tools) and Event-B verification tools (for instance ProB model checker and Atelier B prover).

References

1. Behjati, R., Yue, T., Nejati, S., Briand, L., Selic, B.: An AADL-based SysML profile for architecture level systems engineering: approach, metamodels, and experiments. ModelME! report, 2001-03 (2011)
2. OMG systems modeling language (2005). http://www.omgsysml.org/
3. Feiler, P.H., Gluch, D.P., Hudak, J.J.: The architecture analysis & design language (AADL): an introduction (No. CMU/SEI-2006-TN-011). Software Engineering Institute, Carnegie Mellon University, Pittsburgh, PA (2006)
4. OMG, UML Profile for MARTE: Modeling and Analysis of Real-Time Embedded systems, Beta 2 (2008)
5. Zouaneb, I., Belarbi, M., Chouarfia, A.: Multi approach for real-time systems specification: case study of GPU parallel systems. IJBDI **3**(2), 122–141 (2016)
6. D'Souza, M., Ramesh, S., Satpathy, M.: Architectural semantics of AADL using Event-B. In 2014 International Conference on Contemporary Computing and Informatics (IC3I), pp. 92–97. IEEE (2014)
7. Filali-Amine, M., Lawall, J.: Development of a synchronous subset of AADL. In: Frappier, M., Glässer, U., Khurshid, S., Laleau, R., Reeves, S. (eds.) ABZ 2010. LNCS, vol. 5977, pp. 245–258. Springer, Heidelberg (2010). https://doi.org/10.1007/978-3-642-11811-1_19
8. Van Lamsweerde, A.: Goal-oriented requirements engineering: a guided tour. In: Proceedings Fifth IEEE International Symposium on Requirements Engineering, pp. 249–262. IEEE (2001)
9. Matoussi, A., Gervais, F., Laleau, R.: A goal-based approach to guide the design of an abstract Event-B specification. In: ICECCS 2011, pp. 139–148 (2011)

10. Tueno Fotso, S.J., Mammar, A., Laleau, R., Frappier, M.: Event-B expression and verification of translation rules between SysML/KAOS domain models and B system specifications. In: Butler, M., Raschke, A., Hoang, T.S., Reichl, K. (eds.) ABZ 2018. LNCS, vol. 10817, pp. 55–70. Springer, Cham (2018). https://doi.org/10.1007/978-3-319-91271-4_5

Automatic Generation of DistAlgo Programs from Event-B Models

Alexis Grall[1,2(✉)]

[1] LORIA UMR 7503, Vandœuvre-lès-Nancy, France
alexis.grall@loria.fr
[2] Université de Lorraine, Vandœuvre-lès-Nancy, France

1 Motivations

The development of distributed algorithms offers challenges in verifying that they meet their specifications. The correct-by-construction approach consists in developing a model of the algorithm before transforming this model into a program. This transformation can introduce errors that were not present in the model. Our objective is to develop an automatic transformation of distributed algorithm Event-B [2] models into DistAlgo [7] programs. The Event-B language combines refinement techniques and state based modelling and is adapted to the verification of distributed systems [3,12]. The DistAlgo language is a high-level programming language for distributed algorithms. Its high-levelness makes DistAlgo closer to the mathematical notations of Event-B and improves the clarity of DistAlgo programs. A verified automatic transformation ensures that the properties proved in the model still hold in the program and facilitates the developing process.

2 Related Works

Code generation from Event-B models has been a subject of interest in the B community. The B0 [11] language defines constraints on classical B for code generation and an equivalence between B types and usual programming types such as arrays, integers *etc*. B0 can be translated to Ada, C, C++ using the AtelierB [4] tools. In Event-B, several plugins have been developed for the Rodin [10] software. The EB2ALL [8] framework provides a list of transformations of Event-B models into classical programming languages (C, C++, Java, ...) and this work can be considered as adding a new target programming language but with the target of a distributed program. Automatic generation of distributed programs was proposed in ViSiDiA [1,9] together with Event-B with the plugin B2VISIDIA [12] relating the local Event-B model and a ViSiDiA program. A Tasking Event-B [6] plugin for Rodin extends the Event-B language to provide features for specifying concurrent multi-tasking systems. The plugin enables the decomposition of a model into several machines performing tasks and provides a tool support for translating a tasking specification into Ada code. However, a global state is preferable for verifying global properties on distributed algorithms.

© Springer Nature Switzerland AG 2020
A. Raschke et al. (Eds.): ABZ 2020, LNCS 12071, pp. 414–417, 2020.
https://doi.org/10.1007/978-3-030-48077-6_34

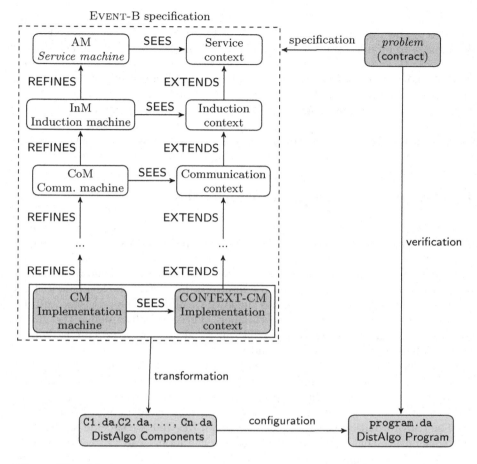

Fig. 1. The global methodology for correct-by-construction distributed algorithms.

3   General Approach

The general methodology, presented in Fig. 1, starts by stating the *problem* to solve by listing the requirements attached to the problem. One has then to specify the abstract EVENT-B machine AM translating the main requirements for the given problem. The process is progressing by a list of formal EVENT-B refined machines leading to a final concrete EVENT-B machine and context, CM and CM-CONTEXT. Finally, the translations of these final context and machine into DistAlgo components and programs are generated in two main steps: the automatic transformation of CM and CM-CONTEXT into a DistAlgo program and the possible tuning of the obtained DistAlgo components using some auxiliary configurations (*e.g.* the number of processes).

The final pair of machine and context, CM and CM-CONTEXT are supposed to be specified in a language LB (Local Event-B) which restricts the Event-B language to models of distributed algorithms that are local, which means that the constants and variables are local to processes (they are associated to processes.) and that events only use local variables and constants of one process. Transformations of this LB model into a DistAlgo program are based on the extraction of information concerning the network and the process classes from the context CONTEXT-CM, and on the analysis of the localization of the different variables and events of the machine CM. Constants whose values are not defined in the context are instantiated during the configuration phase.

4   Development of the Transformation

The LB language [5] defines a distributed algorithm in Event-B as a collection of processes with a specific local algorithm for each process. We derive the general architecture of the distributed algorithm from the context CONTEXT-CM which specifies the topology of the network, the processes, the local constants of the processes and the communication operations. We derive the local algorithms from the machine CM which specifies the local variables of the processes and the variable specifying the state of the channels. These variables define the global state of the distributed algorithm while the local state of a process is defined by the values of its local variables and the states of the channels involving this process. The transition relation between the local states of a process are then defined by the local events (of this process) in the machine CM. An additional variable pc in the machine helps specifying the flow of the algorithm and can be used for constructing control flow graphs describing the local algorithms.

The transformation rules add constraints on LB that are specific to DistAlgo and that ensure that the expressions used in the model can be translated into DistAlgo by specifying a correspondence between Event-B and DistAlgo types. The transformation into DistAlgo generates programs based on the local control flow graphs, therefore giving a tool to help understanding the program.

Refined models of distributed algorithms have been manually and successfully transformed into DistAlgo programs following the given transformation rules. These models describe a simple communication between two processes, a simple communication algorithm between processes in a star network and the election in a connected acyclic network.

We need to ensure that any safety property that was proved with invariants in an Event-B model still holds in the DistAlgo program obtained from the transformation of the model, *i.e.* the soundness of the transformation. We have already shown the soundness of the translation for an earlier version of LB, *i.e.* a program obtained by the transformation refines the machine it is obtained from. We have thus, that for any possible execution of the program, an equivalent behaviour can be observed for the model machine. However, the completeness of the transformation is not guaranteed when no restrictions are imposed on the corresponding programs.

5   Future Work

The implementation of the transformation as a Rodin plugin is currently a work in progress with so far encouraging results on simple cases. The proof of the soundness of the transformation remains to be completed and modified with respect to the current version of the transformation. Other future work includes the extension of LB to enable more possibilities in modelling, such as enabling the broadcast of messages, the use of user defined functions/libraries in the resulting programs or the addition of configurable timeout delays. Additional case studies of parallel computing are to be refined into LB and transformed. These case studies will be an occasion to compare the performance of the automatically generated program with an hand written version of this program. Long term future work includes the derivation of other transformations into other programming languages from the transformation to DistAlgo.

References

1. Abdou, W., Abdallah, N.O., Mosbah, M.: Visidia: A java framework for designing, simulating, and visualizing distributed algorithms. In: 2014 IEEE/ACM 18th International Symposium on Distributed Simulation and Real Time Applications, pp. 43–46. IEEE (2014)
2. Abrial, J.R.: Modeling in Event-B: System and Software Engineering. Cambridge University Press, Cambridge (2010)
3. Abrial, J.R., Cansell, D., Méry, D.: A mechanically proved and incremental development of IEEE 1394 tree identify protocol. Formal Aspects Comput. **14**(3), 215–227 (2003)
4. Atelier, B.: The Industrial Tool to Efficiently Deploy the B Method. http://www.atelierb.eu/en/
5. Cirstea, H., Grall, A., Méry, D.: Generating Distributed Programs from Event-B Models. Technical report, Loria & Inria Grand Est, March 2020. https://hal.inria.fr/hal-02496623
6. Edmunds, A., Butler, M.: Tasking Event-B: an extension to Event-B for generating concurrent code. In: PLACES 2011, February 2011
7. Liu, Y.A., Stoller, S.D., Lin, B., Gorbovitski, M.: From clarity to efficiency for distributed algorithms. In: ACM SIGPLAN Notices, vol. 47, pp. 395–410. ACM (2012)
8. Méry, D., Singh, N.K.: EB2C : A Tool for Event-B to C Conversion Support (2011–2019). http://eb2all.loria.fr
9. Mosbah, M.: VISIDIA (2009). http://visidia.labri.fr
10. Project RODIN: Rigorous open development environment for complex systems (2004). http://rodin-b-sharp.sourceforge.net/. 2004-2007
11. Tatibouët, B., Requet, A., Voisinet, J.-C., Hammad, A.: Java card code generation from B specifications. In: Dong, J.S., Woodcock, J. (eds.) ICFEM 2003. LNCS, vol. 2885, pp. 306–318. Springer, Heidelberg (2003). https://doi.org/10.1007/978-3-540-39893-6_18
12. Tounsi, M., Mosbah, M., Méry, D.: From event-b specifications to programs for distributed algorithms. IJAACS **9**(3/4), 223–242 (2016). https://doi.org/10.1504/IJAACS.2016.079623

Event-B: From Systems to Sub-systems Modeling

Kenza Kraibi[✉]

Institut de Recherche Technologique Railenium, 59300 Famars, France
kenza.kraibi@railenium.eu

1 Introduction

Event-B [3] is a formal method that allows the verification of critical systems properties. This method is based on the refinement reasoning which consists in adding more details step by step from the abstraction. Modeling critical systems in Event-B requires several steps of refinement in order to take into account all the details of the specification. Therefore, the whole system modeling and proof become more difficult because of the huge size of data and system properties like safety properties described in the system specification.

PRESCOM project (Global Safety Proofs for Modular Design/**PRE**uves de **S**écurité globale pour la **CO**nception **M**odulaire) is an IRT Railenium project in partnership with Clearsy Systems Engineering and under the supervision of Gustave Eiffel University (UGE/COSYS/ESTAS) and Polytechnic University of Hauts-de-France (UPHF/LAMIH). As part of this project, the goal of our thesis[1] is to answer the industrial need, i.e. find a solution to the models voluminosity issue in Event-B when we put the whole specification in the model progressively by the refinement mechanism. This conduces to: study what exists in the literature; apply these approaches on a railway case study, analyze the results and identify their limitations. Based on these identified limitations, we propose a new approach of decomposition called the *decomposition by refinement*.

2 Related Work and Analysis

Many approaches have been proposed to deal with the Event-B decomposition issue, among others one finds: the shared variable decomposition and the shared event decomposition. The shared variable decomposition [4], A-style, consists in distributing events of a system in several sub-systems. This approach proposes to manage shared variables between several events in different sub-systems. It is also used for decomposing parallel programs [6]. The shared event decomposition [5], B-style, is based on the variables partition in each sub-system. Each sub-system

---

[1] This thesis is supervised by: Rahma Ben Ayed (IRT Railenium), Joris Rehm (Clearsy), Simon Collart-Dutilleul (UGE/COSYS/ESTAS), Philippe Bon (UGE/COSYS/ESTAS) and Dorian Petit (UPHF/LAMIH).

© Springer Nature Switzerland AG 2020
A. Raschke et al. (Eds.): ABZ 2020, LNCS 12071, pp. 418–422, 2020.
https://doi.org/10.1007/978-3-030-48077-6_35

contains the chosen variables, and the shared events between the resulting sub-systems are defined in two different signatures for each sub-system. In addition to these two approaches, one finds others such as generic instantiation [4], modularization [7], fragmentation and distribution [10].

The aim of this work is to model the behavior of railway signaling systems in Event-B and at the same time manage the complexity of the resulting models. For this reason, we choose to proceed with the study and analysis of A-style and B-style, because the other cited approaches imply some classical-B [1] method semantics or use other languages.

The analysis of A-style and B-style leads to these results: both approaches require several steps of refinement in order to simplify the model decomposition. For A-style, the shared variables should be copied in the sub-systems and shouldn't be refined. The invariants involving the shared variables are not considered in the sub-systems. As for the shared events decomposition, the distribution of the variables is not always possible because of complex actions involving partitioned variables in different sub-systems or complex predicates (invariants and guards). This requires the separation of these variables by several steps of refinements with mathematical proofs. The detailed description of the state of the art, the application on a railway case study, the analysis and the identified limitations have been presented in [8].

3   Proposed Approach

On the basis of the industrial need and the identified limitations, we define a new approach called the *decomposition by refinement* method. The approach consists in the decomposition using the refinement technique for the purpose of keeping the semantic link between the system and the resulting sub-systems. So, a system is decomposed into one or more sub-systems in such a way they are refining this later. This can be applied to a system either in the abstraction level or in a certain level of refinement as shown in Fig. 1. By this way the sub-systems are still preserving the defined system properties in the abstraction through the refinement. Also, we define a new link between the sub-systems named REFSEES. This link will provide to each sub-system the visibility to the other sub-systems: the state of the private variables, the corresponding invariants, the constants, the sets and the properties.

Let us consider the example of Fig. 1 and let M be the system to decompose[2]. M defines the state variables v where $v = (x, y, z)$ for example, the invariants to preserve involving the state variables $I(v)$ and the abstract events ae. The goal is to decompose M into two sub-systems M_a and M_b where: M_a (resp. M_b) is a refinement of M; w_a (resp. w_b) are the state variables refining some state variables of M. For example, w_a are refining x and y, and w_b are refining y and

---

[2] Due to the limited place in this paper, we show a simple example, but we have already performed our approach on interesting case studies from the railway domain [8].

z; $J_a(w_a)$ (resp. $J_b(w_b)$) is the gluing invariant of M_a (resp. M_b); The events re_a (resp. re_b) are the events of M_a (resp. M_b) refining a part of the abstract ones in M.

The clause *REFSEES* in M_a (resp. M_b) allows to see the state of the private variables of M_b (resp. M_a). So, the private variables of M_a (resp. M_b) can be used in the guards of the events of the machine M_b (resp. M_a). More details about *REFSEES* clause are in [8,9].

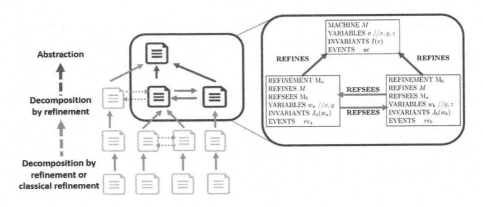

Fig. 1. Proposed approach: decomposition by refinement

Some rules should be considered in order to formalize this approach:

Rule$_1$: Some state variables of the decomposed system M can only be in one of the sub-systems M_a or M_b. But, these variables should all be present at least in one of the sub-systems.

Rule$_2$: The sub-system M_a (resp. M_b) can refer in the guards of their events to the private variables of M_b (resp. M_a).

Rule$_3$: The transition system of the resulting sub-systems M_a and M_b should correspond to one transition system of the behavior of M.

Rule$_4$: M_a and M_b transitions are not synchronized contrary to the decomposition by shared events. So, following what has been presented in [2], we can demonstrate that the theoretical re-composition/combination of the subsystems is a refinement of the system M.

Rule$_5$: For each sub-system, a variant proof obligation rule VAR should be defined because a transition should not be triggered indefinitely. As for the deadlock freedom rules, there are two types: the weak deadlock freedom rule DLF_w and the strong one DLF_s. DLF_w verifies that at least one of the events is triggered. Whereas DLF_s proves that each event is triggered at least one time. This verification should also be done in case of the definition of new events.

4   Conclusion and Future Work

Several approaches have been proposed to deal with the complex and huge system specifications issue in Event-B such as A-style and B-style. The realized analysis and the study conduce to the identification of some limitations of those approaches regarding the industrial need. So, we propose a new approach: the *decomposition by refinement* based on decomposing a system by the refinement technique into several sub-systems. A new clause REFSEES is defined to link the sub-systems to each other which allows the visibility of the state variables. This approach will ensure the preservation of invariants through the refinement technique. Currently, we are working on the definition of the strategy to follow for the application of the approach. This strategy will define: the way to decompose the state variables of the system and its events, and how to define, in each sub-system, new invariants, new state variables and new events. As a short-term perspective, we will demonstrate that the fact of combining -theoretically- the sub-systems constitutes a one refining component of the initial system regarding the theoretical definition of the refinement in B method. As a long-term perspective, new proof obligations will be specified, through the new defined link, to ensure the behavior preservation in each of the resulting sub-systems. for the purpose of its scaling up, the approach will be applied to a railway signaling system case study.

References

1. Abrial, J.R.: The B-Book: Assigning Programs to Meanings. Cambridge University Press, New York (1996)
2. Abrial, J.R.: Event Model Decomposition. Technical report/[ETH, Department of Computer Science 626 (2009)
3. Abrial, J.R.: Modeling in Event-B: System and Software Engineering. Cambridge University Press, New York (2010)
4. Abrial, J.R., Hallerstede, S.: Refinement, decomposition, and instantiation of discrete models: application to event-B. Fundamenta Informaticae **77**(1–2), 1–28 (2007)
5. Butler, M.: Decomposition structures for event-B. In: Leuschel, M., Wehrheim, H. (eds.) IFM 2009. LNCS, vol. 5423, pp. 20–38. Springer, Heidelberg (2009). https://doi.org/10.1007/978-3-642-00255-7_2
6. Hoang, T.S., Abrial, J.-R.: Event-B decomposition for parallel programs. In: Frappier, M., Glässer, U., Khurshid, S., Laleau, R., Reeves, S. (eds.) ABZ 2010. LNCS, vol. 5977, pp. 319–333. Springer, Heidelberg (2010). https://doi.org/10.1007/978-3-642-11811-1_24
7. Hoang, T.S., Iliasov, A., Silva, R.A., Wei, W.: A survey on event-B decomposition. Electron. Commun. EASST **46** (2011)
8. Kraibi., K., Ben Ayed., R., Rehm., J., Collart-Dutilleul., S., Bon., P., Petit., D.: Event-B decomposition analysis for systems behavior modeling. In: Proceedings of the 14th International Conference on Software Technologies, vol. 1: ICSOFT, pp. 278–286. INSTICC, SciTePress (2019). https://doi.org/10.5220/0007929602780286

9. Kraibi., K., Ben Ayed., R., Rehm., J., Collart-Dutilleul., S., Bon., P., Petit., D.: Towards a method for the decomposition by refinement in event-B. In: Refinement Workshop at Formal Methods Congress (Refine@FM), Accepted paper but not yet published (2019)

10. Siala, B., Tahar Bhiri, M., Bodeveix, J.P., Filali, M.: Un processus de Développement Event-B pour des Applications Distribuées. Université de Franche-Comté (2016)

A Framework for Critical Interactive System Formal Modelling and Analysis

Ismaïl Mendil[✉]

IRIT/INPT-ENSEEIHT, 2 rue Charles Camichel, 31071 cedex 7 Toulouse, France
ismail.mendil@toulouse-inp.fr

1 Introduction

The human-computer interface (HCI) is the component of an interactive system that allows users to interact with a system. Interactive system development does not follow the same life cycle as software system development. The essential differences lie in the iterative nature of the interactive system development. Hence, the completion of such systems requires usually several iterations. Throughout iterations, the requirements undergo many changes due to the evolution of customer's needs and user feedback after experiencing the prototypes. Furthermore, the formalization of user interaction requirements into HCI specification is a complex task that needs a thorough observation of user behaviour. The challenge is tougher in the case of critical HCI (cockpits, medical systems, etc.). Indeed, critical HCI requires to be designed and built such that safety is put at the forefront of the requirements dictated by standards and norms.

Formal methods offer both theoretical background and support tool enabling the validation and verification of system specification beforehand, i.e. before the system is put into production. In fact, such methods allow dealing with abstract mathematical models of the system on which mathematical proofs are performed. So, different kinds of properties (e.g. safety and usability) are proved, hence providing a higher level of confidence in the system. Several modelling formalisms and tools have been developed to model, analyse and animate HCI models, however, they do not offer HCI domain knowledge integration at modelling level.

In our Ph.D. project, we aim to complete an operational framework for formal verifying and validating critical interactive systems with a special emphasis on the aeronautic field.

2 Challenges

As part of the ANR FORMEDICIS[1] project, an abstract-level language describing the interaction between the system and the user is being developed: FLUID (Formal Language for User Interface Design). The core raison d'être of this language is to allow the design and analysis of critical interactive systems at a high level of abstraction. This endeavour faces, however, several challenges:

---

[1] ANR (French National Research Agency), https://anr.fr/Projet-ANR-16-CE25-0007.

© Springer Nature Switzerland AG 2020
A. Raschke et al. (Eds.): ABZ 2020, LNCS 12071, pp. 423–426, 2020.
https://doi.org/10.1007/978-3-030-48077-6_36

- Defining of a domain theory to formalize HCI design knowledge.
- Ensuring effective integration of the user's requirements in the design process (Human In The Loop).
- Endowing FLUID language with verification capabilities.
- Supplying FLUID with animation extension for the sake of validation.
- Assessing and evaluating the FLUID expressivity and its easiness of use.
- Contributing to the certification of critical interactive systems.

3   FLUID Language

3.1   FLUID Language Main Features

In this section, we sketch the FLUID language definition. It allows describing and verifying the system behaviour. Moreover, it addresses the lack of integration of domain-specific knowledge in HCI development at modelling level. FLUID models are built out of the INTERACTION component which consists of three parts: contextual part (DECLARATION), behaviour part (STATE and EVENT) and properties part(ASSUMPTIONS, EXPECTATIONS and REQUIREMENTS) (Fig. 1).

FLUID models are state-based with interleaving asynchronous event-driven semantics. It allows the designer to express the user requirements in different ways either as general properties (e.g. usability, safety) abstracted away from requirements, or individual scenarios conveying some expected storyboard (SCENARIOS.NOMINAL) or, at the contrary, precise forbidden scenarios (SCENARIOS.NONNOMINAL).

Variables and events in FLUID models are annotated with tags whose definitions lie in domain theory. Hence, we split the domain-specific knowledge and constraints apart from the system model. In addition, proof obligations are generated to integrate the HCI domain constraints into the verification and validation process.

```
INTERACTION component_name
  DECLARATION
    TYPE T_i
    CONSTANT C_i
  STATE
    var_i@{tag_ij} : T_i
  EVENTS
    INIT =
      statement_i
    EVENT event_name_i @ {tag_ij}[arg_ik] =
      WHERE
        G(T_i, C_i, var_i, arg_ik,
                 var_i@tag_ij, arg_ik@tag_ikl)
      THEN
        var_i:| BA(T_i, C_i, var_i, arg_ik,
        var_i@tag_ij, arg_ik@tag_ikl, var_i', var_i@tag_ij')
      END
  ASSUMPTIONS
    A(T_i, C_i)
  EXPECTATIONS
    E(T_i, C_i)
  REQUIREMENTS
    PROPERTIES
      P(T_i, C_i, var_i, var_i@tag_ij)
    SCENARIOS
      NOMINAL
        SC(T_i, C_i, var_i, var_i@tag_ij)
      NONNOMINAL
        NSC(T_i, C_i, var_i, var_i@tag_ij)
END component_name
```

Fig. 1. FLUID basic component template

3.2   Analysing FLUID Models: Event-B and ICO

In our work, two formalisms and two tools are devised to support FLUID language. For the purpose of verifying FLUID models, Event-B[2] will be used. It

---

[2] http://wiki.event-b.org/.

is a formal method for system-level modelling and analysis. Basically, it relies on set theory and first-order logic to formalize systems into models and uses refinement to represent systems at different abstraction levels. Furthermore, it leverages mathematical proof theory for discharging proof obligations [1]. The Event-B language defines two main building blocks to model, in principle, any system: firstly, a context construct for describing the static characteristics of a system through carrier sets, constants, axioms and theorems, secondly, a machine building block for expressing the dynamic aspects through variables, invariants, theorems, variants and events. Rodin[3] [2] is an Integrated Development Environment (IDE) for Event-B modelling language based on Eclipse. It has many features making the modelling and V&V process easier and more efficient. Rodin features project management, incremental model development, proof assistance, model checking, animation and automatic code generation.

For the sake of animation and validation of FLUID models, ICO [6] associated to its PetShop [7] tool will be used as a basis. ICO formal modelling is dedicated to expressing and describing interactive systems. Interactive Cooperative Objects follows the object oriented-paradigm so it incorporates concepts such as dynamic instantiation, classification, encapsulation, inheritance and client/server relationships. It provides means to model the static side of the interactive system inspired by the object-orientation, and it uses the Petri Nets notation to express the behavioural side. Moreover, the ICO notation is fully supported by the PetShop CASE tool.

4   Methodology and Approach

In order to bridge the gap and build a framework resolving the identified challenges listed in Sect. 2, we envision to meet the objectives we present hereafter:

- Modelling HCI domain using Event-B theory extension as supported by Rodin Theory Plugin[4]. Indeed, HCI domain theories connected to design models via tags and the automatic proof obligation generation shall be viewed as a major feature of our approach.
- Developing a transformation schema to embed FLUID models in Event-B [1] [4] to incrementally design interactive system models [3]. The schema needs to integrate the specification and verification of scenarios and properties. This dual specification of requirements is a key feature of our framework addressing the HCI development process, in particular handling of storyboards.
- Defining and implementing transformation rules in order to dive FLUID and/or Event-B models into ICO for the sake of animation using PetShop.
- For the purpose of assessment and evaluation of the framework, case studies will be developed and extended to comply with the HCI domain-specific constraints. In addition, formal methods (Event-B, ICO, LIDL [5]) will be compared, on the basis of different criteria (expressivity, conciseness, etc.), to our framework through development of case studies.

---

[3] http://www.event-b.org/install.html.
[4] http://wiki.event-b.org/index.php/Theory_Plug-in.

5   What's Next?

This article framed the problem tackled in our Ph.D. research project and identified its coarse-grained challenges: mainly, the need for a comprehensive abstract-level HCI modelling framework covering the development cycle of critical HCI. Then, the high-level objectives are set down and enumerated.

Within our Ph.D. work, the next steps are twofold. Firstly, we plan to develop an HCI domain theory to enrich models at the early stages of system development. This theory must encompass the domain knowledge and it connects to the system model via the tag notation. Next, the transformation schema needs to be formalized and implemented, allowing the translation from FLUID into Event-B and ICO. Then, in order to demonstrate the effectiveness, cases studies are planned to be developed based on the novel framework. Finally, a development process will be defined around the novel framework so to facilitate and structure HCI developments. Ultimately, we believe that a successful landing of our project would provide more confidence to tackle the bigger problem of extending our framework to other system development fields.

References

1. Abrial, J.R.: Modeling in Event-B - System and Software Engineering. Cambridge University Press, Cambridge (2010)
2. Abrial, J.-R., et al.: Rodin: an open toolset for modelling and reasoning in Event-B. Int. J. Softw. Tools Technol. Transf. **12**(6), 447–466 (2010)
3. Aït-Ameur, Y., et al.: Encoding a process algebra using the Event B method. STTT **11**, 239–253 (2009). https://doi.org/10.1007/s10009-009-0109-2
4. Geniet, R., Singh, N.K.: Refinement based formal development of human-machine interface. In: Mazzara, M., Ober, I., Salaün, G. (eds.) STAF 2018. LNCS, vol. 11176, pp. 240–256. Springer, Cham (2018). https://doi.org/10.1007/978-3-030-04771-9_19
5. Lecrubier, V.: A formal language for designing, specifying and verifying critical embedded human machine interfaces. Ph.D. thesis. ISAE - Universitéde Toulouse, June 2016. https://hal.archives-ouvertes.fr/tel-01455466
6. Navarre, D., et al.: ICOs a model-based user interface description technique dedicated to interactive systems addressing usability, reliability and scalability. ACM Trans. Comput. Hum. Interact. **16**, 18:1–18:56 (2009)
7. Palanque, P., Ladry, J.-F., Navarre, D., Barboni, E.: High-fidelity prototyping of interactive systems can be formal too. In: Jacko, J.A. (ed.) HCI 2009. LNCS, vol. 5610, pp. 667–676. Springer, Heidelberg (2009). https://doi.org/10.1007/978-3-642-02574-7_75

Author Index

Printed in the United States
By Bookmasters